Advancements in Textile Coloration

Mohammad Shahid · Saptarshi Maiti ·
Shafat Ahmad Khan · Ravindra V. Adivarekar
Editors

Advancements in Textile Coloration

Techniques, Technologies, and Trends

Springer

Editors
Mohammad Shahid
Department of Chemistry
Koneru Lakshmaiah Education Foundation (KLEF)
Vaddeswaram, Guntur, Andhra Pradesh, India

Shafat Ahmad Khan
Department of Chemistry
Galgotias University
Greater Noida, Uttar Pradesh, India

Saptarshi Maiti
Department of Fibres and Textile Processing Technology
Institute of Chemical Technology
Mumbai, Maharashtra, India

Ravindra V. Adivarekar
Department of Fibres and Textile Processing Technology
Institute of Chemical Technology
Mumbai, Maharashtra, India

ISBN 978-981-96-5090-3 ISBN 978-981-96-5091-0 (eBook)
https://doi.org/10.1007/978-981-96-5091-0

This Springer imprint is published by the registered company Springer Nature Singapore Pte Ltd.
The registered company address is: 152 Beach Road, #21-01/04 Gateway East, Singapore 189721, Singapore

Preface

The textile industry has transformed remarkably, evolving from conventional dyeing and printing methods to more advanced, sustainable, and efficient coloration techniques. As the global textile sector continues to expand, there is an increasing demand for innovations that enhance coloration performance and minimize environmental and resource-related challenges. Recognizing this need, *Advancements in Textile Coloration: Techniques, Technologies, and Trends* brings together a comprehensive collection of chapters that explore various cutting-edge developments in textile coloration. This book provides a broad overview of the latest techniques and technologies employed in textile dyeing and printing, highlighting sustainable approaches and emerging trends. The content is thoughtfully organized to cover different aspects of textile coloration, ranging from conventional methods to novel alternatives that align with global sustainability goals. Chapter 1 introduces an overview of various textile coloration techniques, setting the foundation for subsequent discussions. The book discusses sustainable practices in Chap. 2, exploring alternative methods to conventional aqueous dyeing. Chapter 3 further investigates the concept of waterless dyeing, discussing its advancements, challenges, and environmental implications. The innovations in air dyeing and dope dyeing technologies are examined in Chap. 4, emphasizing their contributions to sustainable textile coloration. A major shift toward eco-friendly coloration is discussed in Chap. 5, which explores electrochemical dyeing as a sustainable approach. This is followed by Chap. 6, which focuses on dyes and mordants derived from renewable residues, showcasing their progress and applications. The evolution of bio-mordants and their role in sustainable textile coloration are addressed in Chap. 7. The impact of natural dyeing techniques is explored in Chap. 8, which examines naturally tie-dyed textiles in the context of sustainable fashion. The book discusses the fascinating world of microbial colorants in Chap. 9, where their historical significance, current applications, and future potential are discussed. The integration of statistical methodologies to improve sustainable textile coloration is presented in Chap. 10, offering insights into current trends and future recommendations. Chapter 11 then introduces advanced surface modification techniques for sustainable textile coloration, addressing recent technological innovations. Chapters 12 and 13 thoroughly review the evolution of textile printing from conventional methods to modern digital printing practices and their applications in textiles. Finally, Chap. 14 discusses adopting sustainable technologies to mitigate effluent pollution from the textile industry, a growing concern for researchers and practitioners. This book is valuable for researchers, industry professionals, academicians, and students interested in textile coloration and sustainability. By integrating knowledge from multiple perspectives, this book aims to inspire further research and innovation in the field, contributing to a greener and more sustainable textile industry. We sincerely thank all the authors, researchers, and contributors who have made this book possible.

Guntur, India — Mohammad Shahid
Mumbai, India — Saptarshi Maiti
Greater Noida, India — Shafat Ahmad Khan
Mumbai, India — Ravindra V. Adivarekar

Contents

About the Editors

Dr. Mohammad Shahid Ph.D., is an Assistant Professor in the Department of Chemistry, Koneru Lakshmaiah Education Foundation, India. Before joining KLEF, he worked at various esteemed institutions, including Jamia Millia Islamia, New Delhi; Soochow University, China; University of Glasgow, UK; the Institute of Chemical Technology, Mumbai. He earned his Master's degree in Chemistry from Shibli National College, Azamgarh, in 2006 and his Ph.D. in Chemistry from Jamia Millia Islamia, New Delhi, in 2014. Dr. Shahid has received several prestigious fellowships to support his research pursuits, including grants from UGC, India, China Postdoctoral Fellowship, EU Marie Curie Fellowship, and Dr. D. S. Kothari Fellowship. He has contributed to over 50 scientific publications with ~4300 citations and h-index of 26. In addition, has also edited and authored several books with renowned international publishing houses. His primary research interests lie in developing sustainable biomaterials for textile applications.

Dr. Saptarshi Maiti is an Assistant Professor in the Department of Fibres and Textile Processing Technology at the Institute of Chemical Technology (ICT), Mumbai, India. He completed his post-doctoral research at ICT-Mumbai in 2022. Dr. Maiti holds a Ph.D. (Tech.) and M.Tech. in Fibres and Textile Processing Technology from the same department. He has also worked as a Visiting Research Fellow at the School of Materials, University of Manchester, UK, under the UAA-ICT Dhirubhai Ambani Lifetime Achievement Award. His research interests include graphene and its applications in textiles, dendritic polymers, protein extraction from waste resources, natural dyeing, green processing of textiles, and statistical modeling in textiles. Dr. Maiti has authored more than 50 international publications in reputed journals, contributed to around 22 book chapters in textile chemistry and materials science, and holds 1 patent. He is a life member of The Indian Natural Fibre Society (INFS).

Dr. Shafat Ahmad Khan is a Professor at the School of Basic Sciences, Department of Chemistry, Galgotias University, India. He obtained his Ph.D. from Jamia Millia Islamia, New Delhi, India in 2012. Dr. Khan has accumulated more than 12 years of teaching experience, mentoring students, and contributing significantly to the academic community. His research expertise is reflected in his extensive publication record, which includes 29 international research papers in peer-reviewed journals. Additionally, he has authored ten book chapters and has played a key role in the academic publishing landscape by editing two books. His contributions to innovation and applied research are further evident in his five patents, demonstrating his commitment to advancing scientific knowledge.

Dr. Ravindra V. Adivarekar Ph.D., is a Distinguished Professor in Fibre Chemistry at the Institute of Chemical Technology (ICT), Mumbai, where he has dedicated over 25 years of service. He holds multiple degrees, including an Honorary Fellowship of the Textile Association (FTA). Dr. Adivarekar is renowned for his expertise in textile processing and served as the Head of the Department of Fibres & Textile Processing Technology from 2008 to 2021. He currently serves as the Dean of Human Resource Development (HRD) at ICT Mumbai and as the Director of ICT-IOC Bhubaneswar. His extensive research interests span eco-friendly coloration, natural fiber composites, medical textiles, nanotechnology in textile processing, natural dyes, microbial dyes, and the application of enzymes. A committed mentor, Dr. Adivarekar has guided numerous students and managed research projects with total funding exceeding 18 crores. He has contributed significantly to the academic community with over 200 research papers published in national and international journals, more than 70 conference proceedings, and 3 patents. Beyond his academic achievements, Dr. Adivarekar is an esteemed member of several textile-related organizations and has received significant recognition for his contributions to the industry. He actively engages in advisory roles and government committees, addressing environmental issues in textile manufacturing.

Chapter 1
An Overview of Different Techniques in Textile Coloration

Silpinwita Das and Sankar Roy Maulik

Abstract The coloration of textiles was introduced primarily to serve aesthetic purposes along with functional quality. The history of textile coloration is an age-old journey of craftsmanship using natural dyes. As civilization grew, advanced technologies were introduced to the coloration of textiles. The textile industry relies on diverse coloration techniques, each with unique advantages and challenges. Dyeing and printing are the traditional methods of textile coloration, and they have been the backbone of the industry for a long time. Although effective, these conventional methods are often associated with significant water usage, chemical discharge, and energy consumption. To address these environmental issues, waterless dyeing, viz., supercritical carbon dioxide ($scCO_2$) dyeing, plasma and laser technologies, coloration with nanoparticles, etc., are gaining momentum in the field of coloration of textiles. Air dyeing and foam dyeing are energy-efficient processes involving minimum or zero water. Microwave, ultrasonic, plasma, and nanotechnology-based methods have recently been adopted for the efficiency and precision of the dyeing process. Digital printing has revolutionized textile coloration by enabling high-resolution, customizable designs without screens. This has become a more potent threat to traditional block printing, mechanical processes, and hand painting. Pigment-based coloration is another technique involving the application of pigment particles to the fabric surface without the need for complex chemical processes, resulting in a more sustainable and resource-efficient coloration method. These advanced techniques face challenges that persist in their widespread adoption. The initial investment required for digital printing or nanoscale coloration can be a barrier for smaller manufacturers, especially in large-scale production. Adopting new techniques for coloration to meet diverse needs has its unique set of benefits and challenges. This chapter provides an overview of the different techniques for textile coloration, including traditional methods and innovative approaches, highlighting their significance in the context of efficiency, sustainability, and color quality. Advanced technology, environmental pollution reduction, reusing and recycling textile wastes, etc., address the essence of sustainability.

S. Das · S. Roy Maulik (✉)
Department of Silpa-Sadana, Visva-Bharati (A Central University), Sriniketan, India
e-mail: s_r_moulik@yahoo.co.in; sankar.raymaulik@visva-bharati.ac.in

M. Shahid et al. (eds.), *Advancements in Textile Coloration*,
https://doi.org/10.1007/978-981-96-5091-0_1

Keywords Dyeing · Digital · Microwave · Nano · Printing · Technology · Ultrasound · Waterless

1.1 Introduction

Textile industries are among the oldest and most prominent for manufacturing, designing, and distributing yarns, fabrics, and related goods. As one of the largest industries, it tries to use minimum energy, minimizing waste and effluent. The pretreatment, dyeing, printing, and wet processing stages involve many highly toxic chemicals, including heavy metals and significant organic pollutants, which are eventually dumped in the waterbodies and wastelands. Various measures have been taken worldwide to cope with waste management, which involves a variety of challenges as well. This industry is also a water-reliant sector; freshwater remains an integral part of a textile product from the beginning to the end. The sustainable practice of textile coloring gained attention in this scenario because it used less water and minimum energy sources and saved time by introducing more recent, environmentally friendly technological innovations. A lot of water is required in the conventional textile dyeing process and a considerable amount of dye is lost as it stays in the dye bath. The effluent is released into the environment, sometimes without maintaining the safety rules and regulations. Nowadays, greener and more eco-friendly alternatives to conventional dyeing practices are gaining momentum in the coloration of textiles. Microwave, ultrasonic, plasma, and nanotechnology-based processes have recently been adopted to efficiently extract colorants from natural resources and dyeing, using significantly less or no water.

1.2 Microwave Technology

It is primarily used to save conventional energy resources required for extracting the colorants and dyeing the textile materials. This particular radiation falls between infrared and radio waves. It can generate rapid, controlled, and uniform volumetric heat with low electricity by increasing the molecular vibration frequencies without directly affecting the molecular structure of a compound, thus saving time and energy. This process is considered sustainable and cleaner [1]. There is much research on microwave technology for applying natural and synthetic dyes to different textile substrates. The dyeing of acrylic knitted fabric with turmeric in microwave conditions produced uniformity in color and high fluorescence properties [2]. The different color-related properties are enhanced on wool fiber when dyed with natural dyes under microwave treatment [3]. The colorants extracted from *Phlomis floccose* L. and *Artemisia monosperma Del* under microwave conditions are used to dye wool fabric [4]. The color strength and fastness properties are enhanced when the cotton fabric is dyed with pomegranate rind and turmeric under microwave irradiation [5].

The colorants can be extracted from saffron and safflower through a microwave-assisted technique and have a high potential for polyamide (nylon) fabric previously irradiated with microwave [6]. Microwaves enhanced the dye uptake and color fastness of polyester dyed with henna [7]. Similarly, the microwave-assisted technique increased the color yield of natural dyes extracted from lac and produced high surface color strength onto irradiated and unirradiated wool [8]. Microwave irradiated safflower produced darker shades and excellent fastness properties on unirradiated silk [9]. It has been reported that *Juglans Regia* L. bark extracted through microwave irradiation produced good color strength and fastness on wool fabrics [10]. Lanthanum chloride, a new mordanting agent, improved various color properties of microwave-assisted wool fabrics dyed with walnut extract [11]. The microwave treatment is used for surface modification and to enhance the color characteristics of cellulosic fiber dyed with vat dyestuffs [12]. The color strength and fastness of polyester fabric dyed with disperse dye are also improved through microwave treatment [13].

1.3 Ultrasonic Technology

It is effective in extracting colorants and also for the dyeing process. Sound waves with frequencies exceeding the typical audible range for humans (above c 20 kHz) are referred to as ultrasonics [14] and consist of a series of longitudinal waves that vary in density and propagate across a liquid medium [15]. It gives rise to high-temperature and high-pressure conditions by inducing chemical reactions and covering various operational conditions, saving energy and time and thus being considered an efficient heating source. Sonochemistry involves the application of ultrasound radiation in the frequency range of 0.020 MHz to 10 MHz [16]. The dye extraction from *Punica granatum* rind through ultrasound, enzyme, and enzyme-mediated ultrasound is reported in the literature [17]. This technique is also used to extract colorants under optimized conditions from avocado seed [18], *Dillenia indica* leaf [19], and Haar Singhar flower [20]. The ultrasonic method is more effective at low temperatures than conventional heating in a short time, with higher color strength and good colorfastness properties on wool fabric dyed with lac and *Curcuma aromatica* rhizome [21, 22]. Using an ultrasound-assisted dyeing method, mercerized cotton fabric dyed with marigold produced superior color strength, and thus, the process saves water, energy, and salt [23]. *Acacia catechu* and *Tectona grandis* showed enhanced dye uptake with higher color strength after enzyme pretreatment on cotton fabric using ultrasonic dyeing [24]. Using ultrasonic waves, wool fabric dyed with ratanjot root significantly improved the dye uptake, resulting in a darker shade, good color intensity, and fastness properties compared to conventional heating methods [25]. The color strength for the nylon-6 fiber dyed with reactive dyes using the ultrasonic technique is higher than that obtained using conventional heating [26]. Sonicated jute fabric dyed with reactive and basic dyes produced higher color strength than those dyed without sonication [27]. Ultrasonic-assisted wool dyeing with an acid dye increased the dye absorption rate with a good fastness property [28]. Using the ultrasound technique, cellulose nano-fibers dyed with reactive dyes produced a higher color yield [29].

1.4 Supercritical Carbon Dioxide ($scCO_2$)

The worldwide interest in environmentally friendly approaches forces the textile industry to cope with newer and cleaner coloration and finishing methods, using no or significantly less water. Supercritical fluids are non-polar solvents, capable of dissolving only hydrophobic dyes, and are gaining popularity for dyeing synthetic fibers due to their ecological aspect. They need less time and energy than conventional processes [30]. Using high pressure and low temperature simultaneously facilitates the liquefaction of gases. A specific gas cannot be liquefied beyond a critical temperature and requires a minimum critical pressure to achieve liquefaction at the critical temperature. The conjunction of critical temperature and pressure is called the critical point. At the critical point, the liquid and gas phases have the same density, displaying gaseous characteristics while maintaining a typical liquid property and functioning as a single phase. This particular phase is referred to as a supercritical fluid [31]. Carbon dioxide is a supercritical fluid with a low critical temperature (31 °C), pressure (73 bar), non-flammable, toxicity, non-explosive, and affordability. It is ideal for the uniform coloration of textiles with natural and synthetic dyes under specific conditions [32]. Supercritical dyeing is environmentally friendly and requires no subsequent drying steps or solvent removal procedures. The process does not produce effluent, and the unused dye particles can be separated for further use. [33]. The dyeing of hydrophilic fibers with direct, reactive, vat classes of dyestuffs in $scCO_2$ is not preferred as they are not soluble in $scCO_2$. However, $scCO_2$ showed the possibility of swelling hydrophobic fibers by increasing the softening and flexibility of the polymer at a comparatively lower pressure and temperature than that polymer's conventional glass transition temperature [34]. The dyeing of pre-treated cellulosic fiber in $scCO_2$ condition with disperse and natural dyes improves the dye uptake and colorfastness to washing and light [35]. The dyeing of polyester fibers with the disperse class of dye under $scCO_2$ is an effective and reliable method to achieve excellent wash fastness properties [36]. The wool dyed with various classes of disperse dyes in $scCO_2$ showed excellent colorfastness and reduced tensile strength [37]. Polyester, nylon, silk, and wool are dyed with disperse-reactive dye in the $scCO_2$ system, and it is noticed that the dyeing behavior of polyester is increased with pressure [38]. A reverse micellar system with a supercritical carbon dioxide dyeing medium is used to dye silk and wool to produce darker shades than conventional water-soluble acid dyes [39, 40]. The supercritical carbon dioxide dyeing process needs trained operators for the costly machinery system and thus waits for its large-scale successful commercial implementation. In the future, this concept of green dyeing can be considered one of the best alternatives to conventional dyeing.

1.5 Plasma Technology

Plasma, the fourth state of matter, is an ionized gas, and its application is gaining attention in textile industries worldwide as an alternative to wet processing. As temperature rises, molecules become more energetic, transforming solid matter into plasma through liquid and gaseous states. In this stage, at least one free electron of a molecule initiates the ionization process, converting the atoms or molecules into positively charged ions, making plasma electrically conductive. It also contains positive and negative ions, neutral particles, and UV rays [41]. Plasmas are classified into hot/high-temperature and cold/low-temperature plasmas (Fig. 1.1). Cold plasma is categorized into atmospheric, vacuum, or low-pressure plasmas (LPPs). The low-temperature plasmas (LTPs) are particularly useful for textiles, as most fiber-based materials are heat-sensitive polymers. These plasmas modify the surface characteristics to enhance functional properties and improve wettability, capillary action, and dye uptake [42].

Low-pressure plasma can increase the hydrophilicity of polyester and polypropylene fabrics, which makes them more receptive to various chemicals, including dyes. However, they are challenging to use for the continuous processing of textiles [43]. Only a very thin layer surface is exposed in the low-temperature polymer modification technique. The changes induced by plasma treatment depend on the nature of the textile substrate and the gases or mixtures used [43, 44]. Atmospheric plasma technologies have become a cost-effective and efficient dry application alternative, reducing the need for expensive vacuum equipment for low-pressure plasma [45]. The plasma pre-treatment and subsequent bio-surface modification using chitosan enhanced the fastness properties of polyester dyed with *Rubia tinctorum* L. [46]. The combined air/He plasma treatment for polyester fabrics improved the color yield and anti-bleeding performance [47]. A hydrophilic functional group can be incorporated on the hydrophobic surface of polyester by combining the nitrogen plasma technique and ionization treatment. Researchers also identified improved wettability, tensile

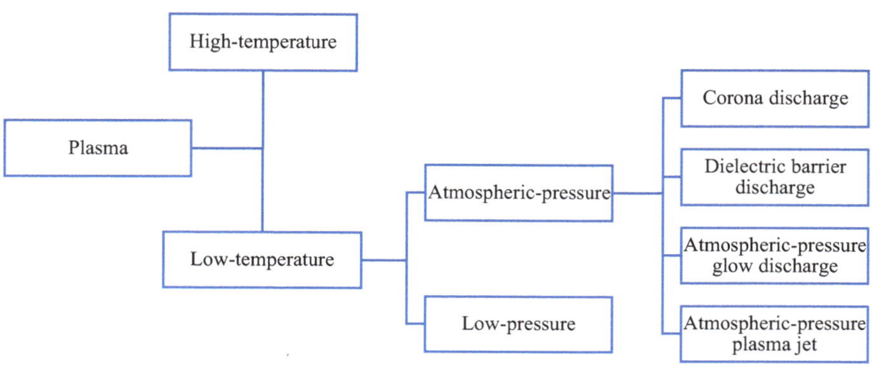

Fig. 1.1 Classification of plasma

strength, and roughness after dyeing and printing textile substrates with madder roots [48]. The enhancement of the surface color strength of polyester fabric dyed with curcumin in atmospheric pressure air plasma has been reported [49]. Dielectric Barrier Discharge (DBD) at atmospheric pressure plasma is used to dye polyester fabric with different natural dyes by improving the wettability of the fiber and hereby increasing the adsorption of natural dyes [50]. Polypropylene fabrics, pretreated with atmospheric pressure plasmas, significantly increased color strength when dyed with water-insoluble vat class of dyestuffs [51]. The treatment with acrylic acid, water, and argon plasma on polyacrylonitrile fabrics increased wettability, soil resistance, and dyeability [52]. The cotton fabric dyed with direct color, followed by post-plasma treatment, improved the fastness and water-repellent properties of the dyed fabric [53]. Low-temperature plasma-treated wool fabrics dyed with cochineal and Chinese cork trees showed increased brightness compared to untreated wool [54]. The wicking property and tensile strength of silk fabric treated with air dielectric barrier discharge plasma have improved due to increasing the hydroxyl group. The plasma-treated silk fabrics' wicking rate and wet-out time are close to 0° contact angle [55]. Mulberry silk fabrics, when treated with a low-temperature DBD plasma utilizing helium and oxygen gases before dyeing with acid dyes, enhancing the dye adsorption capacity [56].

1.6 Laser Technology

Laser, a light amplification device, is widely used in material manufacturing, medical procedures, and precise measurement, but its application in textile processing requires comprehensive support. Laser technology in the textile industry has revolutionized conventional finishing methods, replacing harmful processes like sand-blasting and hand sanding, etc. Automated lasers are widely used for marking the surface of the fabric, controlling the cutting of the fabric, fading effect, etc. Laser marking is a complex technology used in textile production, creating energy by heating polymeric material from microns to hundreds of microns, promoting eco-friendliness. The laser can be classified into a solid-state (Nd: YAG), glass or semi-conductor (AlGaAs, GaAsSb, and GaAlSb), liquid (solutions of dyes in water/alcohol and other solvents), and gaseous laser. Gas can be further sub-grouped into neutral atoms (He–Ne, Cu, or Au vapor), ions (argon $Ar+$ and krypton $Kr+$ ion), molecules (CO_2 or CO), and excimer lasers (XeCl, KrF) [57]. Laser is used to fade the surface color of jeans to achieve the whitening effect without damaging the material. This laser technology also opened a new possibility in 3D printing/painting on fabric surfaces. Natural cellulose and protein fibers need careful handling to achieve optimal results, and the quality may deteriorate due to incorrect laser combinations [58, 59]. Carbon dioxide lasers are mainly used to modify the surface of polyester fabric to improve dye uptake [60]. Laser treatment increased the dye absorption property of the wool and synthetic polymer by creating more functional groups, resulting in better color fastness properties. However, in the case of cotton dyed with direct dye, the depth of shade decreased [61].

1.7 Coloration with Nanoparticles

A nanometer is one billionth of a meter and is described as a system acting on the atomic or molecular level with significantly improved properties due to its nano-size, which has applications in the real world. Nanotechnology is in high demand worldwide because of its unique economic potential, and for commercial purposes, 1–100 nm ranged nanoparticles are commonly used. During the last few decades, a wide range of research has been conducted to pave the road for nanomaterial dyeing processes. This technology enhances the properties of textile materials by dispersing dye particles and reducing agglomeration in exhaust dye baths, resulting in increased uptake and reduced dye requirements [62]. Cochineal (*Dactylopius coccus Costa*) is used as a reducer/capping reagent, substituting the synthetic chemicals for the green synthesis of silver nanoparticles (AgNPs), which has multifunctional properties and opened new areas for functional textiles [63]. Nanoparticles, as mordanting agents, are a new approach in reducing the large quantity requirement of conventional heavy metal-based mordants for dyeing textiles with natural dyes. Zinc oxide nanoparticles are used as a mordant for dyeing silk fabric with an aqueous extract of *Camellia sinensis* and showed improved fastness properties [64]. Three different metal nanoparticle mordants (copper oxide, titanium dioxide, and zinc oxide) enhance the dyeability and color fastness properties of cotton fabrics dyed with *Eucalyptus globulus* bark and *Rubia tinctorum* root [65]. β-cyclodextrin nanoparticles grafted silk yarn using succinic acid and sodium hypophosphite and dyed with natural dyes showed enhanced color strength compared to control samples treated with tannic acid and copper salt [66]. Cellulose nanocrystals as mordant improved the fastness properties of cotton fabric dyed with an aqueous extract of mango, teak, and Indian ash tree leaves [67]. The safflower dye nanoparticles are produced using a planetary ball milling system. Ultrasound-irradiated extraction of this nanoparticle is used to print on cotton, wool, and polyester fabrics, which shows excellent fastness properties [68]. Considerable improvements in the fastness properties of cotton/wool (50/50) blended fabric printed with turmeric, madder, and rhubarb nanoparticles against regular dye particles are also reported [69]. The natural dye Red Poppy (*Papaver rhoeas L.*) is applied on chitosan nanoparticles at pH 9 on ramie fabrics through pigment printing without a binder and fixation process, improving the surface color strength of the printed fabric [70]. The nano pigment ink opens a new application and market in digital printing, exploring the possibilities of combined design patterns with reduced light scattering phenomena and good light purity without requiring pre- or post-treatment of the fabrics [71]. Nowadays, the electrospinning technique is used to produce cellulose nano-fibers, and it is reported that ultrasonic dyeing produces higher color yield than conventional dyeing with improved color fastness properties [72]. Au–Ag mixed nanoparticles-based pigments were developed for ink-jet printing on ceramic wares through a microwave-assisted synthesis route, with a low environmental impact, and can easily be transferred to large-scale production [73].

1.8 Air Dyeing

Air dyeing is an environment-friendly process. The fabrics are dyed in rope form under high pressure, using air as a transporting medium. This dyeing technique reduces water pollution caused by conventional dyeing. The dye liquor is atomized, mixed with high-pressure airflow, and then sprayed on fabric to be dyed. The dye particles are strongly infused with the large contact areas of the fabric, which passes with great speed, making this technology only feasible for synthetic materials like polyester. It can also dye the fabric with two colors on opposite sides [74]. Several air dyeing technologies have been developed in the last decade for cellulosic fibers. As this technique is relatively new, and the machine installation cost is high, developing low-cost, air-dyeing technology for natural and synthetic fabrics could be the future of textile dyeing [75, 76].

1.9 Foam Dyeing

Foam is a micro heterogeneous colloidal system where gases are dispersed within a liquid state of equilibrium or disequilibrium. Foam technology is considered a potential process requiring less water compared to traditional application methods [77]. Developed in late 1930, foam application technology is used to apply dye, pigment, or finishes on textile surfaces. It requires low liquid loading and wet pick-up compared to traditional methods. These innovative technologies develop through various combinations and manipulating essential variables [78, 79]. This process constantly generates foam to facilitate stable bubble formation and is applied to the fabric uniformly. During dyeing, the bubbles containing dye particles and auxiliaries collapse immediately, transferring the dye liquor to the substrate and getting fixed. Foam dyeing can be carried out in batches and as a continuous process. In continuous pad-dyeing processes, foaming of the liquor can adversely impact the quality of the dyeing. Researchers optimized the foam dyeing process for cotton fabrics with reactive dyes, which requires smaller chemical dosages and reduces effluent load apart from achieving excellent wash and rubbing fastness [80]. Foam technology can also help apply multifunctional finishing on both surfaces of the cotton fabric, which is durable for up to 50 washing cycles [81]. Foam technology can also be used to reduce indigo. However, using sodium hydrosulfite and sodium hydroxide as reducing agents is considered harmful to the environment [82]. Due to the low density of the foam solution and the low wet pick-up ratio, the dyes cannot penetrate the fabrics, causing the finishing agents/dyes to stay on the fiber surface. This shortcoming can be an advantage for the wash-out effect on knitted fabric [83]. Acid dyes are applied on nylon carpets by foam technology using rotary screens and appear to be significantly faster in pick-up compared to conventional methods [84]. In roller and screen printing, foam inhibitors are used in the printing paste to improve print quality. Still, more comprehensive research is needed for environment-friendly and low-cost stabilizing agents to achieve level dyeing and printing quality [85, 86].

1.10 Digital Printing

Digital printing on paper, film, plastic, etc., emerged over two decades ago as a prototyping tool. Since then, the improvement in ink chemistry, software, and machinery helped digital inkjet printing to grow and has been extensively used in textile applications. The first digital printing application on carpets was launched in the 1970s, using a pulsed air jet process, with jets of air drawing ink from dye-filled nozzles. The print quality mainly depends on technical aspects of ink chemistry, hardware requirements, and substrate. The application of digital textile printing is moving forward rapidly to serve mass-customized products, superseding the traditional wooden, metal block, industrial flat, rotary screen-printing, and roller printing techniques, adversely affecting the market demand for conventional printing [87]. However, many advancements in digital printing technologies still need to be made to make an impact in a broader commercial context. Digital printing has the potential benefit of zero stockholdings of printed fabrics, and many colors can be printed. Multicolor inks are used in this process and are compatible with any surface to print. High-speed inkjet is used to drop the ink onto the substrate. However, the print quality primarily depends on the image's resolution [88]. A digital textile printing system comprises an inkjet head, ink chemistry, media, and color management software. The main principle of the digital printing process follows the mechanism of a dripping tap, and each drip allows a drop of color to fall onto the surface of the moving textile substrate. The inherent rate of dripping of the ink has led to various innovations in the printing methods. Drop-on-demand and continuous ink jet are the two major printing processes primarily used in digital printing [89]. In digital printing, dye deposits are made on the surfaces of fabrics. In contrast, screen printing impregnates the dye into the fabric, using colors prepared in a color kitchen. The final appearance of a printed surface may resemble the screen-printed fabrics. Digital printing is primarily based on cyan–magenta–yellow–black inks without needing a color kitchen. Digital printing should not compete with rotary screen printing or other conventional mass production techniques, as it is used mainly for correct color with possible visual texture for small batch prototyping and photoshoots, reducing the lead time for product development. Inks contain a colorant (dye or pigment), additives (surfactants and salts), a carrier base (water, solvent, oil, phase change fluid, or UV curable fluid), and different additives. Most inks used for textile printing are water-based as they can evaporate quickly once printed on the substrate and are non-toxic to the environment. Colorants used as ink for textile printing have a higher fixation rate and good washing, light, and rubbing fastness on proper pre-treated and post-treated fabrics. In digital printing, reactive, disperse, and acid dyes are used for cellulose, polyester, protein, and nylon fibers [90]. Surface-modified and micro-encapsulated pigment dispersions printed on a silk fabric produce higher color yield, better tone reproduction, and fastness properties than only micro-encapsulated pigmented inks applied on silk [91]. Recently, several types of research have been done using fungal pigments as environmentally friendly inkjet colorants for digital printing on textiles [92].

1.11 Pigment-Based Coloration

Pigments are insoluble in water or their application medium and contain no available water solubilizing groups; they have no intrinsic affinity for textiles. The use of pigments for block printed textiles by ancient Chinese in the eighteenth century has been reported in the literature [93]. Pigments are used in various applications like cosmetics, ceramics, food, paints, paper, plastics, printing inks, and textiles. Any pigments applied to textiles require additional polymeric materials, known as binders, to hold them on the surface. The pigment and auxiliaries with an environmentally friendly character, the highest color strength, and good fastness properties are the primary requirements of the textile printing industry [94]. Pigments can be applied to the textile substrates by pad-liquor, exhaust, and print-paste techniques. The affinity of pigment emulsion on the textile substrates can be improved by modifying the surface with different aminating agents before pigment application [95, 96].

1.12 Conclusion

The textile industry is considered one of the most polluting industries because of its high energy and water consumption and the use of chemicals and auxiliaries. Thus, eco-friendly and sustainable methods are becoming viable options. Microwave and ultrasonic techniques can reduce the use of natural resources and carbon footprint. Green and water-saving textile dyeing technologies, viz., supercritical carbon dioxide, air, and foam, are getting the attention of textile industrialists. Plasma and laser technology in various textile applications opened a new area of textile processing and coloration. Coloration with nanotechnology can be an effective tool to enhance multiple qualities of textiles, amplifying the dyeing behavior. Air and foam dyeing are relatively new technologies used in innovative textile coloration. Despite their shortcomings, extensive research can open new possibilities. Today, digital printing is extensively used in the textile industry due to its time and cost-saving aspects. However, it should not be used for fraudulent imitation of traditional and conventional prints. These recent innovations can also reduce the effluent load of conventional dyeing processes by implementing them on a large-scale production capability to be economical and environmentally friendly.

Declaration The authors declare that they have no known competing financial interests or personal relationships that could have appeared to influence the work reported in this chapter. AI tools were used in the writing process only to improve the readability and language of the work. After using this tool, the authors carefully reviewed and revised the content to ensure accuracy and quality. The authors take full responsibility for the final version of this publication.

References

1. Hitchcock R (2012) Radio-frequency and microwave radiation. In: Patty's toxicology. John Wiley & Sons, Inc. https://doi.org/10.1002/0471435139.tox101
2. Popescu V, Astanei DG, Burlica R et al (2019) Sustainable and cleaner microwave-assisted dyeing process for obtaining eco-friendly and fluorescent acrylic knitted fabrics. J Clean Prod 232:451–461
3. Adeel S, Hussain M, Rehman FU et al (2019) Microwave-assisted sustainable dyeing of wool fabric using cochineal-based carminic acid as natural colorant. J Nat Fibers 16(7):1026–1034
4. El-Apasery M, Hussein A, Nour El-Din N et al (2021) Microwave-assisted dyeing of wool fabrics with natural dyes as ecofriendly dyeing method: part I. Dyeing performance and fastness properties. Egypt J Chem 64(7):3751–3759
5. Naveed R, Bhatti I, Adeel S et al (2022) Microwave assisted extraction and dyeing of cotton fabric with mixed natural dye from pomegranate rind (*Punica granatum* L.) and turmeric rhizome (*Curcuma longa* L.). J Nat Fibers 19(1):248–255
6. Rehman F, Adeel S, Haddar W (2022) Microwave-assisted exploration of yellow natural dyes for nylon fabric. Sustainability 14(9):5599
7. Arain R, Ahmad F, Khatri Z et al (2018) Microwave-assisted henna organic dyeing of polyester fabric: a green, economical and energy proficient substitute. Nat Prod Res 35(2):327–330
8. Adeel S, Rehman F, Pervaiz M et al (2021) Microwave-assisted green isolation of laccaic acid from lac insect (*Kerria lacca*) for wool dyeing. Prog Color Color Coat 14(4):293–299
9. Adeel S, Naseer K, Javed S et al (2020) Microwave-assisted improvement in dyeing behavior of chemical and bio-mordanted silk fabric using safflower (*Carthamus tinctorius* L.) extract. J Nat Fibers 17(1):55–65
10. Ticha M, Slama N, Dhouibi N et al (2022) Bark residues recovery of *Juglans regia* L. for the dyeing of wool fabrics: development of microwave-assisted extraction and dyeing processes. J Nat Fibers 19(13):5282–5296
11. Wang X, Yi F, Zhang W et al (2022) Optimization of the application of walnut green peel pigment in wool fiber dying and fixing process under microwave-assisted condition. J Nat Fibers 19(13):4854–4867
12. Adeel S, Saeed M, Abdullah A et al (2018) Microwave-assisted modulation of vat dyeing of cellulosic fiber: improvement in color characteristics. J Nat Fibers 15(4):517–526
13. Muneer M, Adeel S, Ayub S et al (2016) Dyeing behaviour of microwave assisted surface modified polyester fabric using disperse orange 25: improvement in colour strength and fastness properties. Oxid Commun 29(2):1430–1439
14. Bensebaa F (2013) Chapter 2—wet production methods. In: Bensebaa F (ed) Interface science and technology, vol 19. Elsevier, pp 85–146
15. Pirkonen P, Ekberg B (2015) Chapter 9—ultrasonic. In: Tarleton S (ed) Progress in filtration and separation. Elsevier, pp 399–421
16. Lowe M (2001) Ultrasonics. In: Braun S (ed) Encyclopedia of vibration. Elsevier, pp 1437–1441
17. Tiwari H, Singh P, Mishra PK et al (2010) Evaluation of various techniques for extraction of natural colorants from pomegranate rind-ultrasound and enzyme assisted extraction. Indian J Fibre Text Res 35(3):272–276
18. Arlene A, Prima K, Utama L et al (2015) The preliminary study of the dye extraction from the avocado seed using ultrasonic-assisted extraction. Procedia Chem 16:334–340

19. Banna B, Mia R, Hasan M et al (2023) Ultrasonic-assisted sustainable extraction and dyeing of organic cotton fabric using natural dyes from *Dillenia indica* leaf. Heliyon 9(8):e18702
20. Nasreen H, Adeel S, Yameen M et al (2023) Green application of ultrasonic waves for extraction of yellow colorant from haar singhar and its colouring behaviour in cotton dyeing. Text Leather Rev 6:18–36
21. Kamel M, El-Shishtawy R, Yussef B et al (2003) Ultrasonic assisted dyeing: III. Dyeing of wool with lac as a natural dye. Dyes Pigm 65(2):103–110
22. Habib N, Adeel S, Ali A et al (2023) Ultrasonic-assisted sustainable pollution free advanced method for isolation of colouring material from Amba Haldi (*Curcuma aromatica*) for wool dyeing. Sci Prog 106(3). https://doi.org/10.1177/00368504231188610
23. Baig U, Khatri A, Ali S et al (2021) Ultrasound-assisted dyeing of cotton fabric with natural dye extracted from Marigold flower. J Text Inst 112(5):801–808
24. Vankar P, Shanker R (2008) Ecofriendly ultrasonic natural dyeing of cotton fabric with enzyme pretreatments. Desalination 230(1–3):62–69
25. Bemcha O, Bains S, Grewal S (2021) Ultrasonic dyeing of wool fabric with aqueous extract of Ratanjot (*Onosma echioides*) natural dye. Environ Conserv J 22(1&2):79–86
26. Kamel M, El-Shishtawy R, Hanna H et al (2003) Ultrasonic-assisted dyeing: I. Nylon dyeability with reactive dyes. Polym Int 52(3):373–380
27. Hassan M, Saifullah K (2018) Ultrasound-assisted pre-treatment and dyeing of jute fabrics with reactive and basic dyes. Ultrason Sonochem 40(Part A):488–496
28. Ferrero F, Periolatto M (2012) Ultrasound for low temperature dyeing of wool with acid dye. Ultrason Sonochem 19(3):601–606
29. Khatri M, Ahmed F, Jatoi A et al (2016) Ultrasonic dyeing of cellulose nanofibers. Ultrason Sonochem 31(3):350–354
30. Silva de Oliveira C, Viera de Oliveira P, Pellenz L et al (2024) Supercritical fluid technology as a sustainable alternative method for textile dyeing: an approach on waste, energy, and CO_2 emission reduction. J Environ Sci 140:123–145
31. Hendrix W (2001) Progress in supercritical CO_2 dyeing. J Ind Text 31(1):43–56
32. Miah L, Ferdous N, Azad M (2013) Textiles material dyeing with supercritical carbon dioxide (CO_2) without using water. Chem Mater Res 3(5):38–40
33. Agrawal B (2015) Supercritical carbon-dioxide assisted dyeing of textiles: an environmental benign waterless dyeing process. Int J Innov Res Creat Technol 1(2):201–206
34. Goñi M, Gañán N, Martini R (2021) Supercritical CO_2-assisted dyeing and functionalization of polymeric materials: a review of recent advances (2015–2020). J CO_2 Utiz 54:101760
35. Zheng H, Zhang J, Yan J et al (2016) An industrial scale multiple supercritical carbon dioxide apparatus and its eco-friendly dyeing production. J CO_2 Utiz 16:272–281
36. Cardozo-Filho L, Mazzer H, Santos J et al (2014) Dyeing of polyethylene terephthalate fibers with a disperse dye in supercritical carbon dioxide. Text Res J 84(12):1279–1287
37. Zheng H, Xu Y, Zhang J et al (2017) An ecofriendly dyeing of wool with supercritical carbon dioxide fluid. J Clean Prod 143:269–277
38. Kraan M, Cid M, Woerlee G et al (2007) Dyeing of natural and synthetic textiles in supercritical carbon dioxide with disperse reactive dyes. J Supercrit Fluids 40(3):470–476
39. Sawada K, Takagi T, Jun J et al (2002) Dyeing natural fibres in supercritical carbon dioxide using a nonionic surfactant reverse micellar system. Color Technol 118(5):233–237
40. Jun J, Sawada K, Ueda M (2004) Application of perfluoropolyether reverse micelles in supercritical CO_2 to dyeing process. Dyes Pigm 61(1):17–22
41. Fridman A (2008) Plasma chemistry. Cambridge University Press, Cambridge
42. Shah J, Shah S (2013) Innovative plasma technology in textile processing: a step towards green environment. J Eng Sci 2(4):34–39
43. Jelil R (2015) A review of low-temperature plasma treatment of textile materials. J Mater Sci 50:5913–5943
44. Hamdy D, Othman H, Hassabo A (2022) A recent uses of plasma in the textile printing. J Text Color Polym Sci 19(1):1–10

45. Zille A (2020) 6—Plasma technology in fashion and textiles. In: Nayak R (ed) Sustainable technologies for fashion and textiles. Woodhead Publishing, Sawston
46. Agnhage T, Perwuelz A, Behary N (2016) Eco-innovative coloration and surface modification of woven polyester fabric using bio-based materials and plasma technology. Ind Crops Prod 86:334–341
47. Zhang C, Zhao M, Wang L et al (2017) Surface modification of polyester fabrics by atmospheric-pressure air/He plasma for color strength and adhesion enhancement. Appl Surf Sci 400:304–311
48. Abdelghaffar F, Abdelghaffar R, Ahmed H (2020) Highly effective surface modification using plasma technologies toward green coloration of polyester fabrics. Environ Sci Pollut Res 27:28949–28961
49. Kerkeni A, Behary N, Perwuelz A et al (2012) Dyeing of woven polyester fabric with curcumin: effect of dye concentrations and surface pre-activation using air atmospheric plasma and ultraviolet excimer treatment. Color Technol 128(3):223–229
50. Dave H, Ledwani L, Chandwani N et al (2012) Surface modification of polyester fabric by nonthermal plasma treatment and its effect on coloration using natural dye. J Polym Mater 30(3):291–304
51. Yaman N, Özdoğan E, Seventekin N (2011) Atmospheric plasma treatment of polypropylene fabric for improved dyeability with insoluble textile dyestuff. Fiber Polym 12:35–41
52. El-Gabry LK (2015) Modification of acrylic fibers improve many physical and chemical properties. In: 3red conference for industrial, textile research, manpower development, manufacturing technologies and management in textile industries NRC, Cairo, Egypt
53. Ghoranneviss M, Shahidi S (2017) Color intensity and wash fastness properties of dyed cotton fabric after plasma treatment. J Text Inst 108(3):445–448
54. Wakida T, Cho S, Choi S et al (1998) Effect of low temperature plasma treatment on color of wool and nylon 6 fabrics dyed with natural dyes. Text Res J 68(11):848–853
55. Rani K, Chandwani N, Kikani P et al (2018) Optimization and surface modification of silk fabric using DBD air plasma for improving wicking properties. J Text Inst 109(3):368–375
56. Palaskar S, Kale R, Deshmukh R (2021) Influence of plasma treatment on dyeing properties of silk weaves. J Nat Fibers 19(13):7087–7099
57. Dutta Majumdar J, Manna I (2015) Laser surface engineering of titanium and its alloys for improved wear, corrosion and high-temperature oxidation resistance. In: Lawrence J, Waugh D (ed) Laser surface engineering. Woodhead Publishing, Sawston
58. Angelova Y (2020) Factors influencing the laser treatment of textile materials: an overview. J Eng Fiber Fabr 15. https://doi.org/10.1177/1558925020952803
59. Angelova Y, Mežinska S, Lazov L (2017) Innovative laser technology in textile industry: marking and engraving. In: Proceedings of the 11th international scientific and practical conference, vol 3. Rezekne: Environment. Technology. Resources, pp 15–21
60. Akiwowo K, Kane F, Tyrer J et al (2014) Digital laser-dyeing for polyester fabrics. J Text Design Res Prac 2(2):133–151
61. Kane F, Shen J, Morgan L et al (2020) Innovative technologies for sustainable textile coloration, patterning, and surface effects. In: Muthu S, Gardetti M (ed) Sustainability in the textile and apparel industries. Springer, Singapore
62. Ahmed N, El-Shishtawy R (2010) The use of new technologies in coloration of textile fibers. J Mater Sci 45:1143–1153
63. Kumar B, Smita K, Angulo Y et al (2016) Green synthesis of silver nanoparticles using natural dyes of cochineal. J Clust Sci 27:703–713
64. Du H, Yue M, Huang X et al (2022) Preparation, application and enhancement dyeing properties of zno nanoparticles in silk fabrics dyed with natural dyes. Nanomater 12(22):3953
65. Bhandari V, Yadav S, Rose N (2021) Optimization of nanoparticle treatments for cotton fabric and dyeing with eco-friendly natural dyes. Pharm Innov J 10(12):95–104
66. Abkenar S, Malek R (2021) Modification of silk yarn with β-cyclodextrin nanoparticles: preparation, characterization, and natural dyeing properties. Starch 73(7–8):2000209

67. Samarawickrama K, Wijayapala U, Wanasekara N et al (2021) Improving dyeing properties of cotton fabrics to natural dyes with cellulose nanocrystals (CNCs). J Nanotechnol 5(1):1–8
68. Bakr M, Taha M, Osman H et al (2021) Novel green printing of cotton, wool and polyester fabrics with natural safflower dye nanoparticles. Egypt J Chem 64(11):6221–6230
69. Maamoun D, Osman H, Nassar S (2014) Cotton/wool printing with natural dyes nano-particles. Int J Environ Sci 9(1):90–99
70. Hebeish A, Shahin A, Rekaby M et al (2015) New environment-friendly approach for textile printing using natural dye loaded chitosan nanoparticles. Egypt J Chem 58(6):659–670
71. Lin C, Lin C (2013) The study in the application of nanotechnology digital printing in clothing pattern design. Key Eng Mater 562–565:674–679
72. Khatri M, Ahmed F, Jatoi A et al (2016) Ultrasonic dyeing of cellulose nanofibers. Ultrason Sonochem 31:350–354
73. Blosi M, Albonetti, S, Gatti, F et al (2012) Au–Ag nanoparticles as red pigment in ceramic inks for digital decoration. Dyes Pigm 94(2):355–362
74. Kumar D, Garje A, Desa K et al (2010) Dyeing without Water. J Textile Assoc 171–174
75. Mahmud I, Kaiser S (2020) Recent progress in waterless textile dyeing. J Textile Sci Eng 10(6). https://doi.org/10.37421/jtese.2020.10.421
76. Abate M, Tadesse M (2021) Airflow, foam, and supercritical carbon dioxide dyeing technologies. In: Haji LR, Shabbir M (ed) Innovative and emerging technologies for textile dyeing and finishing. Wiley, New Jersey
77. Turner J (1980) An introduction to foam finishing. Text Chem Color 12(3)
78. Baker K, Bryant G, Camp JRJ et al (1982) Foam finishing technology. Text Res J 52(6):395–403
79. Ray Maulik S, Pradhan S (2010) Application of foam in textile wet processing. Asian Text J 19(6):69–72
80. Yu H, Wang Y, Zhong Y et al (2014) Foam properties and application in dyeing cotton fabrics with reactive dyes. Color Technol 130(4):266–272
81. Basyigit Z, Kut D, Hauser P (2020). Development of multifunctional cotton fabric via chemical foam application method. Text Res J 90(9–10):991–1001
82. Zhu D, Wan Z, Zhao X et al (2022) Foaming indigo: an efficient technology for yarn dyeing. Dyes Pigm 197:109862
83. Shang S, Hu E, Poon P et al (2011) Foam dyeing for developing the wash-out effect on cotton knitted fabrics with pigment. Res J Text Appar 15(1):44–51
84. Dawson T (1981) Foam dyeing and printing of carpets. J Soc Dyers Colour 97(6):262–274
85. Wang Q, Zhou W, Du S et al (2019) Application of foam dyeing technology on ultra-fine polyamide filament fabrics with acid dye. Text Res J 89(23–24):4808–4816
86. Capponi M, Flister A, Hasler R et al (1982) Foam technology in textile processing. Rev Prog Color Relat Top 12(1):48–57
87. Dawson T (2003) The use of digital systems in textile printing. In: Miles L (ed) Textile printing, society of dyers and colourists, Hampshire
88. Kan C, Yuen C (2012) Digital ink-jet printing on textiles. Res J Text Apparel 16(2):1–24
89. Ray Maulik S, Md Nasiruddin K, Tunnesa Begam R (2019) Creative digital printing on textiles. Asian Dyer 16(2)
90. Tyler D (2005) Textile digital printing technologies. Text Prog 37(4):1–65
91. Leelajariyakul S, Noguchi H, Kiatkamjornwong S (2008) Surface-modified and micro-encapsulated pigmented inks for ink jet printing on textile fabrics. Prog Org Coat 62(2):145–161
92. Gutierrez S, He Y, Cao Y et al (2019) Feasibility and surface evaluation of the pigment from *Scytalidium cuboideum* for inkjet printing on textiles. Coatings 9(4):266
93. Barrett T (2001) Woodblock dyeing and printing technology in china, c. 700 a.D.: the innovations of Ms. Liu, and other evidence. Bulletin of the School of Oriental and African Studies, University of London 64(2):240–247
94. El-Molla M, Schneider R (2006) Development of ecofriendly binders for pigment printing of all types of textile fabrics. Dyes Pigm 71(2):130–137

95. Aspland J (1993) Pigments as textile colorants: pigmenting or pigmentation. Text Chem Color 25(10):31–37
96. Ray Maulik S, Banerjee A (2009) Dyeing of cotton fabric with pigment colour. Asian Dyer 6(1):24–29

Chapter 2
Sustainable Alternatives to Conventional Aqueous Dyeing in Textile Industries

Farzana Faiza, Kazi Rezwan Hossain, Abdullah Sayam, Muntajena Nanjeba, Sadman Shaharier Mallick, Rifat Jahangir, and Md Arif Mahmud

Abstract The textile industry's dyeing process consumes significant water and utilities, leading to environmental concerns due to wastewater disposal. Traditional dyeing methods often result in water pollution, making sustainability a pressing issue. To address this, researchers are developing eco-friendly alternatives. This chapter explores sustainable dyeing techniques, including foam dyeing, supercritical carbon dioxide ($scCO_2$) dyeing, air dyeing, plasma-assisted dyeing, and laser-assisted dyeing. It provides an overview of conventional dyeing methods, highlighting their environmental impact, and examines innovative waterless dyeing solutions, discussing their principles, benefits, and limitations.

Keywords Conventional dyeing · Natural fiber · Synthetic fiber · Sustainable · Color · Efficiency

2.1 Introduction

The textile industry is one of the most polluting industries globally. With the vast globalization of textile industries, pollution has become a significant environmental problem. These environmental issues have raised concern worldwide, and major steps have been taken to reduce pollution as much as possible. The biggest sufferer of the pollution has been water. The textile industry requires an immense quantity of water for dyeing. The textile sector has been identified as responsible for approximately 17–20% of global water pollution [1, 2]. The dyeing portion consumes most of the

F. Faiza · K. R. Hossain · A. Sayam · M. Nanjeba · S. S. Mallick · R. Jahangir
Ahsanullah University of Science and Technology, Dhaka, Bangladesh
e-mail: faiza.te@aust.edu

Md A. Mahmud (✉)
The University of Manchester, Manchester, UK
e-mail: arif.shawon@hotmail.com

M. Shahid et al. (eds.), *Advancements in Textile Coloration*,
https://doi.org/10.1007/978-981-96-5091-0_2

water being used in the textile sector. Dyeing is a technique of coloring textile materials by dipping their materials in a solution encompassing dyes and other coloring chemicals. This process and the pretreatment of textiles before and after treatments consume most of the water. Also, wastewater treatment mainly involves wastewater coming from the pretreatment, dyeing, and finishing section of textiles [3–6]. Water acts as the solvent or the dispersing medium for the dyes used in the textile industries. The dyes need to be passed inside the fibers, and the same theory applies to all other treatments where water mainly works as the medium for carrying the dyes and chemicals. The researchers have found different ways of replacing the water and using other mediums, both in physical and chemical form, to carry the dyes and chemicals into the textile materials, and these are considered the latest technologies of sustainable dyeing systems. These latest waterless dyeing technologies present different techniques that involve different mediums, including plasma, carbon dioxide (CO_2), etc., to complete the dyeing mechanism. These are also helping to reduce wastewater generation, and the overall process is becoming more sustainable when using these new technologies [7–9]. These technologies are still not being used at full scale worldwide for various reasons, including complexity, cost, and many other factors. Still, these present the best solution to the ongoing environmental issues worldwide with textile industries. This chapter will discuss waterless dyeing techniques while examining their principles, benefits, and challenges. It will explore the technological advancements of these techniques as well as try to highlight other important factors that are essential to understanding these new and promising technologies. A brief look at the current industry practices regarding waterless technology will also be discussed in the latter part of the chapter.

2.2 The Conventional Aqueous Dyeing Method

The conventional dyeing process for textile materials contains numerous crucial processes, with water being the principal medium for dye application. Conventional dyeing is a key procedure in the textile industry that uses dyes and chemicals to provide fabrics with color, pattern, and visual appearance [10, 11]. This procedure, although indispensable for textile manufacturing, presents notable environmental and operational difficulties, particularly with water and energy use and chemical handling [12, 13]. Several steps are involved in the traditional dyeing process [14]. Figure 2.1 illustrates the process flowchart of the traditional method of dyeing.

2.2.1 Brief Description of the Conventional Dyeing Steps

- **Preparation of Textile Materials**—This initial phase is crucial since it carefully prepares textile materials, including natural fibers like cotton and wool or synthetic fibers like polyester. The process, often encompassing scouring and bleaching,

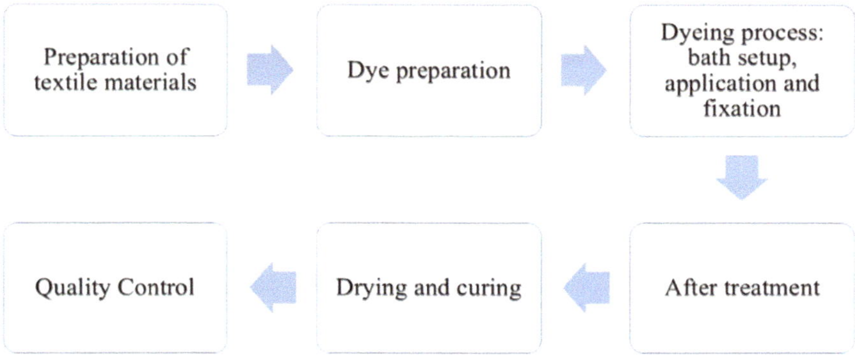

Fig. 2.1 The process flowchart of the traditional aqueous textile dyeing process

aims to eliminate natural oils, waxes, and other contaminants that may obstruct the dyeing method, highlighting the requisite meticulousness [15].

- **Dye Preparation**—The dye is carefully chosen and made into a solution comprising water. The choice of dye is not arbitrary but contingent upon the composition of the textile material. Additionally, cotton employs reactive dyes, wool utilizes acid dyes, and synthetic fibers employ dispersing colors. The dye solution was homogenously blended, guaranteeing uniform coloration and underscoring the importance of this decision in attaining the desired results [16].
- **Dyeing process**—The dyeing process has three primary stages: preparation or bath setup, application, and fixing. Every stage is essential in attaining the intended hue and durability of the dyed textile material [14–17].
  - **Bath Setup**—The material is inundated in a dye bath where water acts as the solvent. The bath parameters, including temperature and pH, are meticulously regulated to maximize dye absorption and align with the specific requirements of the dye and fiber type.
  - **Application**—Dyes are utilized on fabric using different actions, including exhaustion, padding, and continuous dyeing. The exhaustion process comprises the slow dye absorption from the dye bath onto the fibers until equilibrium is reached.
  - **Fixation**—Due to chemical or thermal procedures, the dye molecules are firmly attached to the fiber to guarantee they are permanently connected and will not be removed after washing. For example, a soda ash (sodium carbonate) solution is likely to set the dye permanently after dyeing cotton with reactive dyes.
- **After Treatment**—This stage is crucial as it entails rinsing any unbound dye from the fabric to enhance color durability. Furthermore, it may involve procedures such as applying softeners or finishers to augment specific desirable qualities of the

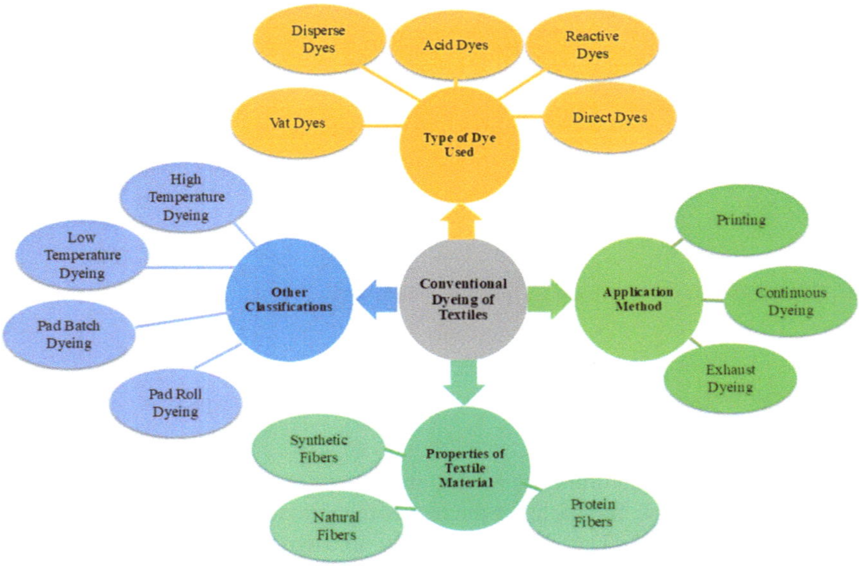

Fig. 2.2 Classification of conventional dyeing in different aspects

fabric. The after-treatment stage is indispensable for ensuring the longevity and quality of the dyed fabric.

- **Drying and Curing**—The textile material experiences a drying procedure following the dyeing and washing process. Curing is a necessary step for certain dyes, involving using heat to enhance the color fully and ensuring the dye is adequately fixed to the fiber.
- **Quality Control**—The last stage of the process entails inspecting the dyed cloth to verify that it adheres to the prescribed requirements regarding color uniformity, resistance to fading, and overall excellence [15, 17, 18].

The classification of conventional textile dyeing can be determined by several measures, such as the specific dye employed, the manner of application, and the characteristics of the textile material being dyed [16]. Figure 2.2 illustrates the classification of conventional aqueous textile dyeing in different aspects.

2.2.2 *Traditional Dyeing Systems*

Traditionally, dyeing involves different dyes applied to the fabrics using various mediums, generally liquids. And the liquid has predominantly been water. All dyes have either been water-soluble or dispersible, essential for dyeing. Direct dyes are acknowledged for their water solubility versatility solution for coloring textile fibers.

They don't entail a mordant and are particularly effective on cellulose fibers like cotton, rayon, and linen. Reactive dyes, enthralling in their chemical connection with textile fibers, are a unique type of dye. They have a special affinity for cellulose fibers and can also dye protein fibers like wool and silk, adding to their intrigue and versatility. Vat dyes need the presence of a reducing agent and an alkaline environment to dissolve and color the fiber, and cellulose fibers are their primary application. Disperse dyes are hydrophobic and are mostly employed for synthetic materials dyeing, such as nylon, polyester, and acetate. Acid dyes are water soluble and are mainly utilized for dyeing protein-based fabrics such as wool, silk, and nylon [14, 16].

Exhaust dyeing is the process of immersing the textile material in a dye bath that contains the dye, allowing the dye to be fully absorbed by the fiber. Continuous dyeing involves the uninterrupted passage of textile material through a dyeing machine, where the dye is applied. Lastly, in printing, dyes are selectively applied to the textile material in predetermined designs using screen printing, roller printing, and digital printing methods [15, 19]. The characteristics of textile materials necessitate the use of various types of dyes. Cellulose fibers originate from botanical sources, including cotton, linen, and rayon. Protein fibers represent textiles originating from animal sources, like wool and silk. Lastly, synthetic fibers refer to textiles produced using artificial polyester, nylon, and acrylic materials [19]. Another type is high-temperature dyeing, which uses water as a medium. High-temperature dyeing is a procedure that involves coloring synthetic fibers by subjecting them to high heat. Low-temperature dyeing is carried out at reduced temperatures and designed explicitly for fragile textiles such as silk. In the pad-batch dyeing technique, the textile material is soaked with dye liquid and then kept in a closed container for a set period. Pad-roll dyeing is a continuous technique in which the textile material is soaked with dye liquid and passed between squeeze rollers to eliminate any surplus dye [10, 11, 20].

Traditional dyeing methods are highly resource-intensive, requiring significant quantities of water and electricity. Processing one kilogram of textile material requires an estimated amount of 100–150 L of water. Due to the large amount of water used and the dye-rich effluents released into the environment, there are significant environmental dangers, such as aquatic life poisoning and water pollution. Energy consumption is an important issue, as a substantial amount of energy is required to heat water for dye baths and then dry the dyed fabrics. The typical dyeing procedures have a significant carbon footprint, which is responsible for releasing greenhouse gases from using fossil fuels [21, 22]. Current studies and technological evolution have prioritized the development of sustainable methods for conventional dyeing of textile materials. Methods such as cold pad-batch dyeing and low liquor ratio dyeing equipment have been formulated to save water and energy usage. These methods strive to augment the dyeing process, reducing the environmental footprint while preserving or enhancing dyeing quality [23]. Environmental apprehensions and the demand for sustainability are anticipated to significantly affect the future of traditional dyeing methods in the textile industry. Innovations in dye chemistry, process automation, and waterless dyeing technologies, such as $scCO_2$ ($scCO_2$) dyeing, have significant promise to transform conventional dyeing practices. $ScCO_2$ dyeing uses carbon dioxide in its supercritical form as the dyeing medium, eliminating the necessity for water and minimizing the environmental effect of the dyeing process. These technologies can mitigate the environmental impact of dyeing processes and enhance their sustainability [22, 24]. The subsequent section of this chapter will address the topic of sustainable alternatives to conventional aqueous dyeing in textile industries.

2.3 Sustainability Issues of Conventional Dyeing

Sustainability in the textile industry represents a non-hazardous, economically viable, non-polluting, and benign environment for all the stakeholders (workers, consumers, communities) [25]. The textile industry is responsible for 20% of worldwide wastewater output and 10% of CO_2 emissions [26]. Among different steps of textile production, dyeing has staggering environmental impacts. The process of dyeing textiles contributes to approximately 17–20% of global water pollution [27]. Different dyes, chemicals, and water are accounted for in conventional dyeing. Various fibrous substances such as cotton, wool, silk, polyester, nylon, and acrylic are categorized into natural, man-made, and mixed. Natural fibers originate from natural sources and are primarily categorized into two kinds: cellulosic fibers (such as cotton, jute, linen, and sisal) and animal fibers (like cashmere, silk, wool, etc.). Man-made fibers can be synthetic polymers (such as acrylic, polyester, and polyamide), regenerated fibers (like acetate, modal, and viscose), and inorganic fibers (including ceramic, glass, and carbon). Each group requires distinct techniques and formulations for dyeing [28, 29].

Most of the widely used chemicals are synthetic and potentially hazardous [30]. Environment-friendly natural dyes account for only 1% of the market share due to restraint of color variation, fastness, and lack of reproducibility [29]. The extensive use of synthetic dyes harms environmental sustainability. Besides, it requires almost 100-150L of water for processing 1 kg of fabric. Subsequently, Manufacturing 4×10^9 pieces of denim fabric requires 2.5×10^9 gallons of water and 1.7×10^9 kg of chemicals, most of which end up as sludge and wastewater [26]. Globally, an estimated 28×10^4 metric tons of dyes are released annually via industrial effluent [31]. Moreover, the wastewater generated during dyeing comprises intricate mixtures of substances, including salt, detergents, peroxides, etc. [32]. The unfixed dyes and salts in the effluent account for higher levels of total dissolved solids (TDS), biological oxygen demand (BOD), and chemical oxygen demand (COD) [33]. If discharged into the atmosphere, the hazardous effluent can penetrate the food chain and be transmitted to individuals through bioaccumulation. [34]. Subsequently, some dyes even require heavy metals like mercury, cadmium, lead, arsenic, or chromium during production and use, which have serious implications for the environment once discharged [35]. Apart from water contamination, further environmental concerns arise from the combustion of fossil fuels to provide energy to dyeing machines, leading to atmospheric emissions [36]. According to the United Nations' Intergovernmental Panel

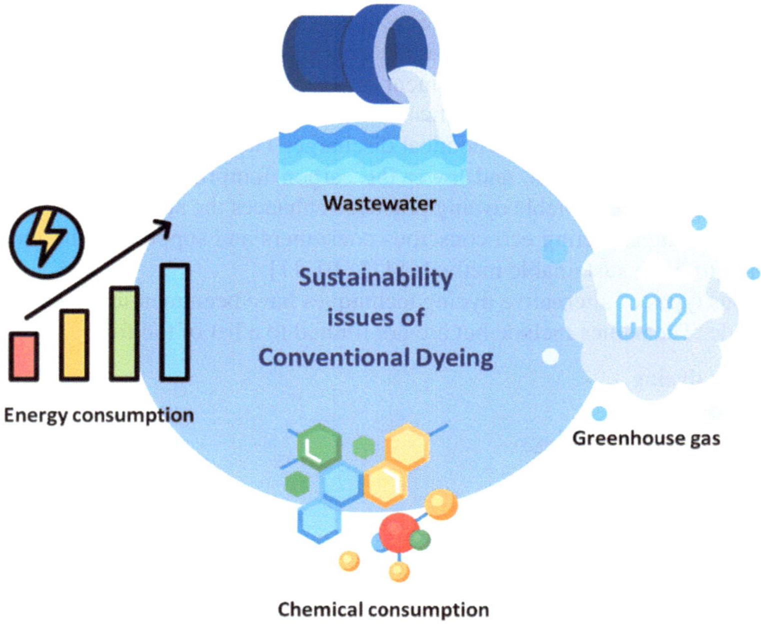

Fig. 2.3 Sustainability issues of conventional dyeing

on Climate Change, the textile industry is responsible for 10% of the total worldwide greenhouse gas (GHG) release, and the textile dyeing step is a main contributor [29]. Additionally, some synthetic dyes have serious consequences for human health, including prickly eyes, allergic skin reactions, wheezing, stuffy noses, shortness of breath, chest tightness, and occupational asthma [27]. So, it can be said that the traditional dyeing process requires a significant amount of water, energy, and chemicals while emitting harmful substances into the air, posing serious threats to both ecology and human health. The following figure represents the sustainability issues of conventional dyeing. Figure 2.3 gives an idea of the sustainability issues of the dyeing industry.

2.4 Alternative Dyeing Techniques: From Lesser Water-Consuming to Waterless Dyeing

In recent years, the textile industry has witnessed a significant shift towards alternative dyeing techniques that aim to reduce water usage and minimize the environmental footprint of textile production. To address the water-intensive nature of traditional dyeing processes, the textile industry has been exploring and adopting lesser water-consuming techniques, focusing on optimizing dye application, improving dye fixation, and enhancing water recycling systems to reduce the overall water footprint of the dyeing process. One of the most significant advancements in sustainable dyeing practices is the development of waterless dyeing technologies. Waterless dyeing technologies offer significant environmental benefits by reducing water use, energy consumption, and chemical discharge. These techniques help conserve resources, reduce pollution, and lower the carbon footprint of textile production. The shift towards sustainable dyeing practices enhances the sustainability profile of textile products, attracting eco-conscious consumers and supporting the industry's transition to more sustainable methods [4, 5, 10, 11].

Various types of alternative dyeing techniques have been mentioned or practiced worldwide. The names include but are not limited to a list of the following:

- Solvent dyeing
- Foam dyeing
- scCO_2 dyeing technology
- Airflow dyeing
- Ozone-assisted dyeing
- Electrochemical dyeing
- Plasma dyeing
- Nanotechnology-based dyeing
- Microwave-assisted dyeing
- Ultrasonic dyeing.

2.4.1 Foam Dyeing

Foam has been increasingly used in textile processing, particularly for dyeing, finishing, and sizing, offering environmental benefits by reducing water and electricity use. The use of foam for dyeing began in the late 1930s when it was discovered to scour and dye clothes simultaneously. This led to the development of the foam dyeing process in the 1970s, with the technique becoming widely used in the textile industry in 1979 [37, 38]. Foam dyeing or finishing adds the dye or finishing chemical to the fabric as foam instead of impregnating the fabric in an aqueous solution. The fundamental idea behind foam dyeing is that a controlled quantity of foam is evenly applied to the fabric utilizing a range of application techniques after foam containing the dye liquid is created in a controlled manner. As the bubbles burst and the dye liquor is transferred to the substrate, the foam is continuously produced and deflated during dyeing. Fixation and washing occur much in the same manner as in the conventional continuous process. The efficient processes of dyeing and dye-fixing are brought about by this constant foaming and draining [39]. Foamed liquor must be applied with the necessary techniques to ensure it is equally dispersed across and along the substrate. The ideal foam should have consistent bubble size, homogeneity, and elegance. To speed up the dye fixation process, the foam must also collapse immediately to allow the dye liquid to permeate the substrate. Batch and continuous methods of foam dyeing are both possible [40]. The foam technique is frequently used with a 20–40% wet pick-up, whereas traditional pad dyeing typically has a wet pick-up in the 60–100% operational range. Using foam to apply chemicals and colorants to textiles is a low-add-on technique used in textile processing that saves a lot of water, shortens the drying time, boosts productivity, reduces waste, and uses less energy because air is used in place of water. By adding air to the formulation in place of some water, foam application reduces the quantity of water added to the textile material. In a foam generator, the air is combined with the chemical formulation to produce foam that typically has the ideal bubble size [39].

Mao et al. [41] examined the states of water present in cotton fibers and the pore volume, pore size distribution, average pore radius, and specific surface area under various moisture level situations. They divided the water found in damp cotton materials into two groups. "Free Water" refers to the water that is found at the fabric's surface and in the spaces between the fibers and threads. Water can flow freely through the fabric assembly, bringing chemicals and dyes with it. Conversely, "Bound Water" refers to the water absorbed in the cotton fibers' pores, where pore walls limit the water's flow. It can transport substances and colors with the right molecular size into the pores. They evaluated the dyeing outcomes of traditional one-bath pad dyeing against low water add-on foam dyeing, considering porosity characteristics and water conditions. They found foam dyeing produced greater color strength and fixation than pad dyeing.

Using a reactive dye and a Neovi-foam technique for low add-on foam dyeing, Yu et al. [42] investigated two-stage pad dyeing of plain woven cotton fabric. Thirty to eighty percent wet pick-up was used for foam dyeing and traditional pad dyeing. The foam was produced by mechanically stirring the mixture for three minutes. The foamability and stability were evaluated by measuring the blow ratio and half-life. Various foaming and stabilizer strengths were combined to create foam coloring. They concluded that foam processing improved the build-up qualities of dyes more than conventional dyeing did by comparing their findings with those of traditional water dyeing systems and analyzing different foaming parameters based on fixing rate and color strength. It was observed that guar gum, dodecanol, and sodium dodecyl sulfate (3:4 in weight) were efficient stabilizers and foaming agents, respectively. When foam stabilizer concentration increased, the dye fixing rate decreased, and the K/S value of the dyed fabric increased. A recent research reported on hydrolyzed keratin foam for foam coloring cotton and wool [43]. The foam generation and durability of hydrolyzed keratin foam are excellent for the reactive dyeing treatment of cotton and the acid dyeing treatment of wool fabrics. It was discovered that acid and reactive dyeing bubble diameters ranged from 0.02 to 0.1 mm and that 40% wet take up of the cloth produced consistent wool and cotton textiles dying. It has been found that the foam dyeing of cotton and wool garments yielded better color strength outcomes than the traditional pad batch procedure. The fastness of rubbing and washing was comparable to traditional pad-dyed textiles.

Foam technique can significantly reduce drying energy costs because it typically uses a very low wet pickup. It has been stated that drying energy costs can be reduced by approximately 50% [44]. Lower drying temperatures and faster processing rates are two more benefits of this energy savings in drying. The decrease in the potential for water use is another major benefit of foam treatment. There was a reported sixty percent reduction in wet pickup compared to pad dyeing and a 30–90% lower water consumption [43]. A 50–60% decrease in the cost of treating wastewater is the outcome of water consumption reduction, reducing the amount of wastewater discharged. The comparison between pad dyeing and foam dyeing cotton showed that foam dyeing produced higher physical qualities, comparable color ability, and significant savings on water, energy, and chemicals [45]. Additional benefits include a simple and quick dying process, reduced energy consumption, better dye diffusion, excellent stability of the colors after diffusion (less color migration), stable dye fixing because of the substrate's quick wetting, and lesser total waste. Furthermore, because existing equipment like dryers, curing chambers or stenters, and standard padders can be used, the foam dyeing method requires very little initial investment. Furthermore, as long as an appropriate foam can be created, all the chemicals used in the traditional textile treatment process can be applied to the foam medium [46].

2.4.2 Supercritical CO_2 ($scCO_2$) Dyeing

The technique of Supercritical Fluid Dyeing (SFD) was pioneered in the late 1980s by Professor Schollmeyer and his research team at the Deutsches Textilforschungszentrum Nord-West e.V. (DTNW) in Krefeld, Germany. Since its introduction, extensive research efforts and experiments have scaled up from the laboratory to the pilot level. Supercritical fluid dyeing offers a compelling alternative to conventional aqueous methods, primarily due to its environmental benefits[47]. Supercritical carbon dioxide is used as the solvent in the SFD dyeing process rather than water.

CO_2 is the preferred and widely adopted supercritical fluid due to its distinct and advantageous attributes. Its affordability, non-flammability, and non-toxic nature make it particularly favorable. Furthermore, its critical temperature, which is at 31.1 °C, and pressure, at 73.7 bar, are lower than other fluids, thus increasing its effectiveness [48]. In the supercritical phase, it displays gas-like viscosity and diffusivity alongside liquid-like density, and its solvent strength can be adjusted by temperature and pressure variations. The $scCO_2$ dyeing process involves key elements, including a dye bath, a tank containing CO_2, a compression system, a dyeing vessel, and circulating pumps. The fabric undergoes treatment within an autoclave, where it undergoes pressurization and heating by prescribed process parameters. After this preparatory phase, the dyeing process is facilitated by utilizing a circulating pump. Following the completion of dyeing, there is a partial reduction in system pressure, accompanied by the rinsing of the dyed fabric with fresh $scCO_2$ to eliminate the residual dye. Following treatment, leftover residual dye and CO_2 are recovered for possible use in later dyeing processes [47]. The initial trials at the UHDE dyeing plant in 1995 demonstrated that dye uptake was significantly higher than results from a static, stirrable dyeing apparatus. In static systems, achieving similar color absorption levels as in the UHDE plant required the addition of 1% (w/v) cosolvents like water, or acetone, alcohols, as specifically shown with CI Disperse Blue 79. This suggests that results from static dyeing systems might not accurately reflect the dyeing behavior seen in dynamic systems with optimal carbon dioxide (CO_2) circulation. The dyeing process greatly influences levelness, dye partition, and exhaustion.

Fluid movement significantly impacts supercritical fluid dyeing processes. Without sufficient movement of CO_2, it may take 2–3 days to reach dyeing equilibrium, highlighting that mass transport of dyes into fibers is a primary resistance factor. However, diffusion in CO_2 is accelerated due to its lower viscosity and the swelling of the polymer caused by CO_2 absorption. The role of CO_2 in enhancing solute mass transport within the swollen polymer matrix is crucial; it acts as a solvent for the solutes, facilitating their diffusion through molecular lubrication. In a static system, cosolvents primarily help by swelling the fiber, thus promoting mass transport. They also increase dye solubility in CO_2, enabling supersaturation and subsequent dye precipitation on the fiber surface when the pressure drops. On the other hand, dyes are constantly delivered to the fiber in a dynamic system, preserving high surface concentrations during the dyeing procedure. Controlled supersaturation via pressure and temperature adjustments can further optimize this, potentially reducing dyeing times and matching the dye uptake of a dynamic system without cosolvents to that of a static system with cosolvents. Ultimately, these findings, corroborated by earlier research by Saus et al. in 1993, illustrate that for optimal color yields in CO_2 dyeing processes, especially with PET fibers, neither cosolvents nor fiber modifications are necessary. This underscores the efficiency of CO_2 in dyeing processes, facilitating quicker, more effective dye uptake and distribution within the fiber [48]. Following extensive testing with the UHDE plant, a refined standard CO_2 dyeing process was established, detailing all essential parameters for scaling up (Fig. 2.4). In some cases, the process specifically addresses the issue of spinning oils in fabrics. Excessive removal of these oils can produce undesirable stiffening of the fabric's texture. Therefore, the initial extraction step using cold CO_2 at the dyeing process's conclusion aims to eliminate unfixed dyes and rapidly lower the polyester (PET) temperature below its glass transition temperature. This careful management prevents the extraction of dyes that have bonded with the fiber. Interestingly, reducing pressure rather than temperature might also be effective for certain dyes. However, this approach is currently limited to dyes that are not ideally suited for CO_2 dyeing. Such dyes have high solubility in CO_2 across various temperatures and pressures, posing challenges in cleaning the equipment and achieving even dye distribution due to their low affinity for PET fibers.

Research in dyeing processes, particularly focusing on polyester with CI Disperse Orange 30, has explored various operational factors' impacts. Studies using factorial design techniques have examined variables like dyeing time, temperature, pressure, and dye-to-fiber ratio. They found that these factors positively influence color strength (K/S), with temperature significantly enhancing wash fastness. Industrial applications using supercritical carbon dioxide have also explored mixing different dyes, such as CI Disperse Orange 30, CI Disperse Red 167, and CI Disperse Blue 79, noting compatible trichromatic dye mixes. These studies indicate that adjusting operational conditions can vary colors effectively without altering dye concentrations. Research

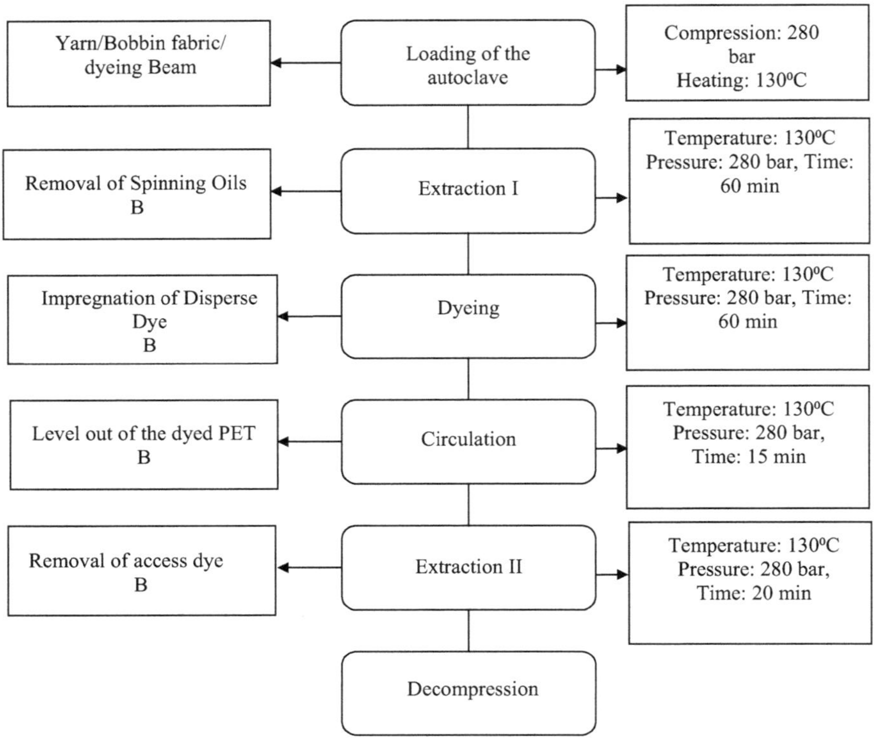

Fig. 2.4 PET dyeing process in supercritical CO_2. Reproduced with permission from Ref. [48], Copyright John Wiley and Sons, 2010

on supercritical processing has shown that altering dye particle size and physicochemical properties can influence dyeing kinetics and distribution, highlighting the role of computational fluid dynamics in optimizing dyeing processes. Research in Supercritical Fluid Dyeing (SFD) focuses on developing new dyes for natural textiles and alternatives for PET fibers, including synthetic dyes like anthraquinone and azo dyes, as well as natural compounds. These efforts aim to enhance dye solubility, reduce costs, and incorporate functional properties such as hypoallergenic, antibacterial, or fluorescent features. $scCO_2$ eliminates the need for pre-treatment and mordants, offering environmental and economic benefits. Promising results with curcumin-based dyeing, including simultaneous extraction and dyeing using turmeric, have shown improved dye exhaustion and fastness compared to aqueous methods. However, challenges like moderate light fastness and its decline at higher dye concentrations highlight the need for further advancements [49].

Understanding dye solubility in $scCO_2$ is crucial for designing efficient dyeing systems. The choice of dye impacts process temperature, pressure, and optimal conditions. Thus, detailed knowledge of solubility in supercritical fluid mixtures is essential. Mathematical models, including Equation of State (EOS) and Density-Based

Models, are commonly used to predict dye solubility. Density-based models, such as the Chrastil model and Kumar and Johnston (K-J) models, offer simplicity and require variables like pressure, temperature, and Super-critical CO_2 density. Thermodynamic models like the Peng-Robinson (P-R) model are also employed, though their use may require estimated values due to limited experimental data. Despite their wide use, EOS models have limitations, including difficulty determining solute-fluid interaction factors. Numerous studies have explored dye solubilities in $scCO_2$, with significant improvements observed at higher pressures. Researchers have developed robust models, such as artificial neural networks (ANN) and semiempirical models, to predict solubility, considering factors like pressure, temperature, and dye properties. Some studies have correlated dye solubility with properties like molecular weight and dye structure, finding consistent relationships [50]. Semi-crystalline polymers like PET fibers experience a plasticization effect during $scCO_2$ dyeing, as CO_2 penetrates amorphous regions, increasing chain mobility and reducing the glass transition temperature by 20–30 °C. This facilitates dye molecule diffusion. Adding cosolvents enhances polymer swelling and CO_2 absorption, improving the plasticizing effect. CO_2 absorption also affects the polymer's physical and mechanical properties, with amorphous polymers showing significant morphological changes while high-crystallinity textiles remain more stable. Heat setting is crucial to preserving fiber structure. Above 100 °C, oligomer migration can impact dye quality, necessitating effective cleaning of dyeing equipment [47]. In $ScCO_2$ dyeing, temperature, pressure, and CO_2 density affect dye solubility, diffusion, and fiber affinity. Efficient dye uptake occurs at high equilibrium partition coefficients (Keq). Disperse dyes are temperature- or density-controlled, with solubility decreasing at lower temperatures or reduced CO_2 density. Proper control prevents dye precipitates that harm colorfastness. Lower molecular weight dyes show higher uptake due to better mobility and pore accessibility [50]. Supercritical dyeing achieves faster mass transfer than aqueous methods due to simultaneous diffusion within the polymer and fluid phase. Diffusion in polymers like PET is enhanced by $scCO_2$, which swells and plasticizes the polymer, improving dye uptake. Factors such as temperature, pressure, cosolvents, and dye structure influence diffusion rates, with smaller molecules moving more easily. The lower activation energy for dye diffusion in $scCO_2$ facilitates easier dye movement. While external mass transfer resistance is minimal, interactions between dye molecules, the polymer, and fluid dynamics play a key role in solute dispersion and uptake [47].

$ScCO_2$ has emerged as a promising method for dyeing natural fibers, although it poses unique challenges compared to dyeing polyester (PET), which is relatively straightforward. The challenge arises from polar dyes typically utilized in conventional water-based dyeing methods exhibiting almost no solubility in $scCO_2$. Consequently, the interaction between the polar natural fibers and dispersed dyes is hindered, as $ScCO_2$ fails to adequately disrupt the hydrogen bonds within the fibers, limiting dye diffusion into their internal structures.

To address this challenge, researchers have proposed two main methods.

- One approach involves reverse micellar techniques, where polar dyes are solubilized in $ScCO_2$ using water/CO_2 microemulsions.
- An alternative approach entails the creation of innovative reactive dyes involving the initial physical treatment of the fiber and utilizing suitable cosolvents. One study has explored the reverse micellar technique for dyeing natural fibers in $ScCO_2$, aiming to overcome these limitations.

The other ways are

- Use of impregnating agent: To enhance the diffusion rate of dyes into fiber pores, impregnating agents like polyethylene oxide or PP glycol are employed. These agents facilitate the dissolution of dispersed dyes in supercritical carbon dioxide ($ScCO_2$) and disrupt hydrogen bonds by swelling the fiber, thereby facilitating dye access. However, drawbacks include needing aqueous mediums during impregnation, necessitating two drying processes, compromised fastness quality, and substantial color strength reduction post-washing.
- Use of cosolvents: Researchers have explored the use of cosolvents to enhance the polarity and dissolving capabilities of supercritical carbon dioxide ($ScCO_2$). A study was conducted on initial investigations into using cosolvents, including alcohol and water, for the dyeing treatment of natural textiles with dispersed dyes in $ScCO_2$. Subsequent studies have further explored the potential of cosolvents in this context. However, a notable drawback of this approach is the challenge of cleaning during the dyeing process, as the dye concentration in the fluid phase increases over time, necessitating dye changes. Additionally, unlike conventional $ScCO_2$ dyeing processes, drying is required post-dyeing to eliminate the cosolvent from the fiber [50].

Dyeing natural fibers with supercritical fluid poses a notable challenge due to the unique properties of supercritical carbon dioxide ($ScCO_2$). Its lack of polarity hinders its effectiveness in coloring hydrophilic textiles while making it a favorable solvent for dyeing hydrophobic fibers. However, natural fibers' hydrophilic nature and polar components lead to poor color retention when dyed with ScCO2, as the interaction between hydrophobic disperse dyes and the fibers is limited. Researchers propose employing modified disperse reactive dyes with particular functional groups to color natural fabrics with $scCO_2$ to get over this restriction. Long et al. demonstrated how reactive dispersion dyes with acyl fluoride functional groups may successfully dye cotton fibers in $scCO_2$, leading to noticeably improved dye fixing. The development of covalent connections between the cotton fibers and the functional groups of the reactive dispersion dyes was credited with this enhancement. According to reports, staining, dryness, and wetness, all have satisfactory color fastness scores of 4–5.

Further research is crucial to fully grasp how dyes dissolve in $scCO_2$ and other cosolvent processes. Exploring the potential of ANNs in predicting dye solubility in $scCO_2$ processes is imperative, given their capability to provide dependable and precise forecasts without the need for complex mathematical models. Leveraging the turning properties of $scCO_2$ is essential; without changing the amount of dye initially applied to the dyeing vessel, a variety of dye hues can be produced by making little changes to the temperature or pressure.

Innovative methods, such as supercritical fluid spray dyeing and incorporating electrophoresis into supercritical fluid dyeing processes, are being investigated as alternatives to current methods. The latter has the potential to facilitate dyeing of natural fibers. Additionally, there's a need for further exploration into enhancing the mechanical properties and colorfastness of synthetic fibers like PP, PLA, and Aramid. Blended fibers, both synthetic and natural fabrics, could be advantageous for dyeing operations. Introducing a dyeing method capable of dyeing blends of PET/cotton fibers simultaneously in one phase would represent a significant innovation in textiles [50].

2.4.3 Air Flow Dyeing

The advancement of eco-friendly technologies in the textile sector is imperative due to the environmental issues associated with traditional dyeing methods, which are characterized by significant water and energy consumption and substantial pollutant discharge. One such innovative technique is Airflow dyeing, where air is employed instead of water to transfer dyes onto fabrics, dramatically reducing the dyeing process's ecological footprint. Introduced by companies like Colorep in California, Airflow dyeing minimizes water use by atomizing dye mixtures conveyed by air, directly contacting the fabric. Other companies like AirDye(R) and ColorZen have extended the application of this technology to synthetic and cotton fibers, respectively, achieving reductions in water and energy use and decreasing greenhouse gas emissions [8]. Despite its environmental benefits, implementing Airflow dyeing technologies is limited by high installation costs and the current applicability to specific fabric types. However, advancing research and development aims to improve these technologies' feasibility and application range.

The Airflow dyeing technology represents an innovative advancement in textile dyeing by employing air circulation rather than water to facilitate the dyeing process. This method shares similarities with conventional jet dyeing techniques [51]. Originating from the innovative work of Wilhelm Christ and THEN Machinen GmbH in 1978, the concept gained traction with the unveiling of the first airflow dyeing machine at ITMA in 1983 in Milan. This process has evolved significantly since then, catering to industry demands and revolutionizing ecological dyeing practices. The THEN Airflow range, renowned for its efficacy, has emerged as a global leader in airflow dyeing machinery. Following a merger in 2010, THEN Machinen GmbH was rebranded as Fong's Europe GmbH, further solidifying its position in the market [52]. The fundamental concept of airflow dyeing resembles that of jet dyeing; however, in airflow dyeing technology, fabric circulation is facilitated by the air, rendering it simpler and comparably swifter than traditional jet dyeing techniques. This approach embraces aerodynamics, employing a flow of air or gas to convey fabric, as opposed to the liquid stream utilized in the conventional jet dyeing apparatus [52]. In these state-of-the-art airflow or air-jet machines, the dyeing process relies on an aerodynamic system, where fabric is carried by a flow of air or gas instead of the traditional liquor stream found in the conventional jet dyeing process. Consequently, these devices are commonly known as air transport dyeing machines, and they are accessible in both high-temperature (HT) and atmospheric variations. In the air jet machines, dyeing cycles are quicker; dyes and chemicals are mixed into the airflow. Once the dyeing process is complete, the used dye solution is drained at a high temperature to speed up cooling. This results in consistent, uniform shades with enhanced colorfastness properties [53].

In a typical airflow dyeing machine, the fabric undergoes a meticulous process to ensure optimal dye penetration and uniform coloration. Firstly, the fabric is loaded in a rope-like formation, minimizing tension and creases. Pressurized air is used to propel the fabric into the nozzle region, priming it for dyeing. Following this, a combination of dye and chemicals is introduced into the flowing air, creating an aerosol mist distributed across the fabric's surface while passing through the dye nozzle. This swift application method guarantees comprehensive dye penetration, yielding outstanding coloration. The fabric remains in constant motion throughout the process, further enhancing uniformity. Excess dye liquor is collected and recycled for future batches, contributing to sustainability efforts. Continuous rinsing is conducted using a spray-type washing system, significantly reducing water usage and rinse time. After rinsing, a bottom valve allows for the discharge of water without additional fabric contact, facilitating the removal of hot bath liquor post-dyeing. This comprehensive approach to airflow dyeing ensures quality results and prioritizes efficiency and environmental conservation.

After Wilhelm Christ's initial patent, advancements in airflow dyeing machines have rapidly progressed. These machines signify the most recent advancement in conventional jet dyeing technology, outperforming overflow and soft flow dyeing techniques. Functioning as batch-type systems, airflow dyeing machines circulate both the dye bath and the material undergoing dyeing, leading to substantial water conservation, chemicals, time, and energy usage compared to conventional jet dyeing methods. All the modern machines are versatile, available in both high-temperature and atmospheric versions, and cater to various fiber types. Manufacturers have focused on enhancing loading capacity, operational speed, and automation features while ensuring gentleness on fabrics. The THEN Airflow Synergy stands out among the latest models, offering low liquor ratios, energy savings, reduced processing times, and adaptability to diverse fiber types and fabric weights. These advancements underscore the industry's commitment to efficiency, sustainability, and superior dyeing quality.

Airflow dyeing machines provide a notable benefit by using minimal water for dissolving dye liquor, as they utilize air or steam propulsion rather than submersion of the fabric. This leads to significantly reduced liquor ratios, with recent progress achieving ratios as low as 1:2 to 1:4 or 5 for synthetic and natural fibers, much lower than conventional jet dyeing techniques. This reduction in water usage has immediate cost benefits and contributes to faster heating and cooling processes and decreased

Table 2.1 Liquor ratio for different types of dyeing

Apparatus	Liquor ratio
Winch beck	1:15–1:40
Overflow	1:12–1:20
Airflow	1:2–1:5
Soft flow	1:4–1:10 (fabric), 1:6–1:20 (carpet)

reliance on salt and auxiliary chemicals. Additionally, lower water volumes reduce dye hydrolysis, thereby enhancing dye yield. Table 2.1 shows the used liquor ratio for different types of dyeing.

Continuous rinsing in airflow dyeing machines enhances efficiency by separating rinse water from the fabric, potentially saving time and facilitating the release of dye bath liquor after high-temperature processing. When coupled with reduced liquor ratios, this continuous rinsing can yield water and heating energy savings of up to 50% compared to traditional hydraulic systems. Moreover, significant reductions in chemical, salt, and dyestuff usage contribute to the sustainability of this technology as a substitute for traditional dyeing techniques.

Comparative data demonstrate the efficiency of airflow dyeing, with significant reductions in auxiliary usage, salt, and steam compared to conventional methods. Although airflow machines consume relatively more electricity, they offer processing benefits such as reduced fabric tension and minimized creasing, ensuring uniform coloration. Additionally, airflow dyeing accommodates a wide range of fibers without requiring modifications to the machine, making it versatile and adaptable. Although the upfront expenses may be greater, the quick return on investment resulting from reduced water, chemicals, and energy expenditures highlights the financial feasibility of airflow dyeing machines. Overall, the driving forces behind adopting such technology remain high productivity and superior dyeing quality, aligning with sustainable practices in the textile industry [52].

2.4.4 Ozone-Assisted Dyeing

Ozone-assisted dyeing is an innovative and sustainable textile technique that utilizes ozone's oxidizing properties to reduce water consumption and environmental impact. This method has gained prominence as the industry shifts toward eco-friendly practices. Recent research highlighted the effectiveness of ozone in decolorizing reactive dyes on cotton and polyester, demonstrating its potential for color stripping and sustainable dyeing [54]. The ozone-assisted dyeing process typically involves ozone generation, application to the textile material, oxidative decolorization of dyes, and subsequent reapplication of the desired color. The optimization of ozone treatment for color-stripping reactive dyes and the incorporation of ozone in electrochemical processes for dye removal showcase the versatility and efficiency of ozone in textile coloration [55]. Environmental studies have focused on the ecological profile of ozone decolorization processes, emphasizing the benefits of ozone-assisted dyeing in reducing water usage, chemical consumption, and the overall ecological footprint. Ozone in textile dyeing conserves water resources and minimizes wastewater generation and chemical pollution, aligning with the industry's sustainability goals [56].

Ozone-assisted dyeing is a cutting-edge technique that leverages the powerful oxidizing properties of ozone to achieve efficient and sustainable coloration of textile materials. This innovative process involves several key steps that contribute to its effectiveness and eco-friendliness.

- **Generation of Ozone**: The first vital move of the ozone-assisted dyeing technique is ozone generation. Ozone gas is produced using specialized ozone generators that convert oxygen molecules (O_2) into ozone molecules (O_3) through an electric discharge within oxygen sourced from pressurized liquid oxygen containers. The transition of ozone from a gaseous state to a liquid state makes it difficult to achieve ozone dissolution in water within the reactor. Different methods of gas dispersion, such as diffusers, static mixers, and injection systems, are commonly utilized in practical applications for this purpose [54, 56]. A pilot ozonation is shown in Fig. 2.5.
- **Application to the textile material**: Once ozone is generated, it is applied to the textile substrate in the dyeing system. This can be achieved through various methods, including bubbling ozone through the dye bath or directly imparting it to the textile material. The interaction between ozone and the textile material

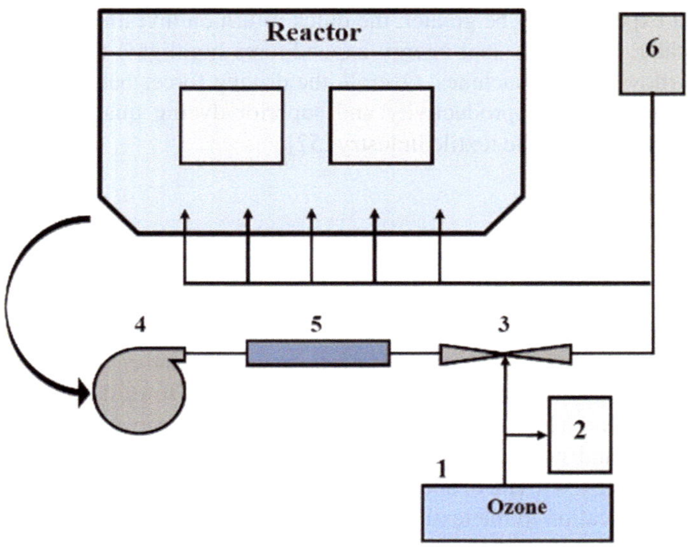

Fig. 2.5 Pilot ozonation (**1**. Ozone Generator; **2**. Analyser Ozone; **3**. Venturi Injection System; **4**. Circulation Pump, **5**. Filter, **6**. Dissolved Ozone Analyzer and pH Meter). Adapted with permission from Ref. [54]

initiates the oxidative decolorization process, where ozone molecules break down the chemical structure of dyes on the fabric.

- **Decolorization of dyes**: Decolorizing dyes on the textile material is a key outcome of the ozone treatment. By effectively breaking down the dye molecules, ozone facilitates the removal of existing colors from the fabric, preparing it for subsequent dyeing processes. This decolorization step is essential for achieving vibrant and consistent color application in the final textile product [55, 56].
- **Reapplication of the desired dye**: Following the decolorization phase, the fabric is ready for reapplication of the desired dye. This step involves conventional dyeing techniques, such as exhaust or continuous dyeing, to impart the desired color to the fabric. The ozone-assisted pretreatment ensures that the new dye penetrates the textile fibers effectively and binds securely, enhancing colorfastness and durability [57, 58].

Ozone-assisted dyeing techniques play a crucial role in the surface characteristics and properties of textile fabrics. This innovative approach to coloration enhances the sustainability of the dyeing process and influences the physical and chemical attributes of treated textiles. One of the notable effects of ozone-assisted dyeing is the improved dye fixation and colorfastness on the fabric surface. The ozone pretreatment helps to break down the chemical structure of existing dyes, allowing for improved penetration and binding of new dye molecules to the textile fibers. This enhanced dye fixation results in improved color retention and resistance to fading, even after repeated washing and exposure to various environmental factors. Studies have demonstrated the effectiveness of ozone treatment in stripping reactive dyes from cotton fabrics and preparing the surface for re-dyeing. The ozone-assisted decolorization process ensures that the new dye can be applied more evenly and consistently, leading to a more vibrant and uniform color appearance on the fabric [59, 60]. In addition to improved coloration, ozone-assisted dyeing can also influence the surface morphology and texture of the textile material. The oxidative action of ozone can slightly change the surface qualities of the fibers, potentially leading to changes in the fabric's hand, drape, and overall feel. However, these changes are typically subtle and can be controlled by optimizing the ozone treatment parameters. Furthermore, ozone-assisted dyeing can affect the fabric surface's hydrophilic properties. The oxidative nature of ozone can increase the wettability and absorbency of the textile, which can be beneficial for certain applications, such as improved dye uptake or enhanced moisture management properties [56]. A few studies have explored the impact of ozone treatment on the hydrophobic and hydrophilic characteristics of cotton fabrics, highlighting the potential to tailor surface properties by optimizing the ozone-assisted dyeing process [55]. Ozone-assisted dyeing techniques have a multifaceted impact on textile fabrics' surface characteristics and properties. From enhanced dye fixation and colorfastness to modifications in surface morphology and wettability, this innovative approach to coloration offers a range of benefits that enhance the overall performance and sustainability of the final textile product [61]. Ozone-assisted dyeing has found diverse applications in textiles, offering an eco-friendly approach to coloration. The utilization of ozone in textile finishing has been reported in various domains, showcasing the versatility and effectiveness of this innovative technique. One significant application of ozone-assisted dyeing is the reduction clearing of dispersed dyed polyester. Ozone has been effectively utilized to remove color from dispersed dyed polyester fabrics, demonstrating its capability to remove existing colors and prepare the fabric for re-dyeing processes. Another notable application is ozone-assisted dyeing, which is applied to the color stripping of reactive dyes from cotton fabrics. The application of ozone treatment for decolorizing reactive-dyed textiles emphasizes the effectiveness of ozone in color-stripping processes. Studies have also shown that ozone treatment can effectively decolorize reactive dyes on cotton fabrics, providing an eco-friendly method for color removal and fabric preparation [55, 62]. Moreover, ozone-assisted dyeing has been utilized in the discharge printing of reactive-dyed cotton fabrics. Ozone treatment has been applied to selectively extract the dye from reactive-dyed cotton fabrics, enabling the creation of intricate patterns and designs through discharge printing techniques. Ozone has been used in the bleaching of raw cotton textiles, showcasing its potential to enhance the whiteness and brightness of cotton fabrics through an environmentally friendly bleaching process. The potential of an ozone treatment to lower the color concentration in wastewater from reactive dyes showcases the potential of ozone-assisted dyeing in wastewater treatment and color removal [63]. The applications of ozone-assisted dyeing are diverse and impactful, offering a sustainable and eco-friendly solution for coloration and fabric treatment in the textile industry. By harnessing the oxidizing power of ozone, textile manufacturers can achieve vibrant colors, efficient color stripping, and enhanced fabric properties while minimizing water consumption, chemical usage, and environmental impact [64]. Ozone-assisted dyeing in the textile industry has significant environmental benefits, making it a more sustainable alternative to traditional dyeing methods. One of the primary advantages of ozone-assisted dyeing is its water-saving potential. Traditional dyeing processes are notorious for their high water consumption, often requiring large volumes of water for dyeing, rinsing, and washing. Ozone-assisted dyeing significantly reduces the amount of water required, eliminating the need for extensive rinsing and washing steps. This water conservation helps to alleviate the strain on water resources and reduce the industry's overall water footprint. In addition to water savings, ozone-assisted dyeing also minimizes the use of harsh chemicals and auxiliaries. The process relies on the oxidizing power of ozone, reducing the need for traditional chemical-intensive dyeing methods. This, in turn, reduces the generation of hazardous wastewater and chemical pollution, as fewer harmful substances are discharged into the environment [63]. The improved dye fixation achieved through ozone-assisted dyeing also contributes to its environmental benefits. Enhancing the binding of dyes to the textile fibers reduces dye loss during subsequent washing and finishing steps, leading to less dye-contaminated wastewater and a lower overall chemical footprint. Furthermore, ozone-assisted dyeing has been found to be more sustainable compared to traditional dyeing methods. The optimization of the ozone generation process can result in reduced energy consumption, further lowering the carbon footprint associated with textile production [65]. The environmental profile of ozone-assisted dyeing has been highlighted in several studies for its advantages in water conservation, wastewater reduction, and chemical footprint minimization. These studies demonstrate the significant potential of this technique in contributing to the textile industry's sustainability goals. The adoption of ozone-assisted dyeing has a profound significance on the environmental performance of the textile field. By reducing water usage, chemical consumption, and energy requirements while improving dye fixation and waste reduction, this innovative technique represents a significant step towards a more sustainable and eco-friendly future for textile coloration [56].

2.4.4.1 Challenges Associated with Scaling up Ozone-Assisted Dyeing

Scaling up ozone-assisted dyeing presents multiple challenges, including process optimization, infrastructure requirements, energy consumption, wastewater management, regulatory compliance, and cost considerations. Figure 2.6 outlines these key challenges and their interconnected aspects. Addressing these challenges through research, innovation, and collaboration with experts in ozone technology and textile processing is essential for successfully scaling up ozone-assisted dyeing for commercial production [66].

2.4.5 Electrochemical Dyeing

Textile industries consume a lot of water and chemicals to make clothes, which creates a big problem. When they throw away the water they use, it can harm the environment [67]. Even a tiny bit of dye in the water can block sunlight from reaching plants and animals in rivers and lakes, hurting them. Some dyes and the things they break down into can even be poisonous and cause cancer. To clean up the water, factories can use different methods like filters, chemicals, and bacteria. However, these methods encounter challenges [68]. They can be expensive, not work well enough, or cause more pollution. Regular filters and bacteria don't always eliminate all the color and tiny bits of bad stuff in the water. So, scientists are trying to find better ways to clean up the water from textile factories [69]. To fix the problems of using too much water and energy and causing a lot of pollution when dyeing a fabric with different dyes, we need to find new ways like using electricity, ozone, sunlight, sound waves, special chemicals, etc. However, among all these methods, fabric dyeing that consumes less energy and water and generates less waste is considered the most efficient [70]. People who work in textile factories around the globe are very interested in searching for ways to dye the fabric without using a lot of salt. In recent years, considerable focus has been given to dyeing a fabric with different dyes (especially reactive) without using much salt. The key to dyeing a fabric without salt is figuring out how to make the dye stick to the fabric even without using salt [71]. There are a few ways to do this: changing the way the fabric and the dye interact, making special dyes that work better without salt, or using a special liquid that can make the fabric and the dye stick

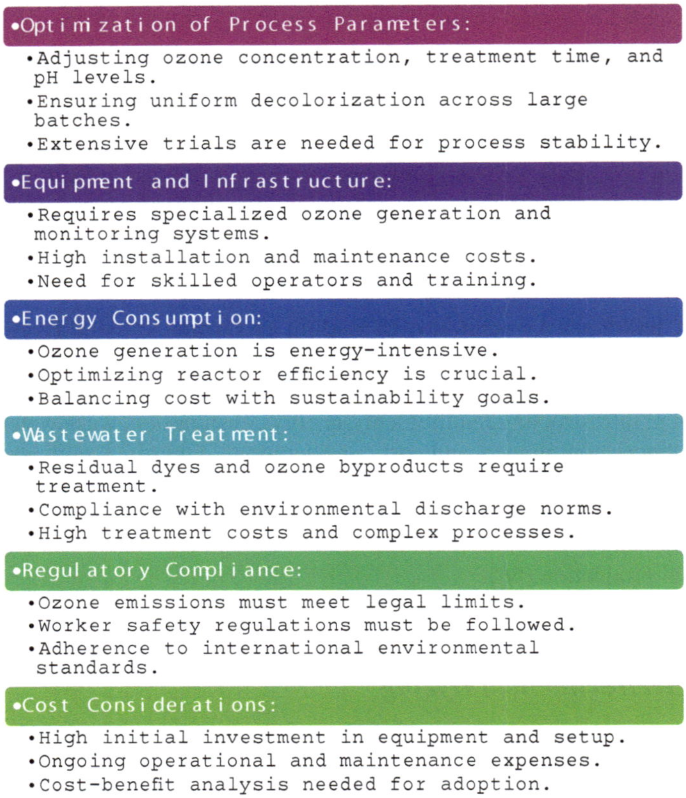

Fig. 2.6 Challenges in scaling up ozone-assisted dyeing

together better. This is important because it can aid in reducing both the consumption of water and energy used, as well as the pollution caused by dyeing different fabrics [72, 73].

Over thirty years ago, a person named Daruwalla tried to change the way vat dyes were used in fabrics. He didn't use electrochemical dyeing, but he did try using electricity to help dye fabrics [74]. He wanted to see if he could use electricity to make a chemical called sodium dithionite stronger. Sodium dithionite is needed to help vat dyes change color in fabrics. When electricity is added, sodium dithionite turns into something even stronger, which allows dye fabrics better. This stronger form of sodium dithionite is made at the cathode, a special part of the electric setup. The cathode changes sodium dithionite into a new form that bears a high reducing power that can help dye the fabrics even more. This change happens because of certain conditions like the amount of sodium dithionite, the pH level, and the temperature [74–76]. Another scientist named Cermak explained that this change happens because sodium dithionite breaks down into a free radical ion. This ion has more power to help dye the fabrics than regular sodium dithionite. However, there arose an issue. After using this method to dye fabrics, the chemicals couldn't be used again. So, scientists thought of a new idea. They came up with a way to use electricity to dye fabrics and still be able to use the chemicals again. This is called electrochemical dyeing. This technique involves the electrochemical reduction of dyes that can be reduced. This reduction occurs in the presence of a cathode, comprised of an electrically conductive base and a cathodically polarized layer that develops in situ, enabling the reduction of organic molecules [77]. All of the traditional hydrogenation metals are ideally useful, but the metals in the Periodic Table's first, second, and eighth transition groups are particularly valuable. Particularly desirable elements include Ag, Cd, Cu, Co, Ir, Fe, Ni, Os, Pb, Pt, Pd, Ru, Re, Rh, and Zn. There are two ways to do this dyeing. They are direct and indirect electrochemical dyeing [74, 78].

When organic dyes undergo direct reduction upon connection with the electrode exterior, it is termed direct electrochemical dyeing. However, in practical application, a traditional reducing agent partially reduces the dyestuff initially, followed by a complete reduction through an electrochemical process. This technique improves the reliability of reduced dye. To start the process, a minute quantity of a soluble reducing agent is essential to generate an initial leuco dye amount via a traditional reaction. Once initiated, the process becomes self-sustaining. This approach proves effective, particularly with sulfur dyes. However, due to limited contact between the cathode and dye particles, a higher dye concentration is required than traditional reduction processes to achieve desired shades. In this system, the dye bath's capacity is restricted, necessitating direct contact between dyestuff particles and the electrode surface for reduction. Later, the previously reduced dyestuff molecule is oxidized by ambient oxygen in the dye solution. The surface area of the cathode must be substantial as the dye requires reduction directly on the cathode surface, posing a constraint. Graphite granules were also used as an electrode material to raise the planar location in a set and fluidized bed reactor to investigate how a direct electrochemical reduction for vat dyes is feasible commercially. The system was calibrated using an indigo concentration of 10 g/l. To attain the concentration of vat and other dye types, they recommend a particular treatment that involves introducing a quinine-like functional group, which can be used on the graphite exterior [79–83].

In 1993, Bechtold obtained a patent for the reduction process for the indirect electrochemical dye and has since introduced other related innovations. In contrast to direct electrochemical reduction, where the dye undergoes immediate reduction at the electrode area, this method conventionally reduces the dye by adding a reducing agent, which then undergoes oxidation after dye reduction. The cathode surface is then reduced by the oxidized reducing agent, making it more suitable for dye reduction. This cycle continues throughout the dyeing process. In electrochemistry, a mediator or reversible redox system is an agent undergoing oxidation and reduction stages. In the indirect electrochemical dyeing system, the mediator, when reduced by exposure to the cathode, facilitates dye reduction instead of close contact between the dyestuff and the cathode, as observed in direct electrochemical reduction. Consequently, this process is termed indirect electrochemical dyeing. The primary objective of the reversible redox system in the dye liquor is to maintain a sustained reduction potential, eliminating the need for a traditional reducing agent and preventing the formation of breakdown products from reducing agents. After the dyeing process, any unexhausted dye precipitates through air oxidation and can be filtered by refining, simplifying the electrochemical dyeing process. Once the dye has been removed, the liquor containing the alkali, mediator, and ligand can be reprocessed and used in further dyeing procedures. This aspect significantly contributes to the process's cost-effectiveness and environmental sustainability. After removing the remaining dye for the first time, indirect electrochemical reduction lowers the value of reduction chemicals and permits full recycling of the dye liquor. Consequently, the dye bath can be disposed of without rinsing water [84–86]. With a steady current, the electrolysis is performed in a Galvano static environment. The fabric sample that has been prepared is immersed in the dye bath. The overall dyeing process uses the exhaustion method, stirring continuously for 30 min. The temperatures for electrolysis and electrochemical dyeing are about 300 ± 2 K. Following the dyeing process, the sample was cleaned with cold water and left outside to allow the dye molecules to oxidize or fix. After the fabric was boiled, it was rinsed with cold water and let air dry [87]. In vat dyeing, the reducing agent provides electrons to change the dye particles so they stick to the fabric. When using electrochemical methods, electrons move from an electrode to the dye in the fabric. Sometimes, this transfer happens directly, but for certain vat dyes, a unique system (mediator) facilitates the movement of the electrons from the cathode to the dye. This can be either a direct or indirect electrochemical reduction with a mediator. Electrochemical Dyeing with vat dyes—package dyeing, developed by researchers, is a technique of electrochemical dyeing in a specialized machine known as a package dyeing unit. This machine speeds up the process by dyeing large amounts of fabric at once. Instead of adding a reducing agent to the dye bath, the fabric is dipped into a special electrochemical cell. This cell is connected to the dyeing machine and utilizes electricity to change the color of the fabric. A setup for package dyeing of 110 kg yarn is shown in Fig. 2.7. A 500 A cell is connected to a dyeing machine, and the dyeing bath is reduced using electricity. Dye particles are added to the bath, where they are reduced with the help of the mediator. By controlling the electrical current, the dyer can adjust the redox potential (a measure of reduction of dye) in the dye bath. Once the dye reduction is complete, the electrical current is lowered to maintain the redox potential. Typical currents at this stage are around 50–100 A.

The mediator is present both in the dye bath and the rinse, which are collected in a tank for later use. After removing the oxidized dye using nano-filtration, the mediator and dye bath can be used again. This process helps reduce waste and saves resources. The dyeing process involves several chemical steps, including oxidation, soaping, and rinsing, to ensure the color sticks well to the fabric. Overall, electrochemical dyeing with vat dyes offers a more efficient and environmentally friendly way to dye the fabric in large quantities [88, 89]. One significant factor affecting how well vat dyes are reduced is pH. Multiple forms of Vat dyes can be found based on the dye bath pH. In traditional reduction, this parameter needs to be continuously maintained—the breakdown of dithionite results in a pH value drop, which may create disruptions. Compared to the conventional method, the electrochemical approach produces a dye bath pH that is more stable. Therefore, adding alkali during electrolysis is not

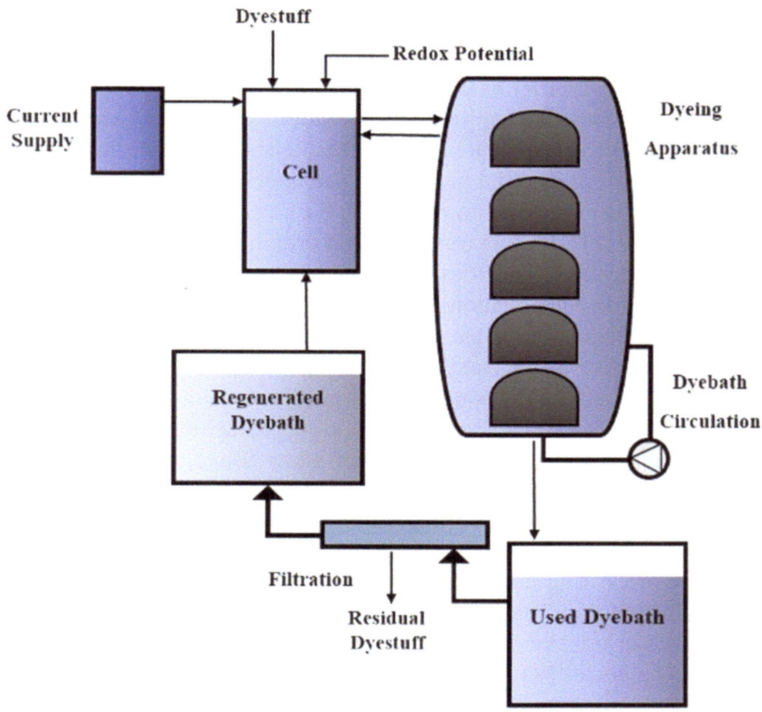

Fig. 2.7 Illustration of package dyeing equipment setup, circulation through the cell, and regeneration of the used dye bath through nanofiltration

necessary [88]. It is really important to control the elements of the bath. ECD helps us avoid the common and challenging occurrences of over- or partial reduction when reduction is carried out by dithionite. It also helps us avoid needlessly extending the reduction period. Also, since complete control over the dyeing parameters is available, the finished fabric is of higher quality [90]. The dye liquor recycling loop in an electrochemical dyeing experiment runs at least nine or ten times. After dyeing, the same catholyte solution is air oxidized and filtered (a 10–15% new electrolyte) to eliminate the insoluble color molecule. The conversion and current efficiency are essentially constant throughout this recycling process. All ten dyeing recycles showed a reasonably constant dye intensity, as shown by K/S values [91].

2.4.6 Plasma Dyeing

The textile industry produces fibers, yarns, and fabrics from different types of materials, natural and synthetic. These fibers undergo spinning processes to create threads, cables, or ropes. Knitting, weaving, or other methods can turn these into fabrics.

Textile materials are widely used in clothing, upholstery, and everyday items [92]. Advancements in textile technology allow for the combination of materials like polymers to enhance the properties of textiles for diverse applications. These applications involve textile functions like wetting, biocompatibility, and electrical conductivity. Many of these properties originate from the textile surface. The surface properties of textiles are crucial in material processing and application. Plasma treatment has emerged as a promising method to modify textile surfaces, opening up new opportunities in the field [93]. Plasma treatment is a promising technique for modifying textile surfaces, enhancing properties such as wetting, biocompatibility, and electrical conductivity for diverse applications.

Recent advancements have demonstrated enormous promise for plasma treatment as a substitute for various textile processing steps. This method offers considerable cost savings, conserves water, and promotes sustainability. Traditional wet processing techniques such as desizing, scouring, bleaching, dyeing or printing, and finishing require the essential amount of water usage, frequently using a lot of energy and producing pollutants through effluents. There is a rising concern for more sustainable, innovative approaches in textile processing to address these environmental concerns. Plasma technology emerges as a promising solution, characterized by its dry processing nature, energy efficiency, minimal chemical usage, and no downstream pollution. By harnessing the capabilities of plasma treatment, the textile industry can transition towards greener and more environmentally friendly manufacturing processes [94, 95].

Plasma, often termed the fourth phase of matter, emerges when the energy of gas particles reaches a level equivalent to the ionization energy of the gas. This heightened energy state triggers a rapid cascade of ionization within the gas, leading to plasma formation. When high electric fields energize neutral gas molecules, certain electrons break loose, turning the gas into a blend of electrons, ionized molecules and atoms, photons, and remaining neutral species. In this phase, plasma exhibits a high level of chemical activity and a propensity for interacting with organic substrates, making it a versatile tool for surface modification even at room temperature. Passing an electrical current through a dielectric gas or fluid creates plasma production, which starts the surface engineering process [96, 97]. Plasma, created through gas heating or electrical stimulation, contains energetic ions, electrons, and reactive particles that can modify polymeric surfaces by breaking and reforming chemical bonds, offering efficient and environmentally friendly surface modifications with minimal gas or vapor consumption [98]. Plasma treatment yields various effects on substrate surfaces, such as grafting, etching, cleaning, activation, and polymerization, with outcomes influenced by gas type and process parameters like frequency, flow rate, duration, power, and pressure. Despite surface modification, the bulk properties of materials remain unchanged [99]. Basically, exciting vapor or gas with electromagnetic or electric fields creates plasma, a cluster of particles consisting of free radicals, ultraviolet radiation, positive ions, neutral species, and electrons. As electrons and free radicals collide with the surface of the material, covalent bonds rupture, and free radicals are created. Afterward, the stimulated material surface merges with excited gas species, providing chemically reactive groups that create covalent interactions with the substrate surface. By the selection of appropriate gases or vapors, the desired surface chemistry can be achieved [95]. Researchers have looked into how cotton structure, characteristics, and dyeing behavior are affected by plasma treatment. They have improved the surface of fibers regarding appearance, texture, and functional performance by using both polymer and non-polymer-producing plasmas. The versatility of plasma surface activation offers various opportunities for its use in textile processing, either independently or as a pre-treatment to enhance the effectiveness of the following processes, referred to as plasma-assisted processing [100, 101].

M. Faraday first studied plasma in the 1880s, and I. Langmuir proposed the concept of plasma in 1926. Initially, plasma was mostly used in the microelectronics industry in the 1960s. But by the 1980s, it had started being used to treat surfaces of metals and plastics. In the textile sector, low-pressure plasma treatments emerged in the 1980s, showing promising results in enhancing the functional properties of fibrous materials. These treatments offered potential improvements in various aspects of plasma-treated textiles. Currently, scientists are studying how to use plasma on textiles at normal air pressure in a way that could be used in factories. This shows that more people are interested in using plasma technology in the textile field.

Plasma processing offers versatility by allowing one system to modify various types of fibers (natural synthetic, cellulose, protein) and textile forms. By carefully selecting gases (O_2, N_2, H_2, etc.) and controlling plasma operating conditions (power, treatment time, gas flow rate, pressure), intrinsic effects on textiles can be achieved. These effects include enhancements in hydrophilicity, hydrophobicity/oleophilicity, chemical reactivity, material mechanical and chemical resistance, and adhesion of coatings and matrices [102, 103].

Since fiber surface modification significantly impacts the qualities needed for textile products to be used in their final form, it is important for numerous procedures and applications. The fiber surface largely determines properties such as dyeability, wettability, resistance to felting, pilling, and electrostatic properties. Plasma treatment is a convenient method for enhancing these properties, as it targets the fiber surface while leaving bulk properties unchanged [104]. A significant advantage of plasma processing is its dry treatment nature, which conserves energy and ensures cleanliness throughout the process. That is why it offers several environmental advantages compared to traditional processing methods. It decreases the need for chemicals in processing, enhances the exhaustion of chemicals from baths, and reduces biochemical oxygen demand (BOD) or chemical oxygen demand (COD) in effluents. Furthermore, plasma treatment shortens wet processing time, lowers processing temperature requirements, and contributes to overall energy savings [99].

2.4.6.1 Types of Plasma

Plasma is categorized into two main types: hot and cold. Cold plasma is preferred in textile applications because high temperatures can damage fabrics. Also known as non-thermal plasma, it operates at low temperatures, making it safe for treating textiles without causing harm.

- Hot plasma: In hot plasma, electrons, and ionic species have very high temperatures. This type of plasma is suitable for certain applications but may not be ideal for textiles due to the risk of damage.
- Cold plasma: It exists at ambient temperature, making it safe for surface modification, polymerization, etching, functionalization, and other treatments. Its energy and ability to penetrate depend upon the pressure at which it functions.

Plasma treatment varies depending on factors like the voltage, the pressure of the chamber, and the frequency of the power generator. These factors determine how the plasma interacts with the material being treated, influencing the outcomes of the treatment process [101].

The power source for plasma treatment can be either DC (direct current) or AC (alternating current).

- *DC Power Source*: With direct current (DC), power is transferred from the main source to the electrodes. However, arrangements like these are not preferred as they may cause electrode degradation over time.
- *AC Power Source*: AC power sources offer a more continuous alternating current, which helps minimize the charging of the material being treated. AC power can be obtained in three frequency levels: radio frequency (13.56 or 27.12 MHz), low frequency (50–450 kHz), and microwave (915 MHz or 2.45 GHz). Each frequency range has advantages and applications in plasma treatment processes [98].

Table 2.2 shows the categorization of different types of plasma.

There are four main types of atmospheric-pressure plasma systems.

Corona Discharge

The earliest known plasma treatment for modifying polymer surfaces is corona discharge. This system consists of two parallel electrodes—one sharp, like a needle, and the other flat, like a plate. Plasma forms approximately 0.5 mm from the needle tip, but its distribution is uneven. Charged particles move toward the grounded electrode, but corona discharge is not commonly used for textile applications due to its inconsistency and limited effectiveness. The plasma density drops quickly with distance from the generation point, so the gap between the electrodes is kept small,

Table 2.2 Different types of plasma [105]

Depending on pressure	Depending on the temperature of electrons and ions	Depending on the frequency of power supply
Low pressure (0.01 kpa) Atmospheric pressure (100 kpa)	Hot plasma (above 10,000 degrees) Cold plasma (below 100 degrees)	Low frequency (50–450 kHz) Radio frequency (13.56 or 27.12 MHz) Microwave frequency (915 MHz or 2.45 GHz

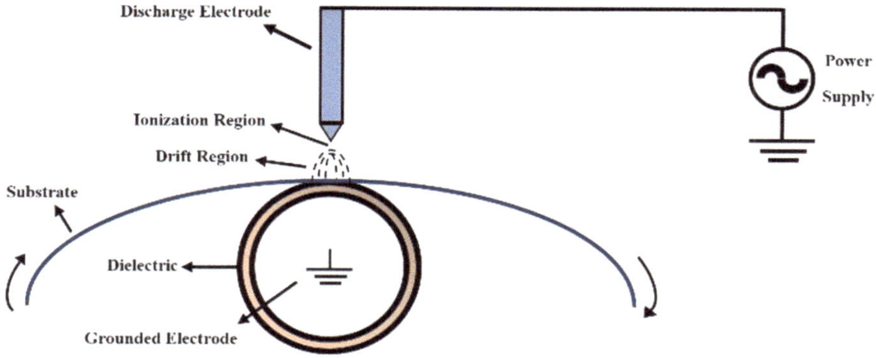

Fig. 2.8 Schematic diagram of corona discharge. Adapted with permission from Ref. [98], Copyright Elsevier 2020

making it unsuitable for thicker fabrics. Thicker fabrics can't be treated effectively using corona discharge because the plasma isn't dense enough across larger distances [106]. Figure 2.8 shows the schematic diagram of corona discharge.

It operates in the 10–30 kHz frequency range and creates small discharges called filaments. These filaments are about 100 μm in diameter and last for 100 ns. One electrode is sharply curved, and the other is less so. This technique is an older plasma method used in textile sectors but is not strong enough for deep fabric treatment. It mainly works well with loose fibers, and its effectiveness decreases over time [107].

Dielectric Barrier Discharge (DBD)

In Dielectric Barrier Discharge (DBD), two parallel flat plates with a tiny space between them are used. They're connected to a power source that applies about a 20 kV potential difference between them. The plates are covered with a dielectric substance, such as glass or ceramic, to stop the arcing and short circuits. This setup is termed as dielectric barrier discharge [106]. Figure 2.9 shows the schematic diagram of dielectric barrier discharge.

DBD can occur in two forms: homogeneous and filamentary. By making careful changes to how the process works, a homogeneous DBD plasma can be attained, which is ideal for uniformly treating textile surfaces. DBD systems use alternating current (AC) with a frequency between 1 and 20 kHz. DBD is considered more uniform than corona discharge because it's consistent across the electrodes [108].

Atmospheric-Pressure Glow Discharge (APGD)

The Atmospheric Pressure Glow Discharge (APGD) system operates with a short voltage of around 200 V and a high frequency ranging from 2 to 60 MHz. This voltage

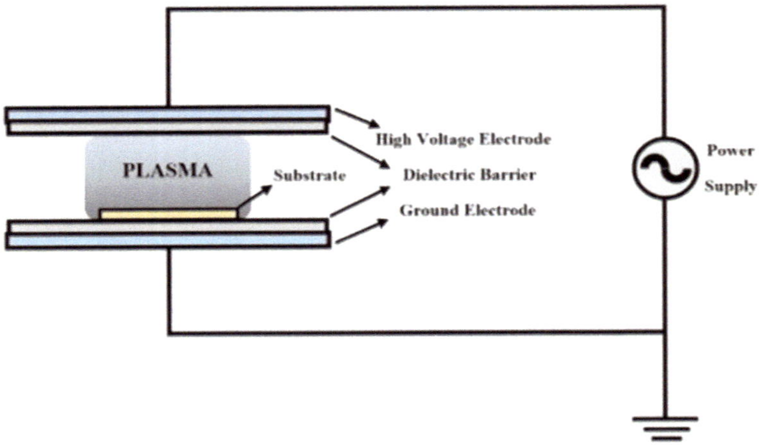

Fig. 2.9 Schematic diagram of dielectric barrier discharge

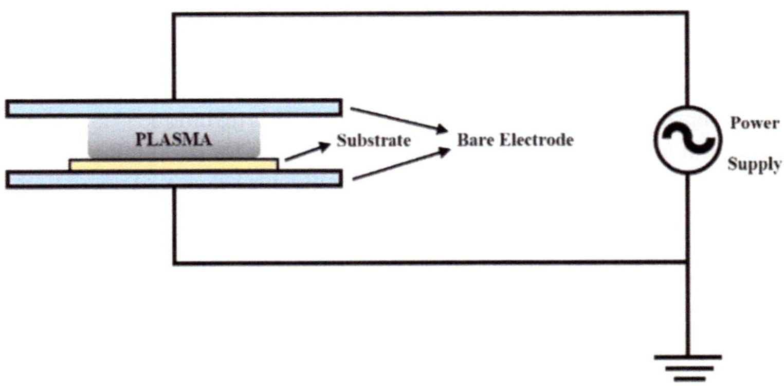

Fig. 2.10 Schematic diagram of atmospheric pressure glow discharge

is applied between two metal electrodes, which can be planar or curved, with a gap typically measuring a few millimeters. This setup promotes the generation of a sharp, uniform, and bright glow between the electrodes, creating a homogeneous discharge. While in APGD systems, helium is the recommended processing gas to avoid arcing between electrodes, its high cost poses a drawback, leading to the occasional use of argon or nitrogen [106]. A schematic diagram of atmospheric pressure glow discharge is shown in Fig. 2.10.

In this system, the Radio Frequency (RF) source connects the two parallel plate electrodes, which are separated by only a very small millimeter difference. This configuration ensures a more consistent and stable discharge compared to DBD. Unlike DBD, APGD doesn't require electrodes to be coated or covered with a dielectric layer, simplifying the setup. APGD offers several advantages over Dielectric

Barrier Discharge (DBD). APGD operates at lower voltages and higher frequencies compared to DBD, making it a relatively uniform, stable, and simple electrode setup, making it a promising option for various plasma processing applications [109].

Atmospheric-Pressure Plasma Jet

The plasma jet has two electrodes arranged, one inside the other. The gas used in the process goes through these electrodes. Applying RF power (between 100 and 250 V) to the inner electrode changes the gas into plasma. This plasma, or ionized gas, comes out through a nozzle and goes towards the surface of the material. Using this spraying technique, surfaces are coated with a fine spray of a selected material. Then, these particles are heated until they melt, and then they are sprayed onto the surface at high speed. When they hit the surface, they cool down and become solid again, creating a protective layer. To make this happen, an electric arc heats gas in a nozzle to extremely high temperatures, over 20,000 °C. This makes the gas expand a lot and shoot out of the nozzle very fast. The powder material for coating is mixed with another gas and sprayed into this super-hot gas jet. The intense heat melts the powder, turning it into tiny droplets shot onto the surface with the fast-moving jet. Plasma spraying works best for materials that can melt without getting damaged. It's commonly used in industries like aerospace and machinery to add strong protective coatings to surfaces [98].

2.4.6.2 Fundamentals of Plasma Dyeing

There are some major effects which will be discussed below shortly.

Surface Activation by Plasma

Plasma surface activation, also called chemical grafting, often follows plasma cleaning. During secondary plasma treatment, organic contaminants on the material are reacted by oxygen atoms, removing them and forming volatile byproducts like water and carbon dioxide. This cleansing process makes polymer surfaces receptive to further reactions, forming functional groups like carboxyl or hydroxyl and carbonyl. This enhances the surface energy of polymers like PP, PE, PET, or PBT, but the effect is temporary and gradually fades over time due to oxygen desorption. The duration of this effect changes according to conditions like substrate type and plasma intensity, with more intense treatments leading to longer-lasting modifications [110].

Etching by Plasma

To etch efficiently, a direct plasma is usually necessary. In this setup, charged particles like ions and electrons bombard the substrate, causing chemical and physical effects. While this method is common in many industries, it's not widely used for textiles and nonwovens. However, there's potential, especially for fabrics. In the textile market, achieving deep, dark colors is a challenge. One solution is to reduce the reflection of textiles' surface after dyeing. Plasma etching can help by creating controlled nano- or micro-roughness on the textiles' surface, which enhances the scattered reflectance and reduces the specular component. As a result, the dyed textile appears darker and more intense after plasma etching. However, etching involves removing hundreds of nanometers from the surface, slowing the process. Additionally, this approach is suitable for very high-end textiles due to its complexity and cost. Despite these limitations, plasma etching offers a promising solution for achieving desired colors in textiles, particularly for those aiming for deep, rich hues [111].

Thin Film Deposition by Plasma Polymerization

Low-pressure vacuum plasma technology is vital in depositing thin film coatings through plasma polymerization. In this process, special gases that can turn into polymers are used. These gases break down into radicals, which then react on the surface of the substrate. The type of gases used greatly affects the properties of the coating formed. Typically, these coatings are quite thin, ranging from 10 to 50 nm, equivalent to 5–30 molecular layers [98]. Initially, plasma polymerization found its use in the medical device industry. However, its applications have expanded to various industrial sectors, including technical textiles and nonwovens. The coatings applied in these sectors can generally be classified as permanently attracting water or repelling water and oil. These coatings often result in unique products that are hard to achieve using other methods. Overall, plasma polymerization offers a versatile and effective way to create specialized coatings with distinct properties, contributing significantly to advancements in various industries [97].

2.4.6.3 Application of Plasma in Textile Industry

In the textile industry, researchers have been working hard since the early 1980s to explore the benefits of using low-pressure plasma treatments on various fibrous materials. These studies have shown promising results, indicating major advancements in different functional attributes of textiles treated with plasma. Many low-pressure plasma machines have been developed over the years for in-line or batch textile processing, primarily as prototypes. These machines have existed for over 15 years and have become increasingly favored in the textile sector because of their numerous advantages over traditional wet processing techniques [94]. One major advantage of plasma treatment is that it doesn't change the fundamental properties of the textile substrate. Instead, it focuses on modifying surface properties, sometimes using environmentally friendly methods. Plasma technology offers a revolutionary approach to textile wet processing, from pretreatments to finishing, by allowing precise tailoring of surface properties. Plasma technology, known for its precise thermal, chemical, and physical effects, operates at low temperatures to avoid sample damage, exists in a non-equilibrium state for innovative material research, and is environmentally friendly with no harsh chemicals or wet processes [112]. Previous studies have explored various plasma technology applications in textile pretreatment and finishing. For example, researchers have successfully enhanced the antibacterial capability of cellulose textile materials by coating them with copper using a DC magnetron sputtering device. Plasma treatment has been demonstrated to enhance the adsorption of silver particles onto cotton fabrics, making them more effective at combating bacteria. Modifying the surface charge of cellulose fibers with nitrogen plasma treatment can greatly enhance the incorporation of silver particles, resulting in stronger antibacterial activity. Moreover, plasma technology can be used to modify polypropylene fibers, which increase the adhesion of copper particles to their surfaces and enhance their antibacterial activity. This highlights the versatility of plasma technology in improving the functional properties of textiles and making them more effective in various applications [113].

2.4.6.4 Application of Plasma Technology in Textile Dyeing

Dyeing textiles is a crucial process, and plasma treatments have significantly enhanced the dyeability and printability of synthetic and natural fibers. These treatments offer various benefits, such as improving capillarity, increasing surface area, reducing external crystallinity, and creating reactive sites on fibers. Additionally, plasma treatments can alter fabric surfaces to enhance dye absorption and facilitate diffraction-based color production [97].

Dye Ability of Natural Fibers

Plasma treatment has shown remarkable effects on the dyeability of natural fibers like cotton. Treatments with air or argon gas have been found to increase water absorbency in cotton, thereby accelerating dyeing processes and enhancing dye uptake. Several factors contribute to this improvement, including changes in surface area, surface erosion, removal of impurities, and chemical alterations on fiber surfaces. Plasma treatments involving oxygen and air alter the surface properties of cotton, encouraging the absorption of direct dyes.

Similarly, plasma treatment enhances the dyeability of wool fibers, increasing the wetability and wicking properties of wool fabric. Oxygen plasma treatment modifies wool fabric surfaces, oxidizing disulfide linkages to create sulphonate groups that perform as active sites for reactive dyes. The etching of hydrophobic epicuticles and increased surface area further enhance fiber wet ability, facilitating dye absorption [96].

Dye Ability of Synthetic Fibers

In synthetic fibers, plasma treatment induces etching and introduces polar groups, improving dyeability. For instance, in polyester, polyamide, and polypropylene fabrics, in situ, polymerization of acrylic acid enhances dyeability. Plasma treatment of micro denier polyester produces cationic dye-able fibers, offering the potential for low-energy and environmentally friendly dyeing technologies.

Polyamide fabrics treated with tetrafluoromethane plasma exhibit improved dyeing properties with acid and dispersed dyes. While plasma treatment marginally reduces the exhaustion rate, it does not decrease dye uptake. Disperse dyes show markedly improved dyeing properties on plasma-treated nylon fabric, along with a slight enhancement in colorfastness compared to untreated fabric.

Overall, plasma treatments play a significant role in enhancing the dyeability of textiles, offering a scope of benefits for both natural and synthetic fibers. These treatments hold promise for improving dye uptake, facilitating color production, and advancing environmentally friendly dyeing technologies across various textile industries [97].

Key Features of Plasma Dyeing

Plasma technology is a powerful tool used to make textile materials better in many ways. It can change both the surface and inside of textiles, making them stronger, more durable, and more useful. This technology has led to notable enhancements in a diverse array of textile items, spanning from common fabrics to advanced composite materials. One important improvement plasma technology brings is in the way textiles interact with water, known as wettability. By treating fibers like polyester, polypropylene, and wool with plasma, their ability to absorb and retain water droplets or moisture on their surface can be enhanced. This means they stay dry or absorb water more efficiently, depending on what's needed [114]. Additionally, plasma treatment can give textiles a hydrophobic or water-repellent finish. For example, when cotton fibers are treated with plasma gas like hexamethyldisiloxane (HMDSO), they become smoother, and water rolls off them more easily. This hydrophobic treatment makes textiles more resistant to water and stains. Plasma technology also helps improve adhesion, which is how well things stick together. It increases the ability of chemicals and dyes to stick to textile materials, ensuring they stay on the fabric and don't wash or wear off easily [97]. Another significant benefit of plasma treatment is its impact on product quality. For instance, wool garments can become fuzzy over time due to the structure of the fibers. Traditional methods to prevent this can make the fabric feel bad and harm the environment. However, using oxygen plasma can prevent felting in wool without causing these negative effects [96]. Moreover, different plasma gases can give textiles special abilities, such as protecting against UV rays, fighting bacteria, or even having medical uses. These functionalities can add value to textile products and make them more versatile in various applications. Plasma technology has also been shown to improve the dyeing process for polymers, enhance colorfastness (how well the colors resist fading), and increase the resistance of fabrics to washing. Unlike wet processes that perforate deep into fibers, plasma solely affects the exterior layer of textiles, typically around 100 nm deep. This implies that it can be utilized on both natural and synthetic fibers, along with non-woven fabrics, without causing harm to their internal compositions. Overall, plasma technology offers a wide range of benefits for improving textile properties and performance [115].

2.4.7 Microwave-Assisted Dyeing

Electromagnetic radiation includes radio waves, microwaves, gamma rays, ultraviolet rays, and X-rays. This type of radiation can propagate through space in the form of waves or discrete energy packets called photons. They have varying wavelengths and frequencies. The electromagnetic spectrum consists of a wide range of wavelengths. Microwaves are a component of the electromagnetic spectrum that is higher in frequency and distinguishable from radio waves by the technology used to access them. Microwave frequencies range from around 300 MHz to over 300 GHz, with wavelengths that range from 1 m to 0.001 m [116].

Microwaves behave similarly to light waves as they move through materials. It transmits with minimal loss through other materials. Microwave absorbers include water, carbon, food, and some chemical solvents. Per mole of microwave, photons bear 0.125 kJ energy. A chemical bond requires much more energy than this to be formed. Thereby, the electronic configurations of atoms and the molecular structure of the substance are unaffected by microwave radiation [117].

The textile industry is researching the use of microwave radiation for surface modification, pressing, finishing, drying, heating, condensation, and dyeing [118]. They have been used as a substitute for traditional heating methods in different operations. It heats materials quickly and evenly throughout processing. Microwave energy provides several benefits, including shorter processing times, faster heating of materials, and energy conservation. It causes dye molecules to quickly come into contact with the fiber [119]. Microwaves promote organic reactions as a safe and efficient way to speed up many chemical reactions. Microwaves assist in reactions that are conducted for extended periods at higher temperatures by lowering the duration of the reaction and energy consumption. Instead of conventional heating methods, microwave heating can be used, which is a cleaner process and eco-friendly [117].

2.4.7.1 Introduction of Microwave-Assisted Dyeing

During the Second World War, microwaves played a significant role in radar systems. The advantages of Microwaves were first used to heat things in 1946 after radio-frequency heating had already been established. Ciba-Geigy was among the first to patent dyeing methods by microwave heating and print fibers using reactive dyes in 1966. The idea of using microwaves for textile finishing processes first originated in the 1970s. In the experiment, cellulosic fabrics were treated with Durable Press (DP) finishing chemicals, followed by curing in a microwave oven [120]. Microwave heating works at the molecular level, unlike traditional heating technologies that rely on convection and thermal conductivity. Microwave radiation relies on the dielectric heating process. The dielectric property influences the dyeing by dipolar dye rotation. When exposed to an oscillating field of microwave, particles attempt to align with it. Resisting forces constrain particle mobility, resulting in random motion. Therefore, energy is lost due to friction and dielectric loss. Insufficient heating occurs when the dipole does not have adequate time to reorient or when it prematurely reorients during the chemical reaction. In the presence of a microwave field oscillating at approximately 2450 MHz, the dye solution demonstrates two polar components. As a result, dye and water molecules start to vibrate. Resistive heating is one way that ionic conduction happens. Ionic stimulation through the dye solution allows fiber molecules to engage with dye molecules. The mordant improves dye absorption and color fastness for microwave-dyed materials [121]. Using a traditional heating source results in slow and ineffective energy transfer due to the thermal conductivity of the materials. This causes the reaction chamber to be higher in temperature than solutions. Microwave radiation is directly coupled with the reaction mixture's solvents, reagents, or catalysts, resulting in effective internal heating, in other words, core volumetric heating. Microwave heating is characterized by a volumetric process, which enables it to be more rapid compared to conventional heating, which typically involves a surface-based mechanism [120]. Figure 2.11 shows the difference between conventional heating and microwave heating.

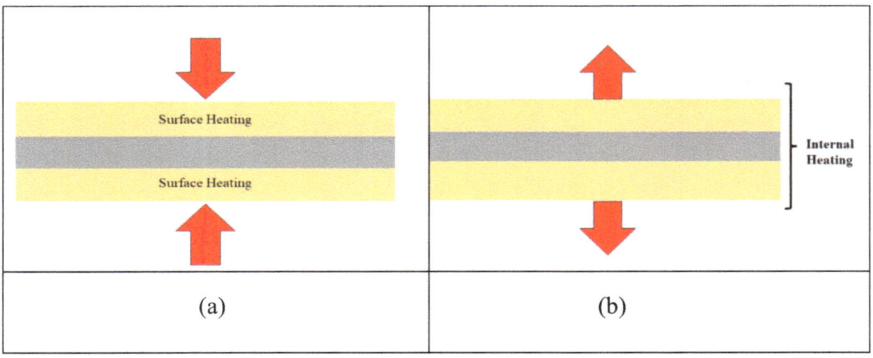

Fig. 2.11 Difference between **a** conventional heating and **b** microwave heating

2.4.7.2 Fundamentals of Dyeing Through Microwave Heating

Polyester Dyeing

Saleh Mohammed Al-Mousawi et al. studied the application of dispersed dyes to polyester fibers using conventional and microwave dyeing processes. Their research revealed that heating through microwave radiation enhanced the dyeing process of polyester by increasing dye exhaustion and improving dyeing rates. Fabrics dyed through microwave heating demonstrated outstanding wash and perspiration fastness [122]. In a separate study, Manik J. Kale and Narendra Bhat examined the impact of microwaves in the pretreatment process utilizing dimethylformamide. Their research indicated that this treatment enhanced crystal distribution without notably altering the melting and crystallization temperatures. Surface cavitation and erosion were observed through scanning electron microscopy after the treatment, leading to increased dye uptake and color strength in the fabric. Additionally, light fastness tests indicated positive results, suggesting that microwave treatment in dimethylformamide enhanced the dyeing properties of polyester fabric [123].

Cotton Dyeing

Microwave enhances the speed and uniformity of cotton fabric dyeing, surpassing traditional methods. Microwave technology allows for rapid dyeing and solid color due to the quick conversion of microwave energy into heat energy, achieving maximum kinetic energy for dye molecules and fiber expansion. This results in efficient dyeing and solid color in a short time frame. Additionally, microwave technology provides good levelness and color fastness in the cotton fabric after dyeing, making it a feasible, low-energy consumption, and environmentally friendly option [124]. Microwave radiation enhances the color qualities of cotton fabric treated with Vat Blue 4 by increasing dye absorption and generating a strong bond between the dye molecules and fabric. Modifying the cotton fabric surface during microwave treatment for an optimal time of 1 min leads to better color strength. Additionally, microwave treatment activates dye molecules, facilitating significant absorption of dye into the cotton fabric, resulting in good color strength and better fastness properties [125].

Wool Dyeing

The intricate structure of wool presents challenges for dyeing with reactive dyes. The traditional dyeing method involves prolonged dyeing periods at elevated temperatures, potentially leading to premature dye reactions before achieving uniform coloration. Microwave pretreatment enhances the dyeing capability of wool fabric by disrupting hydrogen bonds, separating molecular chains, and creating more space, size, pores, and voids within the wool structure, thereby improving its dye absorption properties [126].

Nylon Dyeing

Haggag et al. conducted a study comparing the dyeing of polyamide and nylon 6,6 fabrics using dispersed dyes with microwave technology against traditional heating methods. Microwave irradiation was found to save time, energy, and water while improving color strength and fastness in dyed materials. Optimal conditions for dyeing polyamide fabric included a power not exceeding 200 W, a liquor ratio of 1:5, and a 5% dye conc. Microwave irradiation significantly influenced the dyeing process of polyamide fabric by reducing the dyeing time by approximately 40 min, eliminating the need for a carrier and dispersing agent, and proving to be a time-, energy-, and cost-efficient method [127].

2.4.7.3 Key Features of Microwave

Its advantages include the following points:

- Microwave energy can be controlled easily.
- Uniform heating is possible by microwave, and dyeing is also enhanced.
- In textiles, non-contact microwave drying technology can lessen the damage done by material finishing, lessen drying strain, and enhance product quality.
- Microwaves can boost output speeds, create high power intensity, and lower product costs.
- The use of microwaves can help reduce dangerous gas emissions [128].

On the other hand, microwave radiation may be reflected, transmitted, or absorbed. It can cause serious harm to the human body. Microwaves are known to cause hyperthermia, which is the most severe biological impact on humans. These include ocular injury and testis. The health risks associated with microwave exposure depend on factors such as the power, frequency, and duration of the exposure [129]. In addition, the following points can be termed as application limitations at the industry level.

- The cost of equipment is very high.
- Application is limited.
- If the temperature changes, then the dielectric properties will also change. So, the temperature should be carefully controlled [120].

2.4.8 Ultrasonic Dyeing

Ultrasound consists of high-frequency sound waves beyond human hearing, typically exceeding 17 kHz [130]. It is classified as either power or diagnostic based on the frequency range. Power ultrasound enhances chemical and physical processes through cavitation, where small bubbles form and collapse explosively in a liquid [131]. These rapid collapses generate "hot spots" with high pressure, temperatures, shock waves, and shear stress, potentially breaking chemical bonds. In textile dyeing, ultrasound-based processes leverage cavitation to improve coloring by breaking aggregates, accelerating dye diffusion, generating free radicals, and agitating the liquid [132].

2.4.8.1 Fundamentals of Basic Principle of Ultrasound Wave

Solids can transmit ultrasonic waves in longitudinal and transverse directions, but gases and liquids transfer longitudinal ultrasonic waves. Longitudinal vibrations of molecules in liquids cause compression and rarefaction, creating high and low-pressure zones. Low local pressure creates cavities or bubbles that collapse when compressed, resulting in shock waves. Cavitation, or the production and burst of bubbles, is thought to be the primary cause of ultrasound's physical and chemical effects in solid/liquid or liquid/liquid systems [133, 134].

Ultrasonic Wave Generation

Ultrasonic waves can be created using a wide range of methods—for example, whistles, hooters, sirens, piezo-electric and magnetic transducers [135]. Piezoelectric ceramics are the most effective materials for producing ultrasonic waves. In piezoelectric materials, charge carriers move in response to an electric field, which causes a macroscopic change in length. If the applied voltage is alternating, particles in the medium, such as air or liquid, then it begins to vibrate. Pressure variations happen. Rarefaction of the particles causes the pressure to be lowered, whereas compression causes higher pressure. The wavelength refers to the distance between two rarefaction or compression zones. The generated waves propagate across the surrounding medium. The wave's speed changes depending on the density and elastic qualities of the medium [136].

Ultrasound Equipment

Ultrasound equipment comprises a generating device or machine and a converter. The generator is a standard source of power that transforms up to 60 Hz alternating current into electrical energy of higher frequency, which is then delivered to a transducer and converted into mechanical vibrations. Modern equipment utilizes piezoelectric crystals that expand and compress when exposed to alternating voltage. This action causes the transducer to vibrate longitudinally, transmitting sound waves through the medium and inducing cavitation. [137].

2.4.8.2 Ultrasonic-Assisted Dyeing

Cotton Dyeing

Ultrasound can be used to improve the dyeing rate of cotton. Ultrasound reduces dye molecule size and retention forces between the dye and fiber, ultimately increasing the dye diffusion rate. Cotton cavitations involve the removal of entrapped air bubbles. This improves dye-fiber interaction and causes faster dye flow toward the cotton fiber. The ultrasonic dyeing process requires lower temperatures (35–50 °C) and a frequency of roughly 46 kHz to achieve optimal cavitation levels [138]. The diffusion coefficient of direct dye into cotton fabric increases due to ultrasonic. Ultrasonic energy aids in dye diffusion within fibers by providing the vibrating molecules with the critical energy needed to break their static equilibrium and diffuse. Furthermore, ultrasonic treatment does not harm the fabric surface or the primary wall of fibers [139]. In the study comparing traditional and ultrasonic methods for dyeing cellulosic fibers with reactive dyes, the assessment was based on percentage exhaustion, dye transfer to the washing solution, fastness properties, and color value. The research revealed that ultrasonic techniques improved dye fixation and exhaustion percentage for reactive dyes while showing no impact on the fastness qualities of the colored materials [140].

Cellulose Acetate Dyeing

With the help of ultrasound, it is possible to reduce the use of auxiliary chemicals. It was discovered that using ultrasound in conjunction with mechanical agitation produces an excellent coloring effect with the help of auxiliaries. Cellulose acetate is normally dyed using dispersed dyes. Because dispersing chemicals promote dye solubility and lessen dye aggregation, they are necessary to obtain uniform-colored goods. However, these agents often have poor biodegradability. The use of ultrasound can contribute significantly to reducing the organic load of the dyeing effluent, providing an alternative to the reliance on dispersant agents [141].

Wool Dyeing

Conventionally, wool is dyed at 98 °C with acid dye. Dyeing wool at such high temperatures for an extended amount of time consumes a lot of energy and also causes fiber damage. Furthermore, harmful auxiliaries are also used. However, together with ultrasound and mechanical agitation, wool can be dyed at 60–80 °C without the help of harmful auxiliaries [142]. Ultrasound does not damage the fiber surface. Also, Ultrasound improved the cleaning of scoured wool [143].

Polyester Dyeing

Polyester fiber necessitates a higher dyeing temperature of 130 °C compared to other textile materials due to its crystalline structure. Consequently, dyeing polyester involves elevated temperature and pressure within pressurized closed vessels, increasing the cost of the dyeing process. Dyeing at atmospheric pressure mandates the utilization of carriers, which are not environmentally sustainable. Polyester dyeing may be done at temperatures around 100 °C and atmospheric pressure for an extended period of time utilizing ultrasound. Ultrasonic dyeing saves energy and ensures uniform dying results [144].

2.4.8.3 Key Features of Ultrasonic-Assisted Dyeing

Even though ultrasonic application in textile wet processing has many benefits, it still has some limitations.

- This technology is still on a laboratory scale. Industry scale application is difficult.
- A large reactor is necessary for industry applications, which is not commonly available.
- Tailoring ultrasonic equipment for specific applications increases the expenses of integrating this technology into industrial processes.
- In certain scenarios, direct contact between liquids and ultrasonic transducers can lead to potential issues [145].

2.4.9 Other Techniques

2.4.9.1 Solvent Dyeing

A wealth of information is available on solvent dyeing using solvents made from organic compounds like tetrachloroethylene (perchloroethylene) [146]. Throughout the past few decades, solvent-assisted dyeing has frequently employed standard solvents such as benzyl alcohol, trichloroethylene (TCE), and tetrachloroethylene (PCE), also referred to as perchloroethylene [PCE] [147]. Peters and Stevens first studied the solvent-assisted dyeing of cellulosic fibers with direct and reactive dyes and a tiny amount of water in 1957 [148]. In 1969, Milićević reported that 642 patents had been issued within twenty years. While pure perchloroethylene has a higher boiling point like 121.1 °C than water (100 °C), it can be heated up more quickly than water. However, 15.8% water in the system can produce an azeotrope, restricting the boiling temperature to 87.8 °C [146]. Since water was needed for the cellulosic fibers dyeing process and other materials with ionic dyes, most research on perchloroethylene dyeing concentrated on the dyeing of polyester fibers with dispersed dye. The main issue, though, was that the disperse dyes' dissolution parameters were more similar to those of perchloroethylene than they were of polyester, which meant that the disperse dyes retained their preferential dissolution in the solvent during the solvent dyeing process of polyester (PET) [146, 149, 150]. The issue of disperse dyes being excessively soluble in perchloroethylene was addressed in an intriguing way by J&P Coats Ltd.'s improvement of the DACSOL process, which uses a perchloroethylene mixture of silicone-fluid and is intended for batch dyeing polyester yarns with fluorescent brightening agents and disperse dyes. The procedure used perchloroethylene and 75–80% volume of silicone fluid with low viscosity. Due to the dye solubility being reduced and the effective liquor ratio being decreased, this changed the coefficient of partition in favor of the colorant in the fiber, increasing dye bath exhaustion [146, 149].

Using CI Reactive Red-31, Chavan investigated the dyeing process of alkali-swollen cotton in combinations of polar and nonpolar solvents. The polar solvent was water, while the nonpolar solvent was acetone. For the pre-swelling of cotton, a suitable ratio of the mixture of solvents to alkali concentration allowed for a 100% exhaustion rate and 77–98% dye fixing. The dielectric constant of the polar solvent determines the ratio of polar to nonpolar solvent needed to achieve the maximum exhaustion rate and color fixation [151, 152]. A reactive dye was recently attempted to be used in two-stage solvent media containing water and dichloromethane for dyeing cotton [153]. A small quantity of water was used in the experiment, and the outputs showed that the uptake of dye was superior to that of traditional dyeing. Moreover, level dyeing can be supported by the use of a hydrophobic in-nature solvent, such as dichloromethane, which uniformly disperses and conveys a tiny amount of dye-dissolved water to materials. Additionally, several attempts have been made to dye cotton fiber using a water–ethanol system [154, 155].

2.4.9.2 Nanotechnology-Based Dyeing

As previously mentioned, the standard dyeing technique uses a lot of water. Therefore, various effluent management technologies are implemented to achieve zero aqueous discharge. Nevertheless, employing this tactic is costly and laborious. Therefore, reducing effluent generation is better than using waste management techniques [156]. Nanobubbles can be applied in numerous sectors. In the textile industry, the nanobubble method is frequently utilized in denim garment finishing processes with a very low liquid ratio [157]. In the textile processing industry, nanobubbles function to transfer various chemicals across the substrate's surface [158]. A revolutionary development in textile wet processing is nanotechnology. Natural dyes can be used to color fibers with a range of diameters. In a particular investigation, nanofibers were colored during bubble electrospinning using a natural colorant derived from the roselle calyx. This dye was extracted from nature to color polyvinyl alcohol (PVA) nanofibers. The dye could color and permeate the nanofibers effectively [159]. To color bleached cotton fabrics using eleven distinct reactive colorants in the exhaust process with a very minimal liquor ratio of 1:1, Mohsin et al. investigated nanobubble technology. The study demonstrates that three liquor ratios—1:1, 1:5, and 1:10—can be utilized in a nanobubble dyeing process to color cotton fabric. The recipe mixture was applied to the fabric for eight seconds at three bars of pressure in the nanobubble form with a pressure nozzle jet. The rapid rate of liquid flow caused nanobubbles to form, and the researchers measured the bubble's size [160].

The nano-dye method is an additional dyeing technique based on nanotechnology. This environmentally safe and sustainable technique was created by Nano-Dye, LLC, a Florida-based corporation. The additional term for this cutting-edge, environmentally friendly method is cationic textile dyeing. Cotton and blended materials can be treated with this procedure using the exhaust approach. In the traditional reactive dyeing technique, untreated cotton acquires a negative charge when immersed in water. Water immersion exerts a negative control on dyestuff. The negative charges on both of them oppose one another, which makes bonding challenging. However, when untreated cotton is immersed in water during the nano-coloring procedure, its negative charge is converted to a positive charge. Bonding is stronger and easier to understand because of the attraction between the positively charged cotton and the negative dye substance charge. Moreover, removing pollutants from the wastewater treatment units' effluent enables future clean coloration procedures and their cost-effectiveness. This procedure can make use of already-existing dyeing machines and equipment. This process is sustainable due to its lack of salt dyeing, 90% colorant exhaustion rate, and significant water and energy reductions. Because of the nano-dye process's increased dye jet manufacturing per hour, less energy is used, preventing climate change [39].

Dyeing with nano-fibrillated cellulose (NFC) is a sustainable and eco-friendly method. NFC is a safe nanomaterial. It can be made from cellulose sources that are easily accessible. This cellulose primarily comes from pulped-cut wood. NFC hydrogel is a mechanically homogenized steady colloidal dispersion. High force is used to push cellulose granules through a homogenizer. Thus, cellulose fibers are used to generate nanofibrils. Experts refer to this phenomenon as defibrillation. NFC's nanocellulose hydrogel is a potent dye carrier to the material substrate. Initially, NFC nanofibers are treated with color using reactive dyes in the exhaust technique under the right conditions, resulting in colored NFC hydrogel. After that, the formed hydrogels are placed at the proper temperature on the material substrate's surface using a particular technique. This new dyeing method based on NFC uses less water. Because NFC has an enhanced dye fixing rate compared to the typical exhaust technique and a strong affinity to cotton fiber, it has a large surface area and reduces alkali and salt. Should this cost-effective and sustainable method prove profitable, the textile sector would reach new heights [161].

2.5 Current Industrial Practices in Alternative Dyeing

According to the literature, one kilogram of textile material can be colored using different amounts of water, depending on the type of fabric, the dye chemistry, and the dyeing equipment. To color one kilogram of fibers using traditional aqueous dyeing, 100–180 L of water are needed [162].

In a conventional dyeing and finishing mill, 150 m^3 of water is used on average for every ton of textile produced. Over 80% of industrial wastewater is discharged into the environment by textile wet processing plants [163]. After the agriculture industry, the textile sector is the largest polluter of pure water. Today's textile business uses about 8000 different chemicals and 3600 different dyes for diverse procedures like bleaching, dying, printing, and finishing. Many of these substances damage water and soil, threaten aquatic life, and are directly or indirectly hazardous to human health. An average-sized textile mill that can process 8,000 kg of fabric daily uses about 1.6 million gallons of water daily [164]. An estimate states that some 280,000 tons of textile colors are released each year as industrial effluent [165]. Excessive water use during dyeing and finishing results in higher energy expenditures since more energy is needed to heat larger amounts of water to the dyeing temperature [166]. The last two decades of the twentieth century saw a dramatic rise in industrial knowledge of environmental issues. The published research on the textile industry's influence on water became widely known. The tightening of environmental protection laws has put great pressure on the textile industry [167, 168]. The textile industry is always looking for alternate ecological dyeing methods to get around these challenges in terms of the environment while also saving time and water during the dyeing process. Consequently, the most impressive methods that might meet these needs include silicone non-aqueous dyeing, solvent dyeing, reverse micelle dying, and supercritical fluid dying [152]. This section aims to examine several approaches, tactics, and techniques for water conservation practiced in textile dyeing, emphasizing their relative benefits, drawbacks, and potential future applications.

2.5.1 Innovations in Textile Processing Machinery for Alternative Dyeing

The amount of water needed in exhaust dyeing is most affected by the ratio of textile material to liquor. When the liquor ratio is reduced from 1:10 to 1:8, 15% less money is spent on processing, and 20% less water is required. In addition to using less water, low liquor ratio dyeing machines also use less salt and alkali—which are added in grams per liter of total liquor—when coloring textile materials [169]. Additionally, lowering the liquor ratio improves the cellulosic fibers' capacity to absorb color, especially when direct dyes are applied [170]. Commercially available equipment from Alchemie Technology is intended to clean up the dye and treatment processes used in the textile industry. Alchemie Technology is a Cambridge-based innovator that is revolutionizing the textile sector by removing the negative environmental effects of these polluting methods with clean-tech digital manufacturing solutions. Traditionally, the wet procedures used to apply the dyes and finishes have resulted in large water use and substantial carbon emissions. Endeavour's digital dyeing technique reduces water usage by 95% and energy use by a much-needed 85%, while Novara machines concentrate on the treatment aspect. High-performance jetting technology is used in the Endeavour machines made by Alchemie. Nanodroplets of dye are fired onto materials by thousands of nozzles, allowing the dyes to form patterns and block colors. After that, infrared light—possibly produced by solar energy or another renewable energy source—physically repairs them. Similar to this, Novara has apps that are specifically designed to be used wherever they are needed. For instance, the machines treat one side to prevent odors while waterproofing the other [171]. These effective procedures starkly contrast to traditional methods, which, since they rely heavily on mass chemical immersion baths and high-energy rinsing, produce enormous volumes of harmful wastewater. Along with international fashion giant H&M, investors At One Ventures, a specialist in US economic sustainability, have contributed to Alchemie's funding [172].

2.5.2 *Industrial Practices in Waterless Dying*

2.5.2.1 Supercritical Fluid Dying

DyeCoo created a waterless dyeing technique in 2008, utilizing $scCO_2$'s capability as a solvent for a variety of synthetic and natural dyes. Compared to the traditional water dyeing method, this procedure offers several benefits, including a 50% cheaper operating cost and a shorter turnaround time. This approach could also lead to high solubility, zero surface tension, and diffusivity. In addition, the procedure is environmentally friendly and economically viable because of the reusability of CO_2 and dyes. Due to the benefits of DyeCoo's techniques, companies like Nike, Adidas, and IKEA have all noticed and begun using this procedure for product dyeing. Additionally, DyeCoo is the first seller of CO_2-dyeing equipment in the globe. Since 2010, waterless textile dyeing with $scCO_2$ technology has been commercially implemented by DyeCoo in Thai Mills; more recently, mills in Taiwan and India have also adopted this technology [173, 174].

2.5.2.2 Air Dyeing

A fresh, cutting-edge airflow dyeing method has been suggested by California-based sustainable technology startup Colorep [175]. The fabric to be dyed is sprayed with the dyeing fluid after it has been atomized and mixed with high-pressure airflow in an airflow dyeing equipment. Since water serves as a solvent for the dyeing liquid and the dye chemicals come into intimate contact with the fabric, very little water is used. The air-dye process uses air instead of water to transfer dye, and no water or hazardous waste is released. Murugesh provides a detailed explanation of the air-dyeing principle [176]. Two businesses have developed their waterless air dyeing process in the past few years. These are American companies called Colorzen and AirDye(R). Rather than dipping them in the customary dye-and-water bath, fabrics are printed using the AirDye(R) technique. This method applies paper colors that have been specially blended onto polyester fabric. Depending on the fabric, AirDye(R) uses 95% less water, 86% less energy, and 84% fewer greenhouse gas emissions than traditional dyeing. Less hazardous waste is emitted, water usage is reduced, and many other benefits of air dye are detailed in several scientific publications and magazines [176–179]. However, natural fibers like cotton cannot be used for the air-dye process; this technology is limited to synthetic materials like polyester. Currently, AirDye(R) is used by a number of textile converters in North America, Asia, and Europe. To color the cotton fibers, a group of gifted engineers and experts in apparel founded ColorZen(R) technology in 2012. To make the cotton more absorbent, they altered the structure of the fibers while taking pretreatment into account. With the use of innovative technology called ColorZen(R), cotton dyeing can now be done more effectively and sustainably. With no harmful emissions, this cutting-edge technology can use up to 90% less water, 75% less electricity, and 95% less chemicals. Recently, airflow-dyeing machines—which use air instead of water—were designed and produced by Cleantech Solutions International, Inc., based in China. Cleantech Solutions' founder and CEO, Mr. Jianhua Wu, claims that the airflow-dying apparatus atomizes water under intense pressure. As a result, two-thirds of the water used by conventional models can be saved by their products. Additionally, they are studying the waterless dyeing apparatus, which is now in the early stages of trial and R&D [180, 181].

2.5.2.3 We Are SpinDye: An Innovation of Dye that H&M Adopts

Before recycled materials are spun into yarn, "We aRe SpinDye" dyes them using leftover garments or post-consumer water bottles. Their process uses no water at all to melt recycled polyester and color pigments together, saving up to 75% of water. The latest word is that H&M's Conscious Exclusive collection uses We aRe SpinDye®'s dying technique [182].

2.5.2.4 Living Color: Pigment-Producing Bacteria to Dye Textiles

The Netherlands-based biodesign initiative Living Color is investigating the potential of employing microorganisms that produce pigment to color clothing. 2020 saw the launch of the first-ever bacterial-colored sports collection, a collaboration between Living Color and PUMA. Value chain participants seek transparency and innovative textile and dyeing goods as businesses strive for greater sustainability and circularity [182].

2.6 Future Prospects in Sustainable Dyeing

The textile sector is under mounting pressure to address its environmental impact, particularly in dyeing procedures. These procedures, historically associated with significant resource consumption and pollution, are now a pressing concern. Stakeholders are increasingly turning to sustainable dyeing procedures to mitigate environmental damage and encounter the growing consumer demand for eco-friendly products. The textile industry is swiftly adopting these procedures in response to the ever-increasing demand for environmentally friendly and ethically manufactured fabrics. With mounting concerns about the environmental repercussions of conventional dyeing methods, players in the textile sector are actively seeking sustainable alternatives. These strategies not only offer environmental benefits but also economic advantages, fostering innovation and ensuring the enduring viability and preservation of the natural environment [183, 184].

2.6.1 Ecological Footprint

The primary catalyst for implementing sustainable dyeing procedures is their diminished environmental footprint. Conventional dyeing methods frequently need substantial quantities of water, energy, and chemicals, contaminating water bodies and soil, as well as significant carbon emissions. On the other hand, sustainable dyeing procedures strive to reduce the usage of resources and the creation of waste. Natural dyeing, waterless dyeing, and low-impact synthetic dyeing employ fewer chemicals and water, decreasing pollution and reducing energy demands. With the tightening of environmental restrictions and the increasing customer demand for sustainable products, these techniques are expected to become increasingly prevalent [185, 186].

2.6.2 Cost Savings

While textile firms may need to make an initial investment to adopt sustainable dyeing procedures, the long-term financial benefits are significant. These methods can save manufacturing costs by reducing water and energy usage and decreasing reliance on expensive chemicals. Moreover, as consumers increasingly show a willingness to pay a premium for sustainably produced textiles, producers can command higher prices for their products, offsetting the initial investment in sustainable technology. The implementation of sustainable practices often leads to operational efficiencies, such as process optimization and waste reduction, ultimately enhancing overall profitability [23, 186].

2.6.3 Innovation and Collaboration

In the textile sector, sustainable dyeing methods are needed to stimulate innovation and foster collaboration. Manufacturers, academics, and other stakeholders collaborate to create and apply innovative technologies and procedures that reduce environmental harm while upholding product quality and performance. This cooperative approach promotes the exchange of knowledge and speeds up the implementation of environmentally friendly practices throughout the sector. Advancements in biotechnology facilitate the creation of environmentally friendly dyes attained from renewable sources, such as microorganisms and plants. Furthermore, digital technologies, like machine learning and artificial intelligence, are employed to enhance dyeing processes and minimize the use of resources [11, 22, 23].

2.6.4 Consumer Demand

Consumer demand for sustainable products is a powerful force shaping the textile sector. Modern consumers are increasingly conscious of environmental and social issues and expect the businesses they support to align with these values. As a result, textile producers are facing mounting pressure to implement sustainable practices across their supply chains, particularly in dyeing. Companies that fail to meet these standards risk losing market share to competitors prioritizing sustainability. However, those who embrace sustainability have the opportunity to gain a competitive edge and build stronger connections with their customers [187].

2.6.5 Regulatory Compliance

Governments worldwide are enacting rules to diminish the ecological consequences of many businesses, including the textile sector. These rules frequently encompass limitations on the utilization of certain chemicals, constraints on the release of wastewater, and mandates for the prevention and management of pollution. By implementing sustainable dyeing methods, textile manufacturers can guarantee adherence to these requirements and prevent fines and penalties. Furthermore, organizations that actively adopt sustainability practices are in a more advantageous position to adjust to forthcoming legislative changes and reduce related risks [186, 188].

Ultimately, the future outlook for sustainable dyeing methods in the textile sector is relatively favorable. Due to increasing environmental concerns and growing consumer demand for sustainable products, textile producers are rapidly adopting eco-friendly dyeing procedures to minimize their environmental impact, decrease expenses, foster innovation, comply with regulations, and meet customer expectations. Textile firms may ensure a more sustainable and lucrative future by investing in sustainable practices today.

2.7 Conclusion

The waterless dyeing techniques are marked as a pivotal step toward meeting the sustainability goals in the textile industry. Traditional dyeing methods are a significant contributor to huge quantities of water consumption and pollution, while the adoption of innovative waterless alternatives promises the options to mitigate these negative environmental impacts on textile industries. This chapter has tried to highlight different waterless dyeing techniques that have been proposed and in the application stage led by researchers around the world. The key advancements and applications of waterless dyeing technologies have been discussed in detail, which include foam dyeing, supercritical CO_2 dyeing, etc.

Supercritical CO_2 dyeing has been found to be one of the most innovative technologies due to the use of carbon dioxide in a supercritical state to effectively dye the textile material without water, which can contribute to sustainability issues significantly. All the techniques mentioned here have been able to contribute significantly to reducing wastewater and chemical runoff, while these can be economically feasible options as well if proper set-up can be established. It has already been found in different industries in South Asian countries such as Bangladesh, which is one of the leading textile manufacturing countries, that waterless dyeing technology can be established at industrial set-up and run successfully. Companies adopting these methods to gain green industry certification and those industries can contribute to substantial reductions in water usage, energy consumption, and environmental footprint, aligning with global sustainability goals and regulatory standards with proper management of the system. However, the transition to waterless dyeing is quite challenging and requires initial investment costs, technological complexities, and other limitations that have been mentioned in the chapter. However, with proper support, waterless dyeing can still become the biggest invention in the textile industry in recent history.

Declaration The authors declare that they have no known competing financial interests or personal relationships that could have appeared to influence the work reported in this chapter. Generative AI and AI-assisted technologies have been used in the writing process to improve the readability and language of the work only. After using this tool, the authors reviewed and edited the content as needed and took full responsibility for the content of the publication.

References

1. Leal Filho W et al (2022) An overview of the contribution of the textiles sector to climate change. Front Environ Sci 10:973102
2. Herrmann S (2017) A new textiles economy: redesigning fashion's future. Circular Fibres Initiative. Availble at: https://www.ellenmacarthurfoundation.org/a-new-textiles-economy
3. Matthews JM (1907) Theory of dyeing. J Franklin Inst 163(6):455–460
4. Christie RM (2007) Environmental aspects of textile dyeing. Woodhead Publishing
5. Chequer FMD, de Oliveira GAR, Ferraz ERA, Cardoso JC, Zanoni MB, de Oliveira DP (2013) Textile dyes: dyeing process and environmental impact. In: Günay M (ed) Textile dyes: dyeing process and environmental impact. IntechOpen, pp 151–176
6. Amin MdR, Mahmud A, Anannya FR (2021) Assessment of carbon footprint of various cotton knitwear production processes in Bangladesh. AATCC J Res 8(6):47–57
7. Saleem MA, Pei L, Saleem MF, Shahid S, Wang J (2020) Sustainable dyeing of nylon with disperse dyes in Decamethylcyclopentasiloxane waterless dyeing system. J Clean Prod 276:123258. https://doi.org/10.1016/j.jclepro.2020.123258
8. Mahmud I, Kaiser S (2020) Recent progress in waterless textile dyeing. J Text Sci Eng 10(6):1–3
9. Zaidy SSH, Vacchi FI, Umbuzeiro GA, Freeman HS (2019) Approach to waterless dyeing of textile substrates—use of atmospheric plasma. Ind Eng Chem Res 58(40):18478–18487
10. Khalil E, Sarkar J, Rahman MM, Shamsuzzaman MM, Das D (2023) Advanced technology in textile dyeing. In: Rahman MM, Mizanur M, Mashud M (ed) Advanced technology in textiles: fibre to apparel. Springer Nature Singapore, Singapore, pp 97–138
11. Yusuf M, Shahid M (2022) Emerging technologies for textile coloration. CRC Press
12. Uddin F (2021) Environmental hazard in textile dyeing wastewater from local textile industry. Cellulose 28(17):10715–10739
13. Saremi R, Rai S, Sharma S (2022) Chapter 21—Dyeing of fibers and impact on the environment. In: Altalhi T, Inamuddin (eds) Green sustainable process for chemical and environmental engineering and science. Elsevier, pp 517–543
14. Drumond Chequer FM, de Oliveira GAR, Anastacio Ferraz ER, Carvalho J, Boldrin Zanoni MV, de Oliveir DP (2013) Textile dyes: dyeing process and environmental impact. In: Eco-friendly textile dyeing and finishing. IntechOpen
15. Uddin F (2019) Introductory chapter: textile manufacturing processes. In: Uddin F (ed) Textile manufacturing processes. IntechOpen
16. Benkhaya S, El Harfi S, El Harfi A (2017) Classifications, properties and applications of textile dyes: a review. J Envir Eng Sci 3(3):311–320
17. Patra AK, Chakraborty S (2024) Dry processes for chemical processing of textiles. In: Sheikh JN, Teli MD (eds) Sustainable textile chemical processing, CRC Press, pp 389–414
18. Touhidul Islam M, Islam T, Islam T, Reazuddin Repon M, Touhidul Islam M, Reazuddin Repon M (2022) Textile & leather review synthetic dyes for textile colouration: process, factors and environmental impact. Text Leather Rev 5:327–373
19. Vhanbatte SB (2022) 1—Background and type of textile materials processed in textile industry. In: Shanmugasundaram OL (ed) Applications of biotechnology for sustainable textile production. The textile institute book series. Woodhead Publishing, pp 1–22
20. Nayak R, Jajpura L, Khandual A (2023) 1—Traditional fibres for fashion and textiles: Associated problems and future sustainable fibres. In: Nayak R (ed) Sustainable fibres for fashion and textile manufacturing. The textile institute book series. Woodhead Publishing, pp 3–25
21. Lara L, Cabral I, Cunha J (2022) Ecological approaches to textile dyeing: a review. Sustainability 14(14):8353
22. Periyasamy AP, Militky J (2020) Sustainability in textile dyeing: recent developments. In: Muthu SS, Gardetti MA (eds) Sustainability in the textile and apparel industries: production process sustainability. Springer International Publishing, Cham, pp 37–79
23. Mahmud I, Kaiser S (2020) Recent progress in waterless textile dyeing. J Text Sci Eng 10:1–3

24. Khattab TA, Abdelrahman MS, Rehan M (2020) Textile dyeing industry: environmental impacts and remediation. Environ Sci Pollut Res 27(4):3803–3818
25. Jacometti V (2019) Circular economy and waste in the fashion industry. Laws 8(4):27
26. Shamsuzzaman M, Islam MM, Hasan HMRU, Khan AM, Sayem ASM (2023) Mapping environmental sustainability of knitted textile production facilities. J Clean Prod 405:136900. https://doi.org/10.1016/j.jclepro.2023.136900
27. Nayak R, Panwar T, Nguyen LVT (2020) Sustainability in fashion and textiles: a survey from developing country. In: Nayak R (ed) Sustainable technologies for fashion and textiles. pp 3–30. https://doi.org/10.1016/B978-0-08-102867-4.00001-3
28. Arputharaj A, Raja ASM, Saxena S (2016) Developments in sustainable chemical processing of textiles. In: Muthu S, Gardetti M (ed) Green Fashion. Environmental footprints and eco-design of products and processes. Springer, Singapore. pp 217–252. https://doi.org/10.1007/978-981-10-0111-6_9
29. Lara L, Cabral I, Cunha J (2022) Ecological approaches to textile dyeing: a review. Sustainability 14(14):8353
30. Nimkar U (2018) Sustainable chemistry: a solution to the textile industry in a developing world. Curr Opin Green Sustain Chem 9:13–17
31. Zainith S, Sandhya S, Saxena G, Bharagava RN (2016) Microbes an ecofriendly tools for the treatment of industrial waste waters. In: Shankar J, Singh DP (eds) Microbes and environmental management. Studium Press, pp 75–100
32. Maiti S, Panda A, Madiwale P, Adivarekar RV (2022) Textile colouration technologies: today and tomorrow. In: Yusuf M, Shahid M (eds) Emerging technologies for textile colouration. CRC Press, pp 1–12
33. Babu KM (2022) Overview on significant approaches to waterless dyeing technology. In: Yusuf M, Shahid M (eds) Emerging technologies for textile colouration. CRC Press, pp 171–188
34. Panda SKBC, Sen K, Mukhopadhyay S (2021) Sustainable pretreatments in textile wet processing. J Clean Prod 329:129725
35. Singha K, Pandit P, Maity S, Sharma SR (2021) Harmful environmental effects for textile chemical dyeing practice. In: Ibrahim N, Hussain CM (eds) Green chemistry for sustainable textiles. Elsevier, pp 153–164
36. Bilińska L, Gmurek M (2021) Novel trends in AOPs for textile wastewater treatment. Enhanced dye by-products removal by catalytic and synergistic actions. Water Resour Ind 26:100160
37. Zhu B, Song Q, Liu J, Liu J, Gao W, Li L (2016) Effects of foaming parameters on sized-foam properties. Text Res J 86(19):2096–2105
38. Abate, MT, Tadesse MG (2021) Airflow, foam, and supercritical carbon dioxide dyeing technologies. In: Innovative and emerging technologies for textile dyeing and finishing. Scrivener Publishing LLC, pp 137–164
39. Rahman MM, Mashud M, Rahman MM (2023) Advanced technology in textiles: fibre to apparel. Springer Nature Singapore
40. Elbadawi AM, Pearson JS (2009) Foam technology in textile finishing. Textile Prog 33(4):1–31
41. Mao Z, Yu H, Wang Y, Zhang L, Zhong Y, Xu H (2014) States of water and pore size distribution of cotton fibers with different moisture ratios. Ind Eng Chem Res 53(21):8927–8934
42. Yu H, Wang Y, Zhong Y, Mao Z, Tan S (2014) Foam properties and application in dyeing cotton fabrics with reactive dyes. Color Technol 130(4):266–272
43. Bhavsar PS et al (2017) Superheated water hydrolyzed keratin: a new application as a foaming agent in foam dyeing of cotton and wool fabrics. ACS Susta Chem Eng 5(10):9150–9159
44. van der Walt GHJ, van Rensburg NJJ (1986) Low-liquor dyeing and finishing. Textile Prog 14(2):1–50
45. Mohsin M, Sardar S (2019) Multi-criteria decision analysis for textile pad-dyeing and foam-dyeing based on cost, performance, productivity and sustainability. Cellulose 26(6):4143–4157

46. Senthil Kumar P, Yaashikaa PR (2018) Sustainable dyeing techniques. In: Muthu, S. (eds) Sustainable innovations in textile chemical processes. Textile science and clothing technology. Springer, Singapore, pp 1–29
47. Banchero M (2013) Supercritical fluid dyeing of synthetic and natural textiles—a review. Coloration Technol 129(1):2–17
48. Bach E, Cleve E, Schollmeyer E (2002) Past, present and future of supercritical fluid dyeing technology–an overview. Rev Prog Coloration Related Top 32(1):88–102
49. Banchero M (2020) Recent advances in supercritical fluid dyeing. Color Technol 136(4):317–335
50. Amenaghawon AN, Anyalewechi CL, Kusuma HS, Mahfud M (2020) Applications of supercritical carbon dioxide in textile industry. In: Inamuddin, Asiri AM, Isloor AM (eds) Green sustainable process for chemical and environmental engineering and science. Elsevier, pp 329–346
51. von der Eltz HU, Christ W (1989) Process of treating textile material in jet dyeing machines and apparatus for performing same. U.S. Patent No. 4,862,546
52. Abate MT, Tadesse MG (2021) Airflow, foam, and supercritical carbon dioxide dyeing technologies. In: Rather LR, Haji A, Shabbir M (eds) Innovative and emerging technologies for textile dyeing and finishing. Wiley, pp 137–164
53. Nair GP (2011) Methods and machinery for the dyeing process. In: Clark M (ed) Handbook of textile and industrial dyeing. Elsevier, pp 245–300
54. Powar AS, Perwuelz A, Behary N, Hoang L, Aussenac T (2020) Application of ozone treatment for the decolourization of the reactive-dyed fabrics in a pilot-scale process—optimization through response surface methodology. Sustainability 12(2):471
55. Powar A, Perwuelz A, Behary N, Hoang LV, Aussena T (2022) Decolourization of the reactive dyed cotton fabrics with the use of ozone assisted eco-technology. J Fiber Bioeng Inform 15(1):27–33
56. Powar A et al (2021) Environmental profile study of ozone decolourization of reactive dyed cotton textiles by utilizing life cycle assessment. Sustainability 13(3):1225
57. Lou J, Ren L, Yuan J, Xu J, Gu Z, Fan X (2023) Mechanism and application of ozone fading: oxidative decolourisation of disperse dyes and waste-dyed polyester fabrics. Colouration Technol 139(3):338–349
58. Chen TY, Kao CM, Hong A, Lin CE, Liang SH (2009) Application of ozone on the decolourization of reactive dyes—orange-13 and blue-19. Desalination 249(3):1238–1242
59. Zulfiqar A et al (2023) A sustainable approach to dyed cotton fabric stripping using ozone. Sustainability 15(9):7467
60. Eren S, Gümüs B, Eren HA (2016) Colour stripping of reactive-dyed cotton by ozone treatment. Colouration Technol 132(6):466–471
61. Prabaharan M, Nayar RC, Kumar NS, Rao JV (2000) A study on the advanced oxidation of a cotton fabric by ozone. Coloration Technol 116(3):83–86
62. Ma X, Wang X, Yin X, Kan X, Wang Z (2018) Electrochemical stripping of cotton fabrics dyed with reactive black 5 in water and wastewater. Chemosphere 206:17–25
63. Muñoz I, Domènech X, Rieradevall J (2003) Life cycle assessment as a tool for green chemistry: application to Kraft pulp industrial wastewater treatment by different advanced oxidation processes. https://www.psa.es/es/unidades/tsa/docs/Tesis_Ivan_Munoz.pdf
64. Chatha SAS, Asgher M, Ali S, Hussain AI (2012) Biological colour stripping: a novel technology for removal of dye from cellulose fibers. Carbohydr Polym 87(2):1476–1481
65. Arzate S, Pfister S, Oberschelp C, Sánchez-Pérez JA (2019) Environmental impacts of an advanced oxidation process as tertiary treatment in a wastewater treatment plant. Sci Total Environ 694:133572
66. Bin AK, Roustan M (2000) Fundamental and engineering concepts for ozone reactor engineering. In: International Specialised Symposium IOA, Toulouse
67. Robinson T, McMullan G, Marchant R, Nigam P (2001) Remediation of dyes in textile effluent: a critical review on current treatment technologies with a proposed alternative. Bioresour Technol 77(3):247–255

68. Pandey A (2004) Concise encyclopedia of bioresource technology. Food Products Press, New York
69. Panizza M, Cerisola G (2009) Direct and mediated anodic oxidation of organic pollutants. Chem Rev 109(12):6541–6569
70. Martínez-Huitle CA, Brillas E (2009) Decontamination of wastewaters containing synthetic organic dyes by electrochemical methods: a general review. Appl Catal B 87(3–4):105–145
71. Xiong W, Yang Y, Hou Y, Huang G, Li H, Gan G (2021) Electrochemical dyeing of cotton fiber for environmental protection. In: Yang M, Lanzinha JCG, Bao X (eds) Hydraulic and civil engineering technology VI. IOS Press, pp 602–607
72. Lee J et al (2015) Conductive fiber-based ultrasensitive textile pressure sensor for wearable electronics. Adv Mater 27(15):2433–2439
73. An Y, Ma J, Zhu Z, Hu M, Shao J (2020) Study on a water-saving and salt-free reactive dyeing of cotton fabrics in non-aqueous medium of liquid paraffin system. J Textile Inst 111(10):1538–1545
74. Kulandainathan MA, Patil K, Muthukumaran A, Chavan RB (2007) Review of the process development aspects of electrochemical dyeing: its impact and commercial applications. Coloration Technol 123(3)
75. Rajkumar D, Kim JG (2006) Oxidation of various reactive dyes with in situ electro-generated active chlorine for textile dyeing industry wastewater treatment. J Hazard Mater 136(2):203–212
76. Chen G (2004) Electrochemical technologies in wastewater treatment. Sep Purif Technol 38(1):11–41
77. Vajnhandl S, Le Marechal AM (2005) Ultrasound in textile dyeing and the decolouration/mineralization of textile dyes. Dyes Pigm 65(2):89–101
78. Mu B, Li W, Xu H, Emanuel L, Yang Y (2019) Salt-free and environment-friendly reactive dyeing of cotton in cottonseed oil/water system. Cellulose 26:6379–6391
79. Roessler A, Dossenbach O, Meyer U, Marte W, Rys P (2001) Direct electrochemical reduction of indigo. Chimia (Aarau) 55(10):879
80. Roessler A, Crettenand D, Dossenbach O, Marte W, Rys P (2002) Direct electrochemical reduction of indigo. Electrochim Acta 47(12):1989–1995
81. Roessler A, Dossenbach O, Rys P (2002) Electrocatalytic hydrogenation of indigo: Process optimization and scale-up in a flow cell. J Electrochem Soc 150(1):D1
82. Roessler A, Crettenand D, Dossenbach O, Rys P (2003) Electrochemical reduction of indigo in fixed and fluidized beds of graphite granules. J Appl Electrochem 33:901–908
83. Roessler A, Crettenand D (2004) Direct electrochemical reduction of vat dyes in a fixed bed of graphite granules. Dyes Pigm 63(1):29–37
84. Bechtold T, Burtscher E, Amann A, Bobleter O (1992) Reduction of dispersed indigo dye by indirect electrolysis. Angew Chem, Int Ed Engl 31(8):1068–1069
85. Bechtold T, Burtscher E, Turcanu A, Bobleter O (1996) Indirect electrochemical reduction of dispersed indigo dyestuff. J Electrochem Soc 143(8):2411
86. Bechtold T, Gutmann R, Burtscher E, Bobleter O (1997) Cyclic voltammetric study of anthraquinone-2-sulfonate in the presence of acid yellow 9. Electrochim Acta 42(23–24):3483–3487
87. Tarábek J, Jähne E, Rapta P, Ferse D, Adler H-J, Dunsch L (2006) New acetophenone-functionalized thiophene monomer for conducting films on electrodes in chemical ion-sensorics: the synthesis and spectroelectrochemical study. Russ J Electrochem 42:1169–1176
88. Bechtold T, Turcanu A (2001) Electrochemical vat dyeing combination of an electrolyzer with a dyeing apparatus. J Electrochem Soc 149(1):D7
89. Bechtold T, Turcanu A, Schrott W (2008) Electrochemical reduction of CI sulphur black 1—correlation between electrochemical parameters and colour depth in exhaust dyeing. J Appl Electrochem 38(1):25–30
90. Sato M, Tanaka S, Kaeriyama K (1987) Soluble conducting polymers by electrochemical polymerization of thiophenes having long alkyl substituents. Synth Met 18(1–3):229–232

91. Kaeriyama K, Sato M, Tanaka S (1987) Electrochemical preparation of conducting polyalkylthiophene films. Synth Met 18(1):233–236
92. Wei Q, Wang Y, Yang Q, Yu L (2007) Functionalization of textile materials by plasma enhanced modification. J Ind Text 36(4):301–309
93. Pane S, Tedesco R, Greger R (2001) Acrylic fabrics treated with plasma for outdoor applications. J Ind Text 31(2):135–145
94. Shahidi S, Ghoranneviss M, Moazzenchi B (2014) New advances in plasma technology for textile. J Fusion Energy 33(2):97–102
95. Verschuren J, Kiekens P, Leys C (2007) Textile-specific properties that influence plasma treatment, effect creation and effect characterization. Text Res J 77(10):727–733
96. Kan CW, Yuen CWM, Tsoi WYI, Tang TB (2010) Plasma pretreatment for polymer deposition—improving antifelting properties of wool. IEEE Trans Plasma Sci 38(6):1505–1511
97. Shah JN, Shah SR (2013) Innovative plasma technology in textile processing: a step towards green environment. Res J Eng Sci 2(4):34–39
98. Haji A, Naebe M (2020) Cleaner dyeing of textiles using plasma treatment and natural dyes: a review. J Clean Prod 265:121866
99. Allan G, Fotheringham A, Weedall P (2002) The use of plasma and neural modelling to optimise the application of a repellent coating to disposable surgical garments. AUTEX Res J 2(2):64–68
100. Dave H, Ledwani L, Nema SK (2019) 8—nonthermal plasma: a promising green technology to improve environmental performance of textile industries. In: Shahid-ul-Islam, Butola BS (eds) The impact and prospects of green chemistry for textile technology. The textile institute book series. Woodhead Publishing, pp 199–249
101. Kan C, Yuen CM (2007) Plasma technology in wool. Text Prog 39(3):121–187
102. Atmospheric pressure plasma vapor treatment of thermo-sensitive poly(N-isopropylacrylamide) and its application to textile materials. J Fiber Bioeng Inform 4(3):285–290
103. Bhat NV, Netravali AN, Gore AV, Sathianarayanan MP, Arolkar GA, Deshmukh RR (2011) Surface modification of cotton fabrics using plasma technology. Text Res J 81(10):1014–1026
104. Liston EM (1993) Plasmas and surfaces—a practical approach to good composites. In: Akovali G (ed) The interfacial interactions in polymeric composites. Springer Netherlands, Dordrecht, pp 223–268
105. Ahmed NSE, El-Shishtawy RM (2010) The use of new technologies in colouration of textile fibers. J Mater Sci 45(5):1143–1153
106. Deshmukh RR, Bhat NV (2011) Pretreatments of textiles prior to dyeing: plasma processing. In: Hauser PJ (ed) Textile dyeing. IntechOpen
107. Joshi M, Butola BS (2013) 14—application technologies for coating, lamination and finishing of technical textiles. In: Gulrajani ML (ed) Advances in the dyeing and finishing of technical textiles. Woodhead publishing series in textiles. Woodhead Publishing, pp 355–411
108. Samanta K, Jassa M, Agrawal AK (2006) Atmospheric pressure glow discharge plasma and its applications in textile. Ind J Fibre Text Res 31(1):83–98
109. McCoustra MRS, Mather RR (2018) Plasma modification of textiles: understanding the mechanisms involved. Text Prog 50(4):185–229
110. Desai Shrojal RP, Singh M (2004) Surface modification of polyethylene. In: Albertsson AC (ed) Long term properties of polyolefins. Springer Berlin Heidelberg, Berlin, Heidelberg, pp 231–294
111. Valko EI (1952) The principles of textile finishing. Text Res J 22(3):213–222
112. Chinta SK, Landage SM, Kumar SM, Society DKTE (2012) Plasma technology & its application in textile wet processing. 1(5). Available at: https://www.ijert.org/plasma-technology-its-application-in-textile-wet-processing
113. Gorjanc M, Bukošek V, Gorenšek M, Mozetič M (2010) CF4 plasma and silver functionalized cotton. Text Res J 80(20):2204–2213
114. Sinha E, Panigrahi S (2009) Effect of plasma treatment on structure, wettability of jute fiber and flexural strength of its composite. J Compos Mater 43(17):1791–1802

115. Zille A (2020) 6—plasma technology in fashion and textiles. In: Nayak R (ed) Sustainable technologies for fashion and textiles. Woodhead publishing series in textiles. Woodhead Publishing, pp 117–142
116. Pozar DM (2012) Microwave engineering, 4th edn. John Wiley & Sons, Inc., Hoboken, New Jersey
117. Atav R (2013) The use of new technologies in dyeing of proteinous fibers. In: Günay M (ed) Eco-friendly textile dyeing and finishing. InTech, ch. 5, pp 128–129
118. Hasanbeigi A, Price L (2015) A technical review of emerging technologies for energy and water efficiency and pollution reduction in the textile industry. J Clean Prod 95:30–44
119. Ali S, Khatri A, Peerzada MH (2018) Emerging technologies for textile dyeing, 1st edn. Higher Education Commission, Pakistan, Islamabad
120. Elshemy N, Haggag K (2019) New trend in textile coloration using microwave irradiation. J Text Color Polym Sci 16(1): 33-48
121. Maiti S, Panda A, Madiwale P, Adivarekar RV (2022) Textile colouration technologies. In: Yusuf M, Shahid M (ed) Emerging technologies for textile colouration. CRC Press, pp 5–6
122. Al-Mousawi S, El-Apasery M, Elnagdi M (2013) Microwave assisted dyeing of polyester fabrics with disperse dyes. Molecules 18(9):11033–11043
123. Kale MJ, Bhat NV (2011) Effect of microwave pretreatment on the dyeing behaviour of polyester fabric. Coloration Technol 127(6):365–371
124. Huang D (2015) The application advantages of microwave fixation in cotton fabric dyeing. In: Proceedings of the 2015 International Conference on Education Technology, Management and Humanities Science. Atlantis Press, pp 535–539
125. Adeel S et al (2018) Microwave assisted modulation of vat dyeing of cellulosic fiber: improvement in colour characteristics. J Nat Fibers 15(4):517–526
126. Xue Z (2017) Effect of microwave pretreatment on dyeing performance of wool fabric. J Text Eng Fash Technol 1(6):217–222
127. Haggag K, El-Molla M, Ahmed K (2015) Dyeing of nylon 66 fabrics using disperse dyes by microwave irradiation technology. Int Res J Pure Appl Chem 8(2):103–111
128. Oghbaei M, Mirzaee O (2010) Microwave versus conventional sintering: a review of fundamentals, advantages and applications. J Alloys Compd 494(1–2):175–189
129. Elshemy N, Haggag K (2019) New trend in textile colouration using microwave irradiation. J Text Color Polym Sci 16(1):33–48
130. Ultrasound (2024) HarperCollins Publishers. Accessed: 27 Apr 2024. [Online]. Available: https://www.collinsdictionary.com/
131. Franco A, Bartoli C (2019) The ultrasounds as a mean for the enhancement of heat exchanger performances: an analysis of the available data. J Phys Conf Ser 1224(1):012035
132. Ahmed NSE, El-Shishtawy RM (2010) The use of new technologies in colouration of textile fibers. J Mater Sci 45(5):1143–1153
133. Manisha M, Sankhe M, Bardhan M (2004) Energy conservation in wet processing: development of low energy dyeing machine. Man Made Text India 47(6):200–206
134. Daniella P, Farid C (2015) Applications of ultrasound in processing, preservation and extraction. In: Application of ultrasounds for transformation processes of agroalimentary products
135. Jack B (1971) Ultrasonics: methods and applications, 1st edn. Butterworth and Co., London
136. Generating Ultrasound with Piezo Components. physik instrumente. Accessed: Apr. 27, 2024. [Online]. Available: https://www.physikinstrumente.com/
137. Jajpura L, Nayak R (2020) Ultrasound applications in textiles and apparels. In: Rajkishore N (ed) Sustainable technologies for fashion and textiles. Elsevier, pp 143–168
138. Shamshad A, Awais K, Mazhar P (2018) Emerging technologies for textile dyeing, 1st edn. Higher Education Commission, Lahore
139. Thakore KA (1988) Application of ultrasonics in dyeing of cotton fabric with direct dyes. Indian J Textile Res 13:208–212
140. Khatri M, Ahmed F, Jatoi AW, Mahar RB, Khatri Z, Kim IS (2016) Ultrasonic dyeing of cellulose nanofibers. Ultrason Sonochem 31:350–354

141. Udrescu C, Ferrero F, Periolatto M (2014) Ultrasound-assisted dyeing of cellulose acetate. Ultrason Sonochem 21(4):1477–1481
142. Ferrero F, Periolatto M (2012) Ultrasound for low temperature dyeing of wool with acid dye. Ultrason Sonochem 19(3):601–606
143. McNeil SJ, McCall RA (2011) Ultrasound for wool dyeing and finishing. Ultrason Sonochem 18(1):401–406
144. Jajpura L, Nayak R (2020) Ultrasound applications in textiles and apparels. In: Rajkishore N (ed) Sustainable technologies for fashion and textiles, 1st ed. Elsevier, ch. 7, pp 152–153
145. Ashokkumar M (2016) Ultrasonic synthesis of functional materials. SpringerBriefs in Molecular Science. Springer Cham
146. Bohrer E (1965) Automation in bleaching and dyeing with reference to the Thies Duomat. J Soc Dyers Color 81(6):258–261
147. Doherty RE (2000) A history of the production and use of carbon tetrachloride, tetrachloroethylene, trichloroethylene and 1,1,1-trichloroethane in the united states: part 1—historical background; carbon tetrachloride and tetrachloroethylene. Environ Forensics 1(2):69–81
148. Tang AYL, Wai Kan C (2020) Non-aqueous dyeing of cotton fibre with reactive dyes: a review. Coloration Technol 136(3):214–223
149. Chavan RB, Subramanian A (1982) Dyeing of alkali swollen and alkali swollen solvent exchanged cotton with a reactive dye. Text Res J 52(12):733–737. https://doi.org/10.1177/004051758205201201
150. Chattopadhyay DP, Chavan RB, Sharma JK (2007) Salt-free reactive dyeing of cotton. Int J Cloth Sci Technol 19(2):99–108
151. Pei L, Liu J, Cai G, Wang J (2017) Study of hydrolytic kinetics of vinyl sulfone reactive dye in siloxane reverse micro-emulsion. Text Res J 87(19):2368–2378
152. Alebeid OK, Hassan EA, Pei L (2021) Recent advances in non-aqueous dyeing systems. In: Rathe LJ, Shabbir M, Haji A(eds) Sustainable practices in the textile industry. Scrivener Publishing LLC, pp 43–73
153. Lim YJ, Kim TK, Cho KH (2001) Reactive dyeing of cotton in water-organic solvent mixture. Sen'i Gakkaishi 57(1):21-24
154. Dong X, Gu Z, Hang C, Ke G, Jiang L, He J (2019) Study on the salt-free low-alkaline reactive cotton dyeing in high concentration of ethanol in volume. J Clean Prod 226:316–323
155. Xia L et al (2018) Environmentally friendly dyeing of cotton in an ethanol-water mixture with excellent exhaustion. Green Chem 20(19):4473–4483
156. Gulzar AT, Farooq T, Kiran S, Ahmad I, Hameed (2019) Green chemistry in the wet processing of textiles. In: Islm S, Butol BS (eds) The impact and prospects of green chemistry for textile technology. Woodhead Publishing, pp 1–20
157. Sato K (2011) Recent patents on micro- and nano-bubble applications and potential application of a swirl-type generator. Recent Patents Mech Eng 4(3):202–211
158. Garcia B (2015) Reduced water washing of denim garments. In:Paul R (ed) Denim. Woodhead Publishing, pp 405–423
159. Ning CJ, He CH, Xu L, Liu FJ, He JH (2016) Nano-dyeing. Therm Sci 20(3):1003–1005
160. Mohsin M, Sardar S, Hassan M, Akhtar N, Hassan A, Sufyan M (2020) Novel, sustainable and water efficient nano bubble dyeing of cotton fabric. Cellulose 27(10):6055–6064
161. Liyanapathiranage A, Peña MJ, Sharma S, Minko S (2020) Nanocellulose-based sustainable dyeing of cotton textiles with minimized water pollution. ACS Omega 5(16):9196–9203
162. Zheng L, Zhang H, Yan J, Zheng J (2016) An industrial scale multiple supercritical carbon dioxide apparatus and its eco-friendly dyeing production. J CO_2 Utiliz 16:272–281
163. Lu J, Liu X, Liu L, Chen R (2010) Textile wastewater reuse as an alternative water source for dyeing and finishing processes: a case study. Desalination 258(1–3):229–232
164. Kant R (2012) Textile dyeing industry an environmental hazard. Nat Sci 4(1):22–26
165. Pang AZ, Abdullah YL (2013) Current status of textile industry wastewater management and research progress in Malaysia: a review. Clean–Soil, Air, Water 41(8):751–764

166. Hasanbeigi A (2010) Energy-efficiency improvement opportunities for the textile industry. Lawrence Berkeley Natl. Lab. Available at: https://escholarship.org/uc/item/6jw8s2gz
167. Chao C, Guan H, Zhang J, Liu Y, Zhao Y, Zhang B (2018) Immobilization of laccase onto porous polyvinyl alcohol/halloysite hybrid beads for dye removal. Water Sci Technol 77(3):809–818
168. Avnir R, Kaufman D, Reisfeld VR (1985) Organic fluorescent dyes trapped in silica and silicatitania thin films by the sol-gel method. Photophysical, film and cage properties. J Non Cryst Solids 74(2–3):395–406
169. Khatri A, Peerzada MH, Mohsin M, White M (2015) A review on developments in dyeing cotton fabrics with reactive dyes for reducing effluent pollution. J Clean Prod 87(1):50–57
170. Burkinshaw SM, Salihu G (2019) The role of auxiliaries in the immersion dyeing of textile fibres: part 4 theoretical model to describe the role of liquor ratio in dyeing cellulosic fibres with direct dyes in the absence and presence of inorganic electrolyte. Dyes Pigm 161:565–580
171. Digital dyeing and finishing pioneer Alchemie joins the British textile machinery association (BTMA). [Online]. Available: https://www.textileworld.com/textile-world/2023/03/digital-dyeing-and-finishing-pioneer-alchemie-joins-the-british-textile-machinery-association-btma/
172. Wright B. H&M Group invests in green dyeing and finishing specialist. [Online]. Available: https://www.just-style.com/news/hm-group-invests-in-green-dyeing-and-finishing-specialist/?cf-view
173. Abou Elmaaty TH, Sofan M, Elsisi H, Kosbar T, Negm E, Hirogaki K, Tabata I (2019) Optimization of an eco-friendly dyeing process in both laboratory scale and pilot scale supercritical carbon dioxide unit for polypropylene fabrics with special new disperse dyes. J CO_2 Utiliz 33:365–371
174. Agrawal B (2015) Supercritical carbon-dioxide assisted dyeing of textiles: an environmental benign waterless dyeing process. Int J Innov Res Creative Technol 1(2): 201–205
175. Lokesh KV, Dhanabalan V, Sukanya. Air-dyeing technology-a review. Textile Today. [Online]. Available: https://www.textiletoday.com.bd/air-dyeing-technology-a-review
176. Kumar D, Garje A, Desai K, Gupta D (2010) Dyeing without water. J Textile Assoc 71(4):171–174
177. Lokesh KV, Dhanabalan V, Sukanya. Air-dyeing technology-a review. Textile Today. [Online]. Available: https://www.textiletoday.com.bd/air-dyeing-technology-a-review
178. Bomgardner MM (2018) These new textile dyeing methods could make fashion more sustainable. [Online]. Available: https://cen.acs.org/business/consumer-products/new-textile-dyeing-methods-make/96/i29
179. Colouring textiles without the use of water. New Cloth Market. [Online]. Available: https://www.fibre2fashion.com/industry-article/4381/airdye-technology-colouring-textiles-without-the-use-of-water
180. Murugesh Babu K. Air-flow dyeing—an eco-friendly water preserving fabric dyeing technology. [Online]. Available: https://www.textileschool.com/5234/air-flow-dyeing-an-eco-friendly-water-preserving-fabric-dyeing-technology/
181. LYDIA HEIDA. Can waterless dyeing processes clean up the clothing industry? [Online]. Available: https://e360.yale.edu/features/can_waterless_dyeing_processes_clean_up_clothing_industry_pollution
182. Willis S. How sustainable dyeing is changing the textile industry. [Online]. Available: https://www.plugandplaytechcenter.com/resources/how-sustainable-dyeing-changing-textile-industry/
183. Pensupa N et al (2018) Recent trends in sustainable textile waste recycling methods: current situation and future prospects. In: Lin CSK (ed) Chemistry and chemical technologies in waste valorization. Springer International Publishing, Cham, pp 189–228
184. Yadav S, Tiwari KS, Gupta C, Tiwari MK, Khan A, Sonkar SP (2023) A brief review on natural dyes, pigments: recent advances and future perspectives. Results Chem 5:100733
185. Yusuf M, Shabbir M, Mohammad F (2017) Natural colourants: historical, processing and sustainable prospects. Nat Prod Bioprospect 7(1):123–145

186. Varadarajan G, Venkatachalam P (2016) Sustainable textile dyeing processes. Environ Chem Lett 14(1):113–122
187. Mabuza L, Sonnenberg N, Marx-Pienaar N (2023) Natural versus synthetic dyes: Consumers' understanding of apparel colouration and their willingness to adopt sustainable alternatives. Resourc Conserv Recycl Adv 18:200146
188. Plakantonaki S et al (2023) A review of sustainability standards and ecolabeling in the textile industry. Sustainability 15(15):11589

Chapter 3
An Overview of Waterless Dyeing in Textile Industry: Advancements, Challenges, and Environmental Impact Analysis

Rifat Jahan and Md. Rashedul Islam

Abstract The use of non-aqueous media in the chemical fiber dyeing and washing processes has become common due to the current problem of water scarcity. The textile sector uses up to 9 trillion gallons of water to produce 60 billion kilograms of fabric yearly, making it one of the biggest global contributors to environmental hazards. More specifically, large volumes of wastewater are produced by the global textile wet processing sector, and their release into the environment could have a negative impact on drinking water and aquatic life. A lot of unfixed dyes flow into water bodies during the dyeing process, and around 10 and 15% of the dye is lost as waste and finds its way into the environment. In addition, concerns about the environment and public health are raised by the standard textile dyeing procedures, which require a lot of energy, water, and auxiliaries rather than producing a lot of harmful effluents. Using heavy metal salts as mordants in the conventional natural dyeing process is not only hazardous for the environment but also a difficult method of controlling color stability; as a result, the procedure is not appropriate for large-scale manufacturing. Therefore, a waterless, environmentally friendly, sustainable textile dyeing method is appealing. This chapter focuses on the waterless dyeing procedures and several mechanisms of these dyeing methods. This chapter will cover recent advancements in waterless dyeing that can lower unfixed color loads in wastewater using different dyeing techniques, including plasma-induced dyeing, supercritical fluid (carbon dioxide) dyeing, ultrasonic wave-assisted dyeing, solvent dyeing, and many more.

Keywords Waterless dyeing · Dyeing techniques · Eco-friendliness · Ultrasonic dyeing · Supercritical fluid

---

R. Jahan · M. R. Islam (✉)
Department of Wet Process Engineering, Faculty of Textile Chemical Engineering, Bangladesh University of Textiles, Tejgaon, Dhaka, Bangladesh
e-mail: rashedul.islam@wpe.butex.edu.bd

M. Shahid et al. (eds.), *Advancements in Textile Coloration*,
https://doi.org/10.1007/978-981-96-5091-0_3

3.1 Introduction

The textile industry heavily utilizes water resources. In particular, the substantial effluent output, high potential for pollution, and high water consumption are mostly the result of the wet processing of textiles. Textile dyeing is carried out at high ratios of liquor to fabric, such as in the range of 1:5 to 1:20. In addition to consuming during dyeing, textile manufacturing facilities release a massive amount of wastewater that comprises dye and several dyeing auxiliaries [1, 2]. The main component used to color textiles is called textile dyes. It could be man-made or organic. Natural dyes were first used in Egyptian tombs approximately 4,000 years ago. Perkin discovered the first artificial dye in 1856, and since then, it has become common practice to color textiles using natural and synthetic dyes. By 2026, the textile dyes market is projected to grow to USD 10.13 billion worldwide. A huge quantity of fresh water is used as a solvent in traditional textile dying. The textile industry is the third-largest global water consumer, following the agricultural and energy production sectors. 1 kg of textile material is estimated to require approximately 100 -150 L of water on an average, with the dyeing process using around 16% of this total [3, 4]. The total amount of water used at various wet processing stages in the textile industry is depicted in Fig. 3.1.

The world's most precious natural resource is water, which comprises about 2.5% fresh water and over 97% salty. Approximately 68.7% of freshwater is found in snow and ice, while nearly one-third of the 2.5% of freshwater is stored as groundwater. Only about 0.3% of freshwater is readily accessible from rivers as surface water, lakes, marshes, and streams. Therefore, producers should create substitute methods for processing textiles [6] to preserve this finite freshwater supply. Textile fibers are typically modified in several ways to improve comfort, functionality, and appearance. These modifications are incorporated through mechanical, chemical, physical, biological, or combined processes. Large volumes of water are used in most commercially used wet processing techniques for textile fibers. These include scouring, bleaching, dyeing, printing, and finishing treatments. Many chemicals are also used

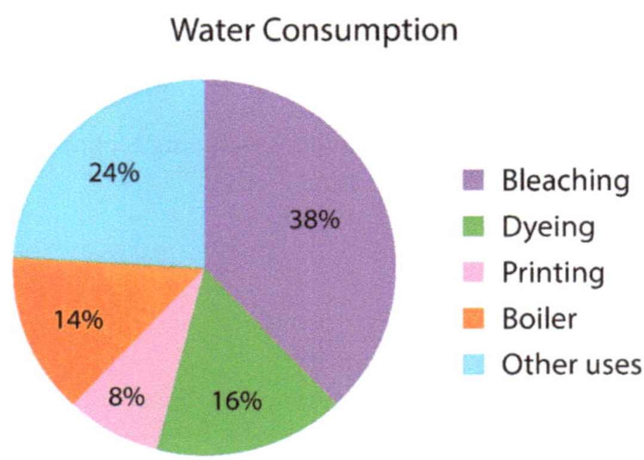

Fig. 3.1 Industrial water consumption in textile wet processing stages [5]

in these dyeing and finishing processes, releasing pollutants in the effluent. Most dyeing enterprises release effluent into the environment without any prior treatment. Human health is impacted directly and indirectly by the toxic and dangerous dyes widely utilized in the dyeing industries. For example, the BOD can be greatly raised by even a basic substance like acetic acid. Thus, it is essential to maintain clean, unpolluted water, particularly when dying textiles, and to create novel techniques such as waterless dyeing. Researchers have tried to lower water use in the textile wet processing industries to create a sustainable strategy for tackling the water shortage and minimizing wastewater treatment. Compared to traditional dyeing methods, non-aqueous or waterless dyeing procedures need less energy, generate less waste, use less water, and emit fewer greenhouse gases. Therefore, non-aqueous dyeing methods have gained priority throughout the past few decades to meet the growing demand for greener technology and move beyond its existing state of sustainability. Eco-friendly waterless dyeing of textile substrates could be achieved through air-dyeing, plasma irradiation, supercritical fluid or foam dyeing, and many more. Based on studies and comments from the scientific community, this chapter attempts to deliver an overview of the present state of development, benefit, and application of waterless dyeing techniques.

3.2 Waterless Dyeing Techniques

3.2.1 Waterless Air-Dyeing Process

The air-dye method places fabrics into dyeing equipment instead of dipping fabrics into the customary dye-and-water bath. This method is usually applied to produce paper-based colors for synthetic fabrics. Compared to conventional dyeing, air-dye technology can reduce water use by up to 95%, depending on the fabric, and lowers energy consumption by 86%. Several scholarly study articles highlight many other advantages of air dye, including reducing water consumption and the emission of hazardous waste into the atmosphere [7, 8]. One of the drawbacks of the air-dye technique is that it cannot be used on natural fiber like cotton, it is only effective on synthetic fabrics like polyester. Air dyeing is used in North America, Asia, and numerous European textile industries. The process flow diagram of the air-dye technique in textile technology is shown in Fig. 3.2.

The overall features of the air-dye technique have been stated in Table 3.1.

The equipment installation cost is high and the air-dye method is comparatively novel. Thus, the researcher's future focus in textile dyeing may be on creating low-cost, straightforward air-dye coloring techniques that produce almost no pollution and require less water for dyeing natural and synthetic fabrics. Recently, airflow-dyeing machines were designed and manufactured by a Chinese company named Cleantech Solutions International, Inc. Mr. Jianhua Wu, the CEO and founder of Cleantech Solutions, claimed that this airflow-dying apparatus could atomize water

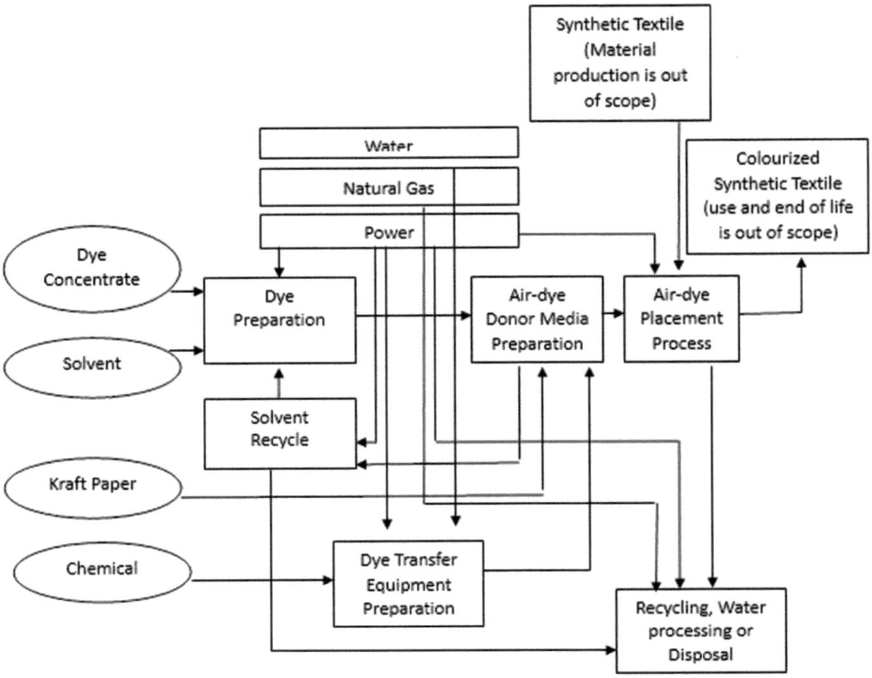

Fig. 3.2 Air-dye process flow diagram [9]

Table 3.1 Some features of the air-dye process [10]

Key-factors	Air-dye process
Power source/energy	Air instead of water
Fiber compatibility	The method functions with synthetic/ man-made fibers. Dying cotton fabrics involves giving the raw cotton a particular treatment
Working process	Atomized dyeing liquor is combined with high-pressure airflow and applied to the fabric through spraying
Water usage	90–95% less water than conventional water-dyeing
Energy consumption	Save 86% of energy than water-dyeing
Efficiency	Overall process time is reduced by approximately 25%
Color fastness of dyed fabric	Good to Better color fastness
Design potential	Dye or print both sides of the fabric simultaneously
Uniformity	Uniform dyeing to all fibers except Wool
Recycle process	Recycles the paper utilized in the process, and the inertness of the dyes allows them to be reverted to their original state for reuse
Environmental sustainability	Minimizes environmental impact and reduces industry's global warming contribution by 84% from
Machine installation	Expensive

under intense pressure. Consequently, the water used by their products can be reduced by two-thirds compared to older versions [9, 11]. A sustainable technology business, Colorep, located in California, has suggested an innovative new airflow dyeing method. The dyeing fluid is suggested to be atomized first in an airflow dyeing machine, then combined with high-pressure airflow, and after that, sprayed onto the fabric, which will be dyed. Water works as the solvent for the dyeing solution, allowing the dye compounds to closely interact with the fabric, thus requiring less water in this procedure. Typically, the air-dye method produces no hazardous waste emissions, doesn't waste water, and uses air rather than water to transfer dye [12, 13].

The Air-dye method significantly lowers the environmental impact of the color application process while enhancing the resulting fabric's performance during usage. For parts of the world without access to the water resources that conventional methods need, air-dye technique offers a significant opportunity to localize manufacturing by eliminating the need for water at the point of color application. The air-dye method also significantly lowers the energy required to apply color because conventional techniques require considerable amounts of energy to boil the water and dry the fabric. Advantages from improved power consumption, power source efficiency, and direct dye application without donor medium are appreciated as air-dye technology develops. The development of air-dye technique is a direct response to the growing public awareness of the negative environmental effects of conventional dye application methods and the advancement of related technologies.

Air-dye fabrics are designed to feel exactly like textiles dyed traditionally. It is possible to dye them all one color, one color on each side, one side with a pattern, and the other with a solid color, or both sides with a design.

Some significant applications of Air-dyed fabrics are listed below:

- **Apparel**: Fabric for clothing, accessories including scarves, aprons, ties, etc., personalized specialty fabrics items, umbrellas, flags, awnings, tents, uniforms for civic, business, and medical purposes, sarongs, swimwear, outerwear, yoga costume, sportswear, activewear, dance costume, men's wear, lingerie, intimate clothing, sleepwear, cosmetic bags, handbags and shopping bags, and more.
- **Industrial Print**: Vinyl transfer, thermo-films, thin paper, melamine, and hard surface laminate.
- **Medical**: Cubicle curtains, hospital gowns.
- **Interior**: Carpet, ceiling tiles, drapes, curtains, blinds, bed linens, wall coverings, tablecloths, napkins, shower curtains, pillows, mattress covers, and furniture, such as couches and chairs.

3.2.2 *Supercritical Carbon-Dioxide ($ScCO_2$) Dyeing Process*

The term "supercritical state" refers to the fourth phase of matter. An idealized pressure–temperature curve of a pure substance is depicted in Fig. 3.3, which indicates that the supercritical state is observed to exist at the temperature and pressure values above the so-called critical point. A substance's isothermal compressibility tends to infinity as it approaches its critical point. The substance's specific volume or density correspondingly alters significantly. A substance that, under normal circumstances, is a gas but has a substantially higher solvent capacity and a density similar to that of a liquid is in the critical region. This tendency is because when density rises, the mean intermolecular distance reduces, increasing the number of interactions between the solute and the solvent. Though supercritical fluids have densities resembling liquids, they also have more characteristics like gases. For instance, diffusivity levels are often high, while viscosity values are comparatively low. Minimal viscosity leads to relatively easier diversion and pumping of supercritical fluids. Mass transport is often facilitated by high diffusivity. Supercritical fluid dyeing was first created by German professor Schollmeyer and his colleagues in the late 1980s. Numerous investigations and trials, ranging from laboratory to bulk production, have been conducted since then [14, 15]. Carbon dioxide (CO_2) is preferred as a supercritical fluid due to its availability and harmless and flammable properties. Furthermore, CO_2 has a considerably lower critical temperature (Tc = 31.1 °C) and pressure (Pc = 74 bar) than many other fluids, making it a more suitable option for dyeing. At this condition, supercritical CO_2 exhibits both liquid and gas behavior. Compared to water, low viscosities and diffusion properties like gas can help minimize dyeing periods, while liquid-like densities help to dissolve hydrophobic dyes [16]. $ScCO_2$ dyeing is a waterless dyeing technique that offers quick dye absorption, a brief dying process, the ability to recycle CO_2 and dye, and zero discharge [17]. Nowadays, $ScCO_2$ dyeing is regarded to be the most affordable, environmentally friendly, and sustainable dying method available.

A schematic diagram of traditional aqueous dyeing and $ScCO_2$ dyeing is depicted in Fig. 3.4. DyeCoo created a waterless dyeing technique in 2008 that uses $ScCO_2$ for various synthetic and natural colors as solvents. Due to the benefits of DyeCoo's

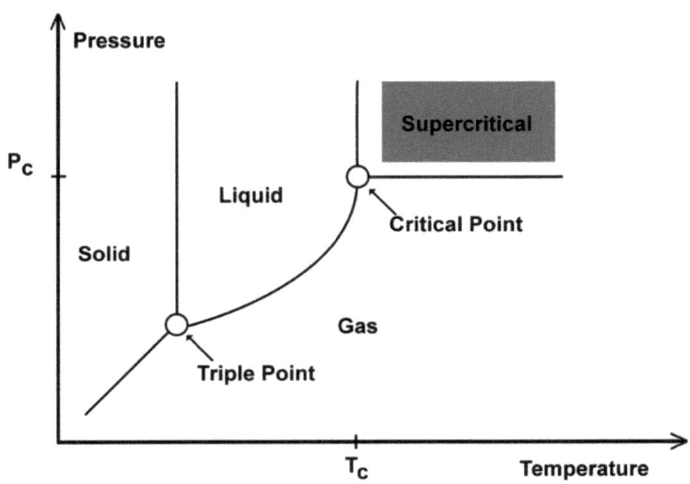

Fig. 3.3 Pressure–temperature curve for a pure substance [17]

techniques, companies like Nike, Adidas, and IKEA have all noticed and begun using this procedure for product dyeing. Additionally, DyeCoo is the first seller of CO_2-dyeing equipment in the globe. Since 2010, Thai mills have begun using DyeCoo's waterless textile dyeing technique using $ScCO_2$, and more recently, Taiwan and India. Yan et al. created a unique dye and colored silk in an eco-friendly way using supercritical carbon dioxide. The synthetic special dye on silk demonstrated excellent adsorption, uptake, and fixing behaviors and the best color fastness, evaluated at 4–5 in the color fading test [18]. In addition, five unique disperse dyes were created by Elmaaty et al. and used to color polypropylene textiles using the supercritical carbon dioxide dyeing method. These dyes demonstrated strong color fastness properties and good color strength (K/S) values on a pilot scale. The dye and the process show that it is possible to dye polypropylene on an industrial scale in $ScCO_2$ [19]. Scientists conducted a second investigation into PET fiber dyeing in a supercritical CO_2 medium. The results showed that equilibrium dye uptake rose at higher pressures and that dye absorption might be improved by adding a small quantity of acetone as a co-solvent [20].

Some significant features of $ScCO_2$ dyeing have been enlisted in Table 3.2.

There are several possible advantages when $ScCO_2$ is used instead of water in the process. The first and most important thing is eliminating the use of water and treating any associated trash. There is a noticeable boost in production due to the substantially shorter dying times compared to the conventional aqueous dyeing procedures, which also eliminate the need for yarn drying. Dryer effluent is naturally eliminated as well since drying is no longer necessary. In $ScCO_2$ dyeing, dye utilization is usually very high; as much as 99% has been recorded in certain cases. Thus, very little waste dye leaves the process and is, in any case, collected for either disposal or reuse. Recycle rates for the $ScCO_2$ utilized during dyeing are usually higher than 90%. Overall, supercritical carbon dioxide offers several advantages, including (a) total removal of water pollution and pre-treatment; (b) lower energy consumption during textile drying; (c) no requirement for auxiliaries; (d) higher fixation and levelness of dyes; (e) savings of time; (f) plentiful in nature and non-hazardous; and (g) ease of recycling [21].

Researchers became interested in dyeing cotton fiber when supercritical carbon dioxide was successfully applied to dye synthetic fibers using disperse dyes. It was found that to achieve good color intensity, cotton has to be modified to an extremely higher degree. A study explained how to dye cotton and wool with disperse dyes by using a supercritical CO_2 medium that first treated the materials with a polyether derivative called Glyezin CD, which is also a swelling agent [22]. In a further study, the hydroxyl groups in cotton were esterified using benzoyl chloride before $ScCO_2$-assisted disperse dyeing. A reaction with sodium benzoyl-thioglycollate and benzoyl chloride was used to modify the cotton [23]. A different investigation detailed pre-treating cotton with polyethylene glycol, a well-known plasticizing chemical, and then dying it in $ScCO_2$ with disperse dye while benzamide was present. The study discovered a good dye uptake, including light and wash fastness, because benzamide partially occluded the cellulose pores, establishing hydrogen bonds with polyethylene glycol and cotton [24]. However, the dyeing of hydrophilic fibers is obviously

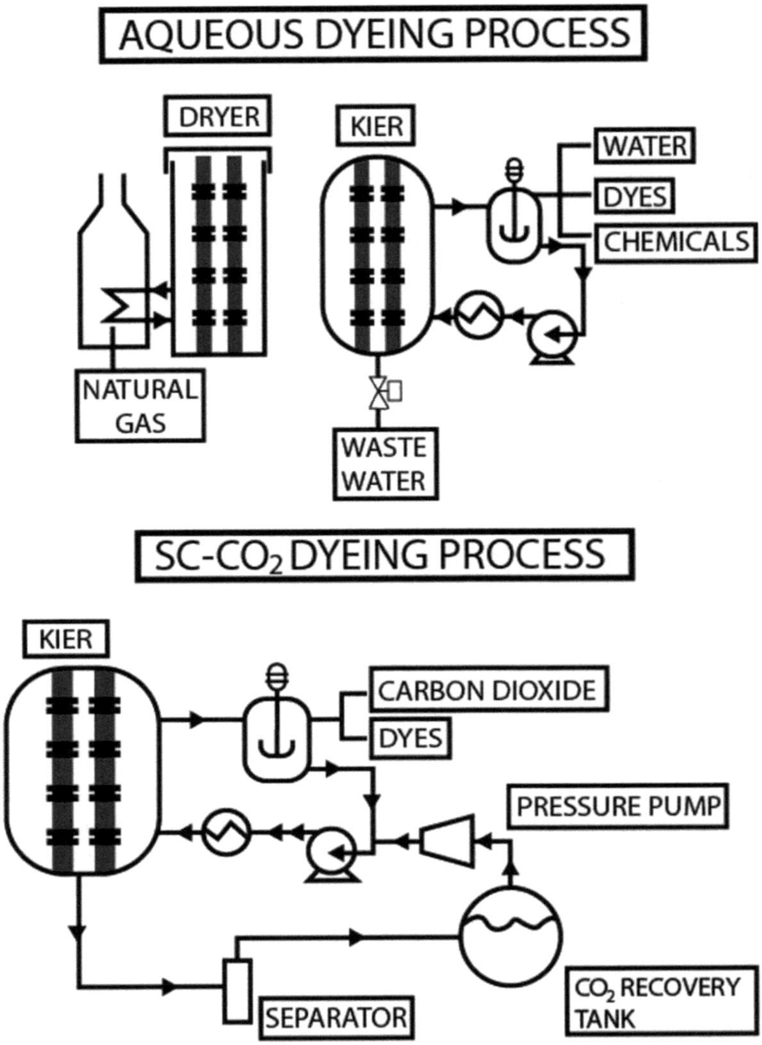

Fig. 3.4 Conventional aqueous and $ScCO_2$ dyeing process [17]

hampered by the low polarity of $ScCO_2$. Due to gravity, non-homogeneous fluid distribution and dye accumulation occur since most $ScCO_2$ dyeing equipment is arranged vertically, resulting in uneven dyeing. Huang et al. reported a pilot scale horizontal apparatus for zipper tape dyeing at a relatively low temperature of 90 ± 2 °C in $ScCO_2$ medium. However, their commercialization would have significant strategic and practical consequences [25].

Table 3.2 Some features of the $ScCO_2$ dyeing process [10]

Key-factors	$ScCO_2$ dyeing process
Power source/energy	$ScCO_2$ as fluid
Co-solvent	Presence of a co-solvent assists in improving the solubility of solutes in $ScCO_2$ even at very low quantity
Fiber compatibility	Suitable for acetate, nylon, and polyester dyeing, although there are some accessible experimental research on natural fibers
Dyeing parameters	Dyeing at high pressure and temperature
Water usage	This is a dry process and eventually eliminates the usage of water
Energy consumption	Very energy efficient
Efficiency	The time required for the solid dyestuff to dissolve will be substantially reduced because of the beneficial diffusion qualities
Color fastness of Dyed fabric	Good to Better color fastness
Uniformity	Any non-uniformity in the fluid medium may impair the consistency of dye distribution resulting in staining
Reuse	After use, CO_2 may be vacuumed away, enabling 95% recovery and reuse
Environmental sustainability	Non-explosive, non-toxic and ecologically safe technique for the environmentally responsible and sustainable textile sector
Machine installation	Expensive

3.2.3 Plasma-Assisted Waterless Dyeing Process

Plasma technology offers a dry, environmentally friendly method of modifying the surface characteristics of different textiles while still maintaining their general properties. Plasma, a partially ionized gas, is frequently regarded as the fourth state of matter. Plasma may be categorized as hot/thermal or cold/non-thermal based on its temperature. The only efficient surface modification method for heat-sensitive textile and polymeric materials is low-temperature plasma (LTP) [26]. A radio frequency (RF) resonant current applied in a coil or an electric field created between two electrodes with a gas between them can both produce an LTP. This can be done in an enclosed area at atmospheric pressure or at a lower pressure. LTPs include several reactive species, such as ions, free radicals, fast-moving electrons, and photons in the shortwave ultraviolet spectrum. All of these chemicals have the ability to initiate physical and chemical processes on the surface of a substrate. These alterations are only a few nanometers deep, indicating that the plasma only impacts the thin outside layer of the substrate [27]. Figure 3.5 shows a textile fiber surface modified by plasma treatment.

Classifying all types of plasma into one category is challenging due to their great diversity. A state of thermal equilibrium among all the species within the gas is a characteristic of thermal plasmas, which can reach temperatures of several thousand degrees Celsius. Stars, the Northern Lights, and other celestial bodies most likely include thermal plasma. The destructive nature of this kind of plasma is too great for

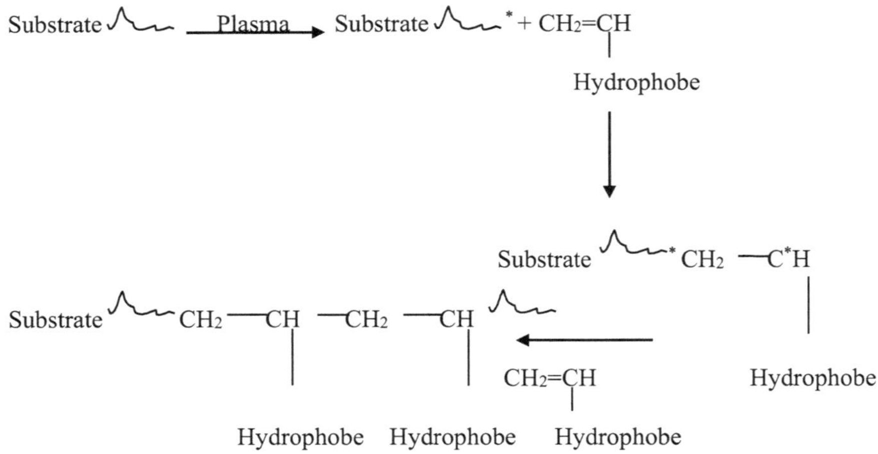

Fig. 3.5 Schematic representation of plasma-induced modified fiber structure

any textile material. Plasma is considered non-thermal when electrons, higher mass particles, such as ions, neutral atoms, or molecules, and pieces of neutral molecules cannot even reach local thermodynamic equilibrium. In this type of plasma, the electrons' temperature is significantly higher than other particles. The electrons can reach temperatures of 104–105 °K (1–10 eV), while the gas remains near room temperature [28]. Non-thermal plasmas, sometimes called cool plasmas, are particularly well suited for processing and surface modification of textile materials because most are heat-sensitive polymers. Such discharges have been found to have the primary benefit of significantly altering the surface's chemistry and morphology, which increases hydrophilicity and increases fibers' susceptibility to various chemical species without changing material properties [29]. There are two types of cold plasmas: vacuum or low-pressure plasmas and atmospheric-pressure plasmas. The following are the four primary categories of atmospheric pressure plasmas used in textiles.

3.2.3.1 Corona Discharge Plasma

The earliest plasma technology used to modify polymer surfaces is called Corona. A corona discharge occurs at atmospheric pressure when two electrodes of varying sizes and shapes are subjected to a high voltage of 10–15 kV at a low frequency. When the distance between the electrodes decreases to approximately 1 mm, the discharge energy density quickly declines, rendering thick materials and rapid, reliable treatments incompatible with very tiny inter-electrode spacing. Corona treatment only has superficial effects on textiles, primarily impacting loose fibers rather than deeply penetrating the fabric. While it can increase the surface area and roughness of fibers, its effects remain limited to the surface [29]. Figure 3.6 shows corona discharge technique in textile dyeing process.

Fig. 3.6 Diagram of a discharge setup showing a power supply connected to an electrode

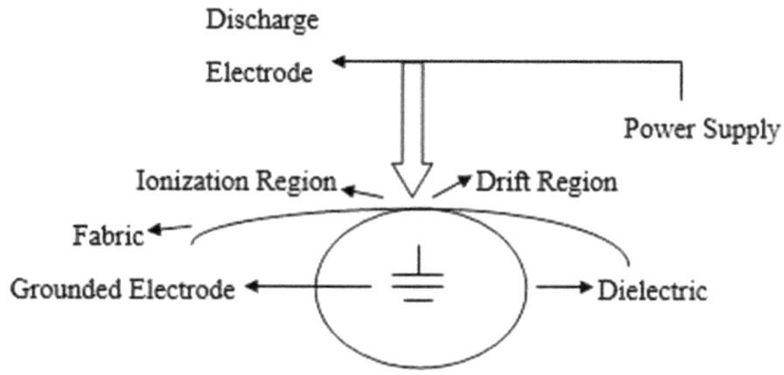

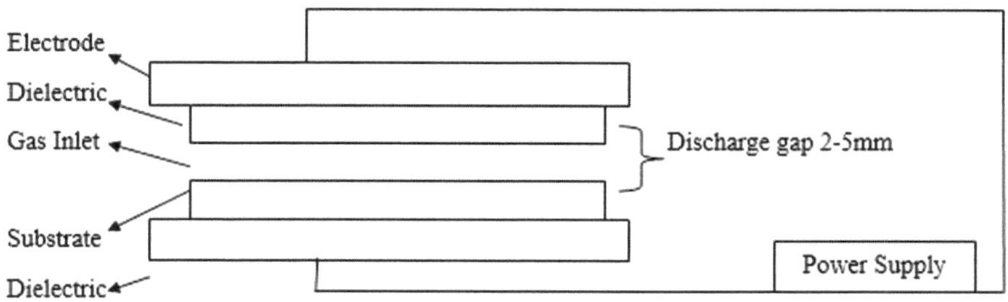

Fig. 3.7 Diagram of a dielectric barrier discharge system

3.2.3.2 Dielectric Barrier Discharge Plasma

The transmitted charge accumulates on the surface of the dielectric layer, which surrounds the minimum of one electrode in the dielectric barrier discharge plasma. This layer serves to both even the distribution of the charge throughout the electrode surface and restricts the amount of charge that may be transferred by each individual micro-discharge. However, it varies somewhat and lasts just a brief while. This can throw off the chain of chemical processes needed to create significant species. The potential for adaptation to very large systems has sparked interest in commercial uses for this highly effective non-thermal atmospheric plasma source [30]. Figure 3.7 displays the application method of dielectric barrier discharge plasma in textile dyeing process.

3.2.3.3 Atmospheric Pressure Glow Discharge Plasma

It is produced at a higher frequency and lower voltage than dielectric barrier discharge plasma. Its features, which guard against surface heating and damage, include homogeneity, a long lifespan, and power densities in low-to-moderate areas. A few millimeters separate two parallel electrodes that are connected to a radiofrequency source. The electrodes must be conductive in order to maintain a DC glow discharge. The simplest situation involves applying a potential between electrodes inside a cell that

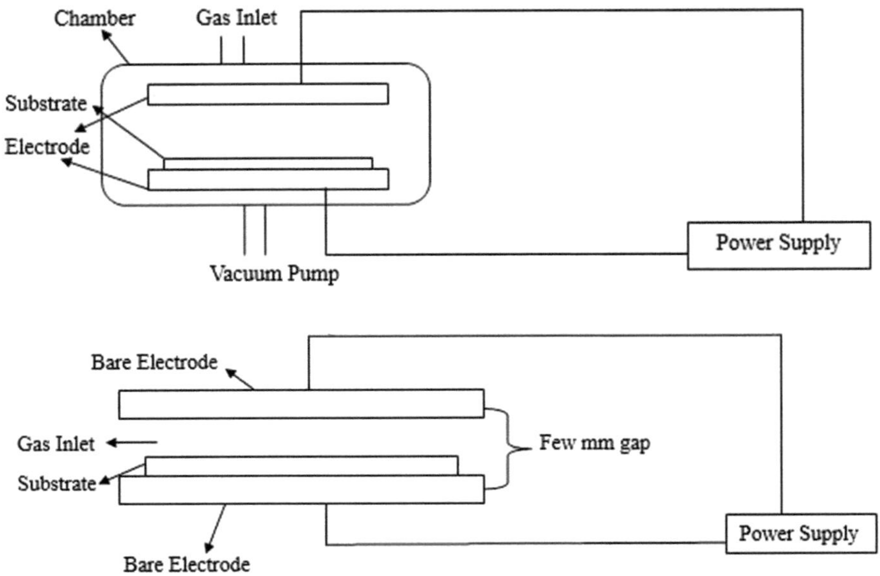

Fig. 3.8 Diagram illustrating a circuit setup for a gas chamber

contains a gas at atmospheric pressure, usually argon or helium, between a few kV and 100 V to cause a discharge [31]. Figure 3.8(a) and 3.8(b) shows low pressure glow discharge plasma as well as atmospheric glow discharge plasma application technique respectively in textile dyeing process.

The benefit of these methods is that, due to the extremely narrow penetration range of plasma treatment, the polymer's surface properties are typically altered without affecting its bulk properties. When the surface-to-volume ratio of a material rapidly increases, as it does in the case of fibers or fibrils, this isn't always the case. However, the understanding of the potential for modifying polymers with glow discharges was applied to technological and natural textile materials to establish plasma therapy for textiles.

3.2.3.4 Atmospheric Pressure Jet Plasma

The ability of atmospheric pressure plasma jets to treat fabrics from one side makes it essential for the plasmas to penetrate to the other side of the fabric to treat fabrics that require treatment on both sides successfully. While it is far gentler than a plasma torch, this room-temperature technique is nevertheless rather effective. Compared to dielectric barrier discharge, plasma jets have the benefit of producing homogenous reactive gasses and being able to function on the surface of any form of object. Nevertheless, the treated material can only be treated on one side facing the plasma jet [32]. A plasma treatment offers a variety of options for treating a polymer surface,

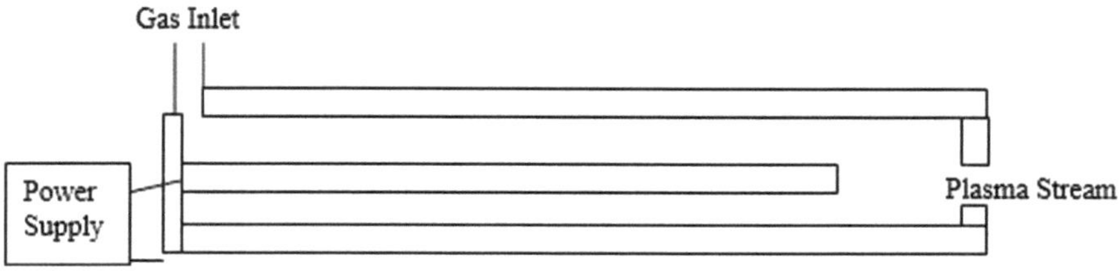

Fig. 3.9 Diagram of a plasma generation circuit

made possible by modifying variables such as gas flows, power, pressure, and treatment duration. Depending on the gas composition and plasma conditions, ions, electrons, fast neutrals, radicals, and vacuum UV radiation impact the polymer treatment, causing activation, etching, and cross-linking. Figure 3.9 shows atmospheric pressure plasma jet application technique in textile dyeing process.

There are four stages that essentially depend on the plasma-assisted textile dyeing conditions. These are surface cleaning, etching, surface activation, and polymerization.

- **Surface cleaning**: The process of eliminating impurities and pollutants from a substrate's surface, including oils, greases, and oxides, is known as surface cleaning. These contaminants are volatilized and removed by plasma cleaning without changing the bulk characteristics of the substrate. The most widely utilized noble gas in plasma treatment is argon due to its chemical inertness with the surface material, low cost, and high ablation efficiency [33].
- **Surface etching**: The method of plasma etching creates volatile chemicals through chemical reactions while physically removing material from the surface of a substrate. The etched surface may still contain pieces of lower molecular weight. Nitrogen or oxygen plasmas and inert gases like argon and helium are frequently used in surface etching. The plasma composition, substrate type, and operating parameters, such as power, gas flow, and substrate placement, all affect the etching rate.
- **Surface activation**: Surface activation introduces new functional groups to a substrate to change its surface energy and impart new properties. By increasing the surface energy, it can enhance their stickiness, printability, or wettability. In non-polymerizing gases, plasma activation occurs. When reactive plasma species come into contact with a treated substance, they break covalent bonds and generate free radicals. Carbonyl, carboxyl, hydroxyl, amino, and other reactive functional groups are produced on the substrate surface as a result of interactions between surface radicals and active plasma species. This activation changes the surface characteristics and chemical reactivity. For instance, oxygen plasma treatment increases the surface energy of a material by introducing hydrophilic and polar groups [34].
- **Polymerization**: With plasma polymerization, electrical monomer discharges can create thin coatings on various surfaces. This is the process of polymerizing an

organic monomer, like hexafluoropropylene or tetrafluoroethylene, in plasma to produce a thin layer of polymer on the surface of the substrate. The process starts with the surface being exposed to plasma radiation, which activates the surface and creates reactive bonding sites in preparation for coating. The plasma deposits and grafts molecular fragments to the surface during the polymerization process [35, 36].

Even though plasma technologies have many potential benefits, environmental friendliness, and application possibilities, their adoption in textile dyeing is still restricted because of three key issues:

- Surface cleanliness: Because plasma treatment primarily affects the textile's outermost layer, contaminations or variations in its surface could have a major detrimental impact.
- Dye penetration: unlike wet treatments, plasma species cannot pierce deeply enough into the three-dimensional textile structure to ensure appropriate treatment. One important factor to consider in this regard is the pressure of plasma treatment.
- Vast surface area: Textiles made up of individual fibers have a vast surface area that is typically orders of magnitude larger than that of flat films [37].

The optimum method for treating textiles is atmospheric non-thermal plasma because most textile materials are heat-sensitive polymers. The use of plasma in textile dyeing is a very advantageous method for increasing dye-ability and lowering water, energy, and time consumption. The industries can demonstrate their high production and high-quality products at a reduced cost. To mitigate the difficulty of producing dye effluent during the standard dyeing process, plasma treatment is typically used as a stage of the pre-treatment process. Although plasma post-treatment can drastically cut dye effluent, little attention has been paid to this, and it is unclear how well plasma fixes natural colors. In research, a plasma-assisted spray-dying procedure with no wastewater discharge was suggested as a clean way to dye wool fabric with onion skin. Four distinct plasma parameters were used, which varied the treatment period (30 s and 60 s) and plasma operating power (40 W and 64 W). The mordanting and dyeing were done in multiple rounds utilizing a sequential spraying-drying technique to produce the same shade as a conventional dyeing procedure done simultaneously [38]. Following plasma post-treatment, the wash and crocking color fastness were improved, and the potential for unfixed colors to adhere to the cloth was verified. Overall, the plasma-assisted spray dyeing method produced equivalent color fastness and better color strength in less time and energy than conventional dyeing, producing no wastewater. This study used air as the gas source in a smart plasma chamber to treat vacuum plasma-assisted dyeing of wool cloth directly, without scouring. Operating at 6.3 A current and 24 V control voltage, the machine had two aluminum electrodes, a 20–50 kHz operating frequency, and 80 W of power. Power and treatment duration were changed throughout the tests, while the operating pressure remained fixed at 0.1 mbar. Samples of fabric were successively treated on both sides with plasma. Using a magnetic stirrer (0.6 W) and 10 ml of water and natural dye extract, 0.1 g of alum was combined to create a mordant solution that was then heated to 40 °C for 10 min. The cloth that had been plasma-treated was then sprayed with approximately 1 ml of dye solution and 1 ml of mordant solution on each side independently. The dyed fabric was rapidly dried using a 2 kW iron set to medium heat (maximum 150 °C) following the spraying process. The drying process in enterprises employing calendar rollers, which apply heat and pressure, was considered when evaluating the ironing method. Following the final ironing stage, the textiles underwent plasma post-treatment under the corresponding plasma pre-treatment conditions. Following the plasma post-treatment, a change was seen in the wool's weak C–H groups (about 1385 cm^{-1}), which may have been connected to dye adhesion. After the post-plasma treatment, the enhanced wash and crock fastness characteristics further supported this.

The effects of different plasma treatments on the dyeing of several natural and synthetic fibers have been stated in Table 3.3 [39].

Table 3.3 Effects of plasma-assisted dyeing of textile fibers

Type	Plasma parameters			Natural dye	Fiber	Effect
	Gas	Power (W)	Duration (s)			
Atmospheric-pressure plasma	Air	50	180	*Berberis Vulgaris*	Wool	Improved colour strength, fastness, and antibacterial properties
	He and He/N_2	5000	7–14	*A. nilotica* and *A. catechu*	Wool	Improved wicking and shade depth
	Air + Argon	50–200	30–300	Almond shell waste	Wool	Increased colour yield, more homogeneous dyeing, reduced amount of mordants
	NH_3	75	900	*Rubia cordifolia*	Polyester	Improved dyeability and fastness properties

(continued)

Table 3.3 (continued)

Type	Plasma parameters			Natural dye	Fiber	Effect
	Gas	Power (W)	Duration (s)			
Low-pressure plasma	O_2, Argon, O_2/ Argon	150	180	*Cuminum Ciminum*	Wool	Improved dyeability
	O_2 + Chitosan	150	150–300	Pomegranate rinds	Cotton	Increased colour strength and fastness
	Water vapour	500	10	*Fallopia japonica*	Cotton	Improved dye uptake and antibacterial activity
	O_2 and Argon	20	300	Lac	Silk	Improved adsorption capacity
	O_2 and N_2	100	600	*Phytolacca Decandra* L	Silk	Improved colour yield
	O_2	300	1200	*Caesalpinia sappan* L. wood	Polyester	Better dye uptake and reduced dyeing time

3.3 Waterless Solvent Dyeing Process

Solvents were first employed in the textile dyeing business to lower wastewater discharges and reduce the effluent load on the environment. Over the course of time, 1,1,1-trichloroethane (TCA), trichloroethylene (TCE), and tetrachloroethylene (PCE) are the solvents that are most frequently used in solvent dyeing. The principal organic solvents utilized in producing the non-aqueous dyeing medium are decamethyl cyclopentasiloxane, heptane, octane, and nonane. The development and expansion of the global synthetic organic chemical industry has been associated with the manufacturing and use of chlorinated solvents throughout history [40]. Tetrachloroethylene has potential use in the textile industry as a carrier for processes such as sizing, de-sizing, scouring, and finishing. Chlorinolysis of light hydrocarbons, oxychlorination of ethylene or C2-chlorinated hydrocarbons, and high-temperature chlorination of ethylene or 1,2-dichloroethane are some of the processes that can be used to make tetrachloroethylene. It could also function as a carrier for silicones, lubricants, and adhesives and a chemical intermediary in synthesizing fluorinated chemicals. However, this solvent was regulated as a volatile organic compound in several countries due to its role as a precursor of ground-level ozone, which resulted in a drop in its utilization. The acetylene synthesis method is the initial step in the manufacturing of trichloroethylene (TCE). Because acetylene is so expensive, the

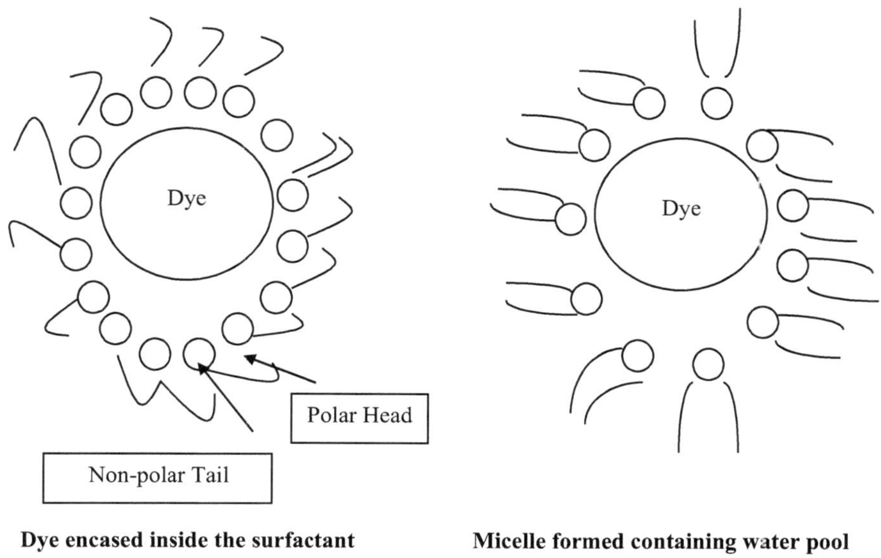

Fig. 3.10 Mechanism of solvent dyeing [44]

industrial process increasingly switches to chlorinating or oxychlorinating ethylene or 1,2-dichloroethane [41]. TCE can be used as a solvent for waterless finishing and to scour textiles when dimethyl formamide-trichloroethylene solvent mixtures are used in the textile industry. Additionally, it may be used as a chemical intermediary in the production of fire-causing substances, polyvinyl chloride, hydrofluorocarbons, and chloroacetic acid. However, the environmental regulations also result in a decrease in its consumption [40, 42, 43]. Regarding 1,1,1-trichloroethane (TCA), it can be produced in several ways, such as by hydrochlorinating 1,1-dichloroethylene in the presence of a $FeCl_3$ catalyst, chlorinating 1,1-dichloroethane, and chlorinating ethane non-catalytically. TCA is a substance used in the textile industry for manufacturing and cleaning leather and suede clothing. It might also be applied as a solvent to textiles to make them oil- and water-resistant. Unfortunately, it is considered an ozone-depleting chemical and a dangerous air pollutant. The reverse micellar system, created using an organic solvent in a non-aqueous medium, is utilized in developing solvent-assisted dyeing systems. In solvent dyeing, reverse micelles have been explored as dyeing media due to their advantages in energy efficiency, environmental sustainability, and safer working conditions. In non-polar liquids, certain surfactants cause the reverse micelles, which are tiny spherical aggregates, to self-assemble. In non-aqueous media, they can create a stable aqueous micro-environment (water pool) by solubilizing a small amount of water in their interior region. The mechanism of solvent dyeing of textiles has been drawn in Fig. 3.10.

Reverse micelles, created by combining surfactant and co-surfactant in non-aqueous solvent media, are often prepared as the carriers for transferring the colors

Table 3.4 Advantages and disadvantages of non-aqueous solvent dyeing process [47]

Dyeing system	Methods	Advantages	Disadvantages
Two-solvent system	Both polar and non-polar solvents apply	Easy application	Harmful to human health and the environment
Dye/solvent suspension system	Dye needs to be ball milled with oleic acid; samples need to be pre-padded with alkali solution	Use of green solvents	Poor suspension of dye molecules without the addition of dispersants
Reverse micellar system	Surfactants are needed for the formation of reverse micelles	Stable encapsulation of hydrophilic dye molecules in non-aqueous solvent media	Phase separation occurs depending
Non-nucleophilic solvent system	Fiber needs to be pre-swelled at a high temperature	Reduction of organic waste and use of electrolytes	High energy consumption for solvent reclamation

from the dye liquid to the fiber's surface in reverse micellar solvent-assisted dyeing. Following their exhaustion from the fiber surface, the dyes are covalently fixed in the fiber matrix. The electrolyte is not added during the entire dyeing process. The final dyeing performance is determined by the dyeing parameters, such as the ratio of solvent to cotton, the ratio of surfactant to co-surfactant, etc. Much recent research has concentrated on using reverse micelles with anionic surfactants like Aerosol [45]. Nevertheless, using anionic surfactants negatively impacted the application of reverse micelles, which had a detrimental influence on the polarity of the water pool and created an uneven micro-environment [46]. Therefore, more focus has been placed on the non-ionic reverse micelles to establish a stable environment in the water pool. Overall, the primary features of solvent-assisted dyeing on textiles are as follows: (a) reduced energy needed for heating and vaporization,(b) faster drying of solvent-impregnated textiles,(c) faster wetting; (d) less swelling and less chance of mechanical distortion; and (e) lower costs. Solvent-assisted systems can prevent color fading and achieve excellent dyeing quality because solvents can quickly and well diffuse dye onto fiber surfaces. It also offers simple and safe operation because solvents can be recycled after dyeing and many more. The advantages and disadvantages of different solvent dyeing techniques are listed in Table 3.4.

3.4 Ultrasound-Assisted Dyeing

Low frequency, sometimes referred to as conventional power ultrasound (20–100 kHz); medium frequency, often referred to as "sonochemical effects" ultrasound (300–1000 kHz); and high frequency, or diagnostic ultrasound, are the three types of sound waves that compose ultrasound (2–10 MHz) [48, 49]. In the second half of the twentieth century, waste materials were dumped into traditionally run wastewater treatment facilities, causing the water environment to rapidly deteriorate due to the widespread usage of complex organic chemicals. Over the past ten years, research and development in cutting-edge technologies have demonstrated the potential of advanced oxidation methods for cleaning up such contaminated water and wastewater systems without producing any hazardous sludge or solid debris [50]. Free radical attacks, especially from hydroxyl radicals, which are far more potent oxidizing agents than any other known oxidant, cause organic materials to disintegrate or mineralize through oxidative deterioration [51]. Numerous processes, including (i) photochemical irradiation with ultraviolet light (combined with a strong oxidizing agent and/or a semiconductor), (ii) Fenton and photo-Fenton catalytic processes, (iii) g-radiolysis, (iv) electron beam irradiation techniques, and (v) sonolysis, can produce free radicals. Even though ultrasonic vibrations create extremely special and "extreme" conditions in liquid media, making it an exceptionally appropriate medium for "high energy chemistry," sonolysis is rarely employed among these. The initial study of ultrasonic-assisted textile dyeing was in 1941 [52], and several academics presented a review of previous work on the use of ultrasound in wet textile processes [53]. Utilizing ultrasonic energy to exhaust color on textiles using reactive dyes and cold pad-batch procedures has also been studied in several works [54, 55]. There, researchers also looked at how ultrasonic energy affected reactive dye wash fastness and the efficiency of dyeing acceleration. A typical sonicator for an ultrasound-assisted textile dyeing process is shown in Fig. 3.11.

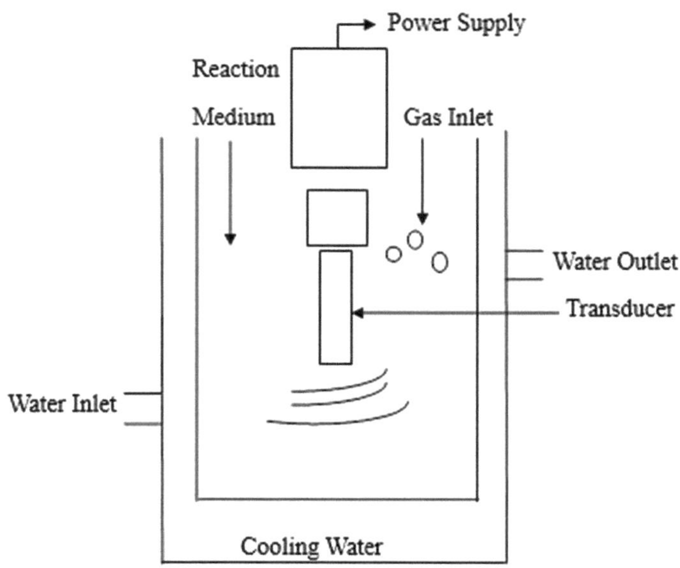

Fig. 3.11 Sonicator to emit low-frequency ultrasound waves into the reaction medium

3.5 Foam Technology in Textile Dyeing

Foam is a collection of gaseous bubbles scattered throughout a liquid. Lamellae, a thin layer of liquid, divide the foam bubbles from one another. One of the two states in which the foam forms is solid or gaseous. Gaseous foam, which has numerous uses in the textile industry, is created when a mixture of liquid and gas is combined. There are two types of gaseous foam: dispersion foam and condensation foam. When a chemical reaction occurs, or the gas inside a liquid undergoes a physical change, condensation foam is created. An improper blow ratio could result from the discharge of the dissolved gas in condensation foam. Foam consequently loses its stability and is not appropriate for textile manufacturing. In the meantime, dispersion foam is created if the gas is added and mixed with liquid from an external source. It's the most prevalent kind of foam used in textile manufacturing. Typically, water is employed as a liquid and air as a gas. The surface-active ingredient in this kind of foam serves as a foaming agent. The blow ratio, stability, viscosity, size, and distribution of the bubbles determine the characteristics of the foam [56, 57]. Figure 3.12 displays the foam dyeing technique of textile fibers.

The following are some potential benefits of employing foam-dying in textile industries:

- Steam savings during the dye-fixation step since there is significantly less water content (pick-up) compared with traditional water dyeing.
- Higher throughput rates during the steam fixation process due to the quicker rates of dye fixation and heating up. An additional benefit is that a slight increase in liquor volume minimizes the effects of "frostiness" or colour "flushing" by causing steam to condense.
- The use of thickening agents is significantly reduced, which lowers costs and effluent load as a result.

The primary advantage of utilizing multiple dye classes in the foam dyeing process is its versatility, allowing it to be applied effectively to both natural and synthetic fabrics with diverse fabric structures. While traditional pad dyeing normally has a wet pick-up of 60–100%, the foam method frequently uses a wet pick-up of 20–40%. Foam processing is a low-add-on technique to apply chemicals and colorants to the textiles. This technique saves a lot of water, reduces drying time, increases efficiency, reduces waste, and uses less energy because air is used in place of water during the drying process. In a foam generator, the chemical composition is combined with air to produce foam that typically has the ideal bubble size. Compared to an unframed liquid, a high volume created by the addition of air to the formulation can be distributed more evenly throughout the textile fabric [58].

Initially, a certain degree of foam containing the colour is made in the foam dyeing procedure. Second, it is applied using suitable, industry-standard techniques to the textile substrates. The foam could be applied to textiles with either direct or indirect applicator systems. Yu et al. investigated a two-stage pad dyeing process for plain-woven cotton fabric, employing a reactive dye and utilizing a low add-on

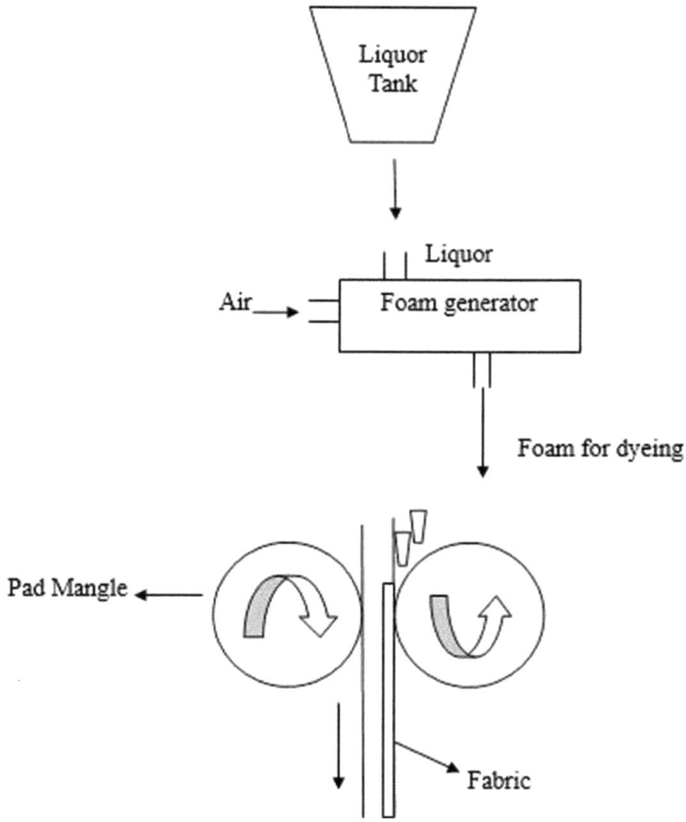

Fig. 3.12 Schematic diagram of foam dyeing on textile fibers

foam dyeing method with the Neovi-foam technique. 30% and 80% percentages wet pick-up were used for foam and traditional pad dyeing, respectively. Here, using several foaming agents with stabilizers, they discovered that foam processing results in better dye build-up properties than traditional dyeing. Additionally, it was stated that guar gum, dodecanol, and sodium dodecyl sulfate were efficient stabilizers and foaming agents, respectively. It resulted a higher color strength (K/S value) of dyed fabric with an increasing stabilizer concentration while lowering the dye fixation rate [59].

In another research, Bhavsar et al. investigated the foam dying process with horizontal padding mangle for cotton and wool fabric using reactive and acid dyestuff, respectively. In the process of dying both materials, they introduced keratin hydrolyzate, a novel foaming agent. For cotton and wool, respectively, the performance of foam dyeing was compared to traditional cold pad batch and pad steam procedures. Aqueous solutions with and without colouring auxiliaries were used to determine the foams with varying characteristics. They observed a blow ratio of roughly 10:1, considerable keratin hydrolyzate surface activity, good foam stability with coloring auxiliaries, and bubble sizes of 0.02–0.1 mm. Hydrolyzed keratin carries dye molecules, and the pH of the dye solution, keratin, and fiber controls the dyeing process. The wool produced a higher level of color strength when foam dying was used instead of traditional dyeing techniques, whereas cotton produced a similar level. The foam-dyed cotton and wool textiles are similar to their conventionally dyed examples in terms of wash and rubbing fastness [60]. In a study, Shang et al. investigated the evolution of pigment-based foam dyeing and its wash-out effect on cotton knit T-shirts. They found optimum results in dyeing performance to the end product [61].

3.6 Environmental Sustainability and Performance Analysis of Waterless Dyeing on Textiles

Wastewater is utilized extensively in the fashion industry to produce textile goods. One of the most polluting businesses in the world, the textile dyeing sector generates trillions of liters of wastewater that is polluted by chemicals. In many developing nations, the textile industry continues to put environmental pressure on water resources, contributing 17–20% of industrial water pollution to dyeing operations. The textile dyeing industry has the potential to drastically reduce water pollution if new waterless dyeing processes are widely implemented [62] A novel dyeing technique that doesn't use water is called waterless dyeing. The technique of waterless dyeing textiles offers a more environmentally friendly manufacturing method and significantly influences the clothes that people wear today. CO_2 is used in this waterless dyeing method to color textile goods. Using CO_2 to color textiles is a more environmentally friendly method as it conserves freshwater, usually used for textile dyeing [63, 64]. Traditional dyeing methods require a lot of water, energy, and chemical additions to establish dye on the fibers. By using CO_2 as an alternative, pollution is reduced, and the amount of chemicals required in textile dyeing is dramatically reduced. People nowadays are more likely to choose textiles made of sustainable and ecologically friendly practices. Consumer awareness and experience with waterless colored clothing and textiles may be improved by research on the color fastness of waterless dyed fabric. In a case study, a pair of leggings consisting of 21% elastane and 79% recycled polyester underwent waterless dyeing where leggings were from one of the well-known international brands of sportswear cloth. The AATCC 15 Colorfastness to Perspiration, AATCC 82 Colorfastness to Crocking, AATCC 61 Colorfastness to Laundering, and AATCC 19 Colorfastness to Flat Abrasion (Frosting) were the standard test techniques used to evaluate the waterless dyed fabric's color fastness [64]. Evaluations utilizing the Gray Scale for Color Change, as per AATCC 15 Colorfastness to Perspiration, revealed considerable shade shifts within a range of 2.3–2.8. The dyed specimen through waterless dyeing showed no appreciable color staining or transfer on other fibers in the multi-fiber test fabric, according to evaluations made using the Gray Scale for Color Staining,all grades were above 3. The Gray Scale for Staining was used to measure the quantity of color transfer from the specimen to the white test square for AATCC 8 Colorfastness to Crocking. The average for the dry test was 3.5, according to the results, while the average for the wet test was 2. The specimen underwent a little color transfer during the dry crocking process and a substantial color transfer during the wet crocking, according to the results. The specimen's shade significantly changes after laundering, according to test findings for AATCC 61 Colorfastness to Laundering, with average ratings of 2.8 for warp and 3.2 for filling. The waterless dyed specimens in the multi-fiber test cloth did not spread discoloration to adjacent fibers,all grades were higher than 3. In the color fastness tests for perspiration and laundering, the waterless colored specimens showed little to no color transfer on the multi-fiber test cloth. This implies that there would be little to no color transfer across materials when they were cleaned together. In conclusion, the test results for AATCC 119 Color Change Due to Flat Abrasion (icing) showed a notable color change on the specimen due to icing, with an average value of 1.25 based on the Grayscale to Color Change. In terms of color fastness against perspiration, crocking, laundering, and frosting, overall test findings show that the waterless dyed fabric did not fulfill the product criteria required for that particular knitted sportswear fabrics category, mostly due to color change and/or staining concerns. Customers wouldn't be happy with the color fastness performance, particularly if the waterless dyed clothing wasn't properly cared for according to the care recommendations. Customers may encounter color or shade changes due to sweating, color transfer from the waterless dyed fabric to a lighter colored fabric through surface contact and rubbing, color or shade changes following multiple launderings, and color changes in flat abrasion-induced localized abrasive areas.

3.7 Challenges and Drawbacks of Waterless Dyeing to End-Use

Waterless dyeing machines, invented by many renowned manufacturers like DyeCoo, ColorZen, and AirDye, do not require the processing of chemicals or the addition of water. In addition to that, waterless dye technologies have been included in Nike and Adidas' product lines. However, the textile industry experts are concerned about difficulties related to cost and limitations. For instance, natural textiles are difficult to dye using waterless dyeing. It is also crucial to look at the waterless dyeing quality and color fastness of the cloth to encourage people to buy eco-friendly textile items in a sustainable manner. Numerous chemical substances and solvents that were previously used have been prohibited or regulated by legislation because of environmental concerns. In recent years, research has been conducted to address the environmental pollution caused by solvent-assisted waterless dyeing. These studies have included low-salt or salt-free dyeing, dyeing with natural dyes and natural mordant using the multiple dip method, dyeing with a water-organic solvent mixture, and recycling wastewater as an alternate water source for dyeing.

The procedures used in ultrasound-assisted dyeing are quite delicate and susceptible to changes in operational conditions. To account for this, it is impossible to govern without a solid grasp of physical and chemical phenomena [48]. In many instances, ultrasonic techniques for textile dyeing are employed in the environment as post-treatment methods to effectively eliminate refractory compounds from wastewater streams. Most of these methods will involve using medium-frequency ultrasound, as these chemicals are typically complex macromolecules possessing hydrophilic qualities. Reactor systems operating on medium frequency radiation are fortunately less problematic to maintain than those using power ultrasound because the latter has issues like cavitational erosion and noise. However, sound-proof materials and appropriate equipment selection, design, and maintenance might help to overcome such issues.

In foam dyeing technology, dyeing using jet dyeing machinery sometimes may cause the pump delivery rate to fall off and eventually can bring the liquor flow and motion of the goods to a standstill, which may result in unlevelled dyeing. In addition, the quality of the dyeing in continuous pad dyeing methods is greatly impacted by the foaming of the liquor. It is caused by the liquor recirculation pumps' operation, or, in multi-bowl machines, by air can trap in the fabric's interstices and may release as the fabric is squeezed in the nip [65].

However, as conventional dyeing methods are harmful to the environment, there is an increasing need for long-lasting, sustainable, and clean substitutes. Therefore, practicing sustainability in textile production is the most important issue in the textile sector. Therefore, dying textiles using ecologically friendly techniques that use less water and energy is a sensible way to satisfy the demand for sustainability. Utilizing state-of-the-art, eco-friendly equipment, modern, sustainable equipment, filtration of runoff water, renewable energy, recycled materials, decent working conditions, supply chain enhancement, green logistics, fewer chemicals, less air pollution, water conservation, and minimal chemical use are all possible components of a sustainable solution. The aforementioned waterless dyeing methods are essential in this regard.

3.8 Conclusion

It is acknowledged that traditional water-based dyeing methods negatively influence the environment and ecology since they require a lot of energy and water and produce a lot of wastewater, raising the treatment cost. Finding alternate methods that can lessen the strain on the environment is crucial. Several notable advantages drive research on transitioning from aqueous-based coloring systems to non-aqueous alternatives: (i) enhanced color yield and improved leveling qualities in dyed samples; (ii) reductions in both energy and water usage; (iii) minimized effluent discharge; (iv) increased feasibility for industrial applications. Furthermore, this chapter covers non-aqueous dyeing procedures and the feasibility of applying these dyeing systems to the coloring of textile fibers, including cotton, polyester, silk, wool, and many more. Although waterless dyeing has versatile advantages, it also has certain disadvantages. Some research includes the fact that most dyeing methods are limited to synthetic materials like polyester and that machine installation costs are high. The dyeing techniques covered in this chapter are still in their infancy, and improvement is always possible. More thoroughly exploring these methods will be a valuable direction for future research.

Declaration The authors declare that they have no known competing financial interests or personal relationships that could have appeared to influence the work reported in this chapter. Generative AI and AI-assisted technologies have been used in the writing process to improve the readability and language of the work only. After using this tool, the authors reviewed and edited the content as needed and take full responsibility for the content of the publication.

References

1. Berradi M et al (2019) Textile finishing dyes and their impact on aquatic environs. Heliyon 5(11):e02711
2. Makene VW et al (2019) Toxicity study of a textile effluent treated with electrohydraulic discharge and coagulant/flocculants. Water Air Soil Pollut 230:167
3. Abner M, Baytar F, Kreiner D (2019) Applying the ESD approach in textile and apparel education. Int J Sustain High Educ 20(1):75–90
4. Chakraborty R, Ahmad F (2022) Economical use of water in cotton knit dyeing industries of Bangladesh. J Clean Prod 340:130825
5. Li Y et al (2017) Decoupling water consumption and environmental impact on textile industry by using water footprint method: a case study in China. Water 9(2):124
6. Khan BU, Tanha JA, Amin J, Fahim TH, Islam MR (2025) Thermal Regulation in Textiles: Phase Change Materials and Sustainability. In Sustainable Finishing Techniques in Textiles, Biranje, S., Shahid, M., Adivarekar, R. V. (eds) Springer Nature Singapore, pp 179–213
7. Reda EM, Ebrahim SA, Mosaad MM (2025) An Overview of Dyeing without Water Techniques. J Text Color Polym Sci 22(1):89–95
8. Abate MT, Tadesse MG (2021) Airflow, Foam, and Supercritical Carbon Dioxide Dyeing Technologies. In: Rather LJ, Haji A, Shabbir M (eds) Innovative and Emerging Technologies for Textile Dyeing and Finishing; Scrivener Publishing LLC, pp 137–164
9. Kumar D, Garje A, Desai K, Gupta D (2010) Dyeing without water. J Text Association, 71:171–174
10. Mahmud I, Kaiser S (2020) Recent progress in waterless textile dyeing. J Text Sci Eng 10(6):1–3
11. Kant R (2012) Textile dyeing industry an environmental hazard. Nat Sci 4(1):22–26
12. Babu KM (2022) Overview on Significant Approaches to Waterless Dyeing Technology. In: Yusuf M, Shahid M (eds) Emerging technologies for textile coloration. CRC Press, pp 171–188
13. Maiti S, Panda A, Madiwale P, Adivarekar RV (2022) Textile Coloration Technologies Today and Tomorrow. In Yusuf M, Shahid M (eds) Emerging technologies for textile coloration. CRC Press, pp 1–12
14. Abate MT et al (2020) Coloration and bio-activation of polyester fabric with curcumin in supercritical CO_2: part II–effect of dye concentration on the colour and functional properties. J Supercrit Fluids 157:104703
15. Bach, E.; Cleve, E.; Schollmeyer, E. (2010) Past, present and future of supercritical fluid dyeing technology – an overview. Rev Prog Color Rel Topics 32:88–102
16. Penthala R et al (2020) Synthesis of azo and anthraquinone dyes and dyeing of nylon-6, 6 in supercritical carbon dioxide. J CO_2 Utilization 38:49–58

17. Hendrix WA (2001) Progress in supercritical CO_2 Dyeing. J Ind Text 31(1):43–56
18. Yan K et al (2020) Development of a special SCFX-AnB3L dye and its application in ecological dyeing of silk with supercritical carbon dioxide. J CO_2 Utilization 35:67–78
19. Abou Elmaaty T et al (2019) Optimization of an eco-friendly dyeing process in both laboratory scale and pilot scale supercritical carbon dioxide unit for polypropylene fabrics with special new disperse dyes. J CO_2 Utilization 33:365–371
20. Chang K-H, Bae H-K, Shim J-J (1996) Dyeing of PET textile fibers and films in supercritical carbon dioxide. Korean J Chem Eng 13:310–316
21. Saus W, Knittel D, Schollmeyer E (1993) Dyeing of textiles in supercritical carbon dioxide. Text Res J 63(3):135–142
22. Gebert B et al (1994) Dyeing natural fibers with disperse dyes in supercritical carbon dioxide. Text Res J 64(7):371–374
23. Özcan AS et al (1998) Dyeing of modified cotton fibres with disperse dyes from supercritical carbon dioxide. J Soc Dyers Colour 114(5–6):169–173
24. Beltrame PL et al (1998) Dyeing of cotton in supercritical carbon dioxide. Dyes Pigm 39(4):335–340
25. Huang T, Kong X, Cui H, Zhang T, Li W, Yu P, Lin J (2019) Waterless dyeing of zipper tape using pilot-scale horizontal supercritical carbon dioxide equipment and its green and efficient production. J Clean Prod 233:1097–1105
26. Jelil RA (2015) A review of low-temperature plasma treatment of textile materials. J Mater Sci 50(18):5913–5943
27. Chan C-M, Ko T-M, Hiraoka H (1996) Polymer surface modification by plasmas and photons. Surf Sci Rep 24(1–2):1–54
28. Zille A, Oliveira FR, Souto AP (2015) Plasma treatment in textile industry. Plasma Processes Polym 12(2):98–131
29. Morent R et al (2008) Non-thermal plasma treatment of textiles. Surf Coat Technol 202(14):3427–3449
30. Borcia G, Anderson CA, Brown NMD (2006) Surface treatment of natural and synthetic textiles using a dielectric barrier discharge. Surf Coat Technol 201(6):3074–3081
31. Vohrer U, Müller M, Oehr C (1998) Glow-discharge treatment for the modification of textiles. Surf Coat Technol 98(1–3):1128–1131
32. Wang CX, Qiu YP (2007) Two sided modification of wool fabrics by atmospheric pressure plasma jet: influence of processing parameters on plasma penetration. Surf Coatings Technol 201(14):6273–6277
33. García JL et al (2010) Cell proliferation of HaCaT keratinocytes on collagen films modified by argon plasma treatment. Molecules 15(4):2845–2856
34. Thurston RM, Clay JD, Schulte MD (2007) Effect of atmospheric plasma treatment on polymer surface energy and adhesion. J Plast Film Sheeting 23(1):63–78
35. Gaur S, Vergason G (2000) Society of vacuum coaters. In: 43 rd annual technical conference proceedings–Denver, pp 15–20
36. Hegemann D (2006) Plasma polymerization and its applications in textiles. Ind J Fiber Text Res 31(1):99–115
37. Mowafi S, Taleb MA (2022) A review of plasma-assisted treatments of textiles for eco-friendlier water-less processing. Egypt J Chem 65(5):737–749
38. Haque ANMA, Naebe M (2023) Zero-water discharge and rapid natural dyeing of wool by plasma-assisted spray-dyeing. J Clean Prod 402:136807
39. Haji A, Naebe M (2020) Cleaner dyeing of textiles using plasma treatment and natural dyes : a review. J Cleaner Prod 265:121866
40. Doherty RE (2000) A history of the production and use of carbon tetrachloride, tetrachloroethylene, trichloroethylene and 1, 1, 1-trichloroethane in the United States: part 1—historical background; carbon tetrachloride and tetrachloroethylene. Environ Forensics 1(2):69–81
41. Doherty RE (2014) History of TCE. In: Gilbert KM, Blossom SJ (eds) Trichloroethylene: toxicity and health risks.Springer, London, pp 1–14

42. Knözinger H, Kochloefl K (2000) Heterogeneous catalysis and solid catalysts. Ullmann's Encyclopedia of Industrial Chemistry. Available at: https://onlinelibrary.wiley.com/doi/book/10.1002/14356007
43. Bakke BA, SP, Waters MA (2007) Uses of and exposure to trichloroethylene in U.S. industry: a systematic literature review. J Occup Environ Hyg 4(5):375–390
44. Sawada K et al (2002) Dyeing natural fibres in supercritical carbon dioxide using a nonionic surfactant reverse micellar system. Color Technol 118(5):233–237
45. De TK, Maitra A (1995) Solution behaviour of Aerosol OT in non-polar solvents. Adv Coll Interface Sci 59:95–193
46. Tang AYL et al (2018) Dyeing properties of cotton with reactive dye in nonane nonaqueous reverse micelle system. ACS Omega 3(3):2812–2819
47. Tang AY, Kan CW (2020) Non-aqueous dyeing of cotton fibre with reactive dyes: a review. Color Technol 136(3):214–223
48. Ince NH et al (2001) Ultrasound as a catalyzer of aqueous reaction systems: the state of the art and environmental applications. Appl Catal B 29(3):167–176
49. Peters D (1996) Ultrasound in materials chemistry. J Mater Chem 6(10):1605–1618
50. Ince NH, Stefan MI, Bolton JR (1997) UV/H_2O_2 degradation and toxicity reduction of textile azo dyes: Remazol Black-B, a case study. J Adv Oxidation Technol 2(3):442–448
51. Legrini O, Oliveros E, Braun AM (1993) Photochemical processes for water treatment. Chem Rev 93(2):671–698
52. Vajnhandl S, Le Marechal AM (2005) Ultrasound in textile dyeing and the decolouration/mineralization of textile dyes. Dyes Pigm 65(2):89–101
53. Thakore K (1990) Application of ultrasound to textile wet processing. Am Dyestuff Reporter 79(10):30–44
54. Khatri Z et al (2011) Cold pad-batch dyeing method for cotton fabric dyeing with reactive dyes using ultrasonic energy. Ultrason Sonochem 18(6):1301–1307
55. Öner E, Başer I, Acar K (1995) Use of ultrasonic energy in reactive dyeing of cellulosic fabrics. J Soc Dyers Colour 111(9):279–281
56. Dawson TL (1981) Foam dyeing and printing of carpets. J Soc Dyers Colour 97(6):262–274
57. Elbadawi AM, Pearson JS (2003) Foam technology in textile finishing. Text Prog 33(4):1–31
58. Garrett PR (1993) Recent developments in the understanding of foam generation and stability. Chem Eng Sci 48(2):367–392
59. Yu H et al (2014) Foam properties and application in dyeing cotton fabrics with reactive dyes. Color Technol 130(4):266–272
60. Bhavsar PS et al (2017) Superheated water hydrolyzed keratin: a new application as a foaming agent in foam dyeing of cotton and wool fabrics. ACS Sustain Chem Eng 5(10):9150–9159
61. Shang S et al (2011) Foam dyeing for developing the wash-out effect on cotton knitted fabrics with pigment. Res J Text Appar 15(1):44–51
62. Hussain T, Wahab A (2018) A critical review of the current water conservation practices in textile wet processing. J Clean Prod 198:806–819
63. Zheng H et al (2018) CO_2 utilization for the waterless dyeing: characterization and properties of disperse red 167 in supercritical fluid. J CO_2 Utiliz 24:266–273
64. Abou Elmaaty T, Abd El-Aziz E (2017) Supercritical carbon dioxide as a green media in textile dyeing: A review. Text Res J 88(10):1184–1212
65. Siddique A, Khawar MT, Xu Z, Kit CC, Mazari A, Naeem S, Naeem J, Saleemi S (2025) Foam dyeing: a review of its potential to reshape sustainable textile processing. Text Res J https://doi.org/10.1177/00405175251324669

Chapter 4
Air Dyeing and Dope Dyeing Technologies: Innovations in Textile Coloration

Mohammad Shohag Babu, Tania Akther, Sk. Mohammad Raafi, and Md. Rashedul Islam

Abstract Conventional dyeing requires significant water and energy consumption and produces contaminated water and other toxic byproducts that deteriorate the environment and public health. To mitigate those issues, the textile industries are concentrating more on alternative green technologies. Air dyeing technology is a green and water-saving technique that uses an airstream rather than a dye bath to convey and color the fabric. Compared to traditional fabric dyeing methods, air dyeing saves 60% water, 50% energy, 40% chemicals, 25% process time, and 10% dyestuff. The air dyeing technology is 25% more economical and colors the fabric with up to 45% elastane content. Dope dyeing incorporates color into the molten polymer during the melt-spinning process before fiber creation, which is another innovative, water-saving approach to dyeing synthetic and regenerated cellulose fiber. Both air-dyed and dope-dyed textiles provide excellent color fastness and uniformity, a smooth and luxurious feel, and less damage during processing. The basic concepts of these technologies, the principles of dyeing, the machinery used and their development, the process of dyeing various fibers, their specific benefits, and the present condition of these technologies are all highlighted here.

Keywords Conventional dyeing · Green technologies · Air dyeing · Water-saving · Dope dyeing

4.1 Introduction

The conventional textile dyeing method consumes an enormous amount of water, chemicals, energy, and dyestuffs, and as a byproduct, wastewater is produced in large quantities. In a conventional dyeing-finishing mill, over 150,000 L of water are used for each ton of material. Several issues with the conventional dyeing process

M. S. Babu · T. Akther · Sk. M. Raafi · Md. R. Islam (✉)
Department of Wet Process Engineering, Faculty of Textile Chemical Engineering, Bangladesh University of Textiles, Tejgaon, Dhaka, Bangladesh
e-mail: rashedul.islam@wpe.butex.edu.bd

M. Shahid et al. (eds.), *Advancements in Textile Coloration*,
https://doi.org/10.1007/978-981-96-5091-0_4

have also been revealed, including a high liquor ratio, a significant sewage outflow, a huge energy consumption, and various defects in dyeing [1]. However, because of environmental consciousness, stringent environmental restrictions, and the desire to cut costs, the textile dyeing and finishing sectors are concentrating on water-saving strategies and sustainable alternatives to traditional processes. Several innovative processing techniques, including air dyeing, supercritical fluid dyeing, foam technology, electron beams, microwaves, ozone, plasma-assisted dyeing, ink-jet printing, and so forth, are examples of these strategies [2–6].

Air dyeing is an environmentally friendly technique that uses little or no water, saves resources, and provides several benefits over conventional dyeing techniques. Compared to conventional fabric dyeing, the air-dyeing approach consumes at least 60% less water, 40% fewer auxiliaries, 10% less dyestuff, 50% less energy, and 25% less process time [1]. The process is less complicated and quicker than the traditional jet dyeing system since only a small amount of water is needed to solubilize the dyes and chemicals, and the material is conveyed with a combination of dye liquor and airstream, which lowers the liquor ratios [7]. Since fabric is carried by air rather than liquor, no wastage of water, no discharge of harmful wastewater, zero hazardous waste material, and significant cost savings on effluent treatment are possible [8].

Dope dyeing is another eco-friendly method of coloration that provides excellent durability and colorfastness. In this method, colorants are introduced to the solution of polymers prior to it being extruded into fibers. As opposed to traditional piece dyeing, at the polymer stage, coloring takes place, resulting in color distributing all over the fiber and lowering carbon emissions while consuming less water and chemicals, having no washing process after dyeing, and producing consistent, vibrant coloration, making the manufacturing process more environmentally friendly [9–11]. This chapter covers dope dyeing and air dyeing fundamentals, their principles, nozzle systems in air dyeing, comparative studies of air dyeing with soft flow and overflow dyeing, the specialty and advancements of machinery in air dyeing, etc. Various fiber processing methods in dope dyeing, the benefits of dope dyeing and air dyeing over conventional dyeing, and their prospects for the future are also discussed.

4.2 Air Dyeing

A significant approach toward lowering the amount of water, energy, and chemicals used in jet-dyeing equipment was substituting an air stream for dyeing liquor in transporting textile material [12]. It utilizes air rather than water to carry the material inside the dye bath making it easier for the dyes to get into the fibrous structure, therefore considered an eco-conscious dyeing technology. Airflow is an essential part of this technology since it is an ideal transport medium without compromising the quality of the fabric. The dye is incorporated into the fibers, and unlike vat-dyed fabrics, air-dyed materials do not fade or lose color so quickly. This technology consumes a small amount of water with the lowest possible material-to-liquid ratio, even at 1:2, whereas a conventional jet dyeing system uses 100–150 L per kilogram of fabric. The annual water consumption for dyeing synthetic fabrics in the traditional system is 2.4 trillion gallons, equivalent to filling 3,700,000 Olympic-sized swimming pools. Additionally, it consumes more energy than 2.8 trillion megajoules annually, sufficient to power almost 12% of American households. Lastly, dyeing textiles generates GHGs (greenhouse gases) in 568 million metric tons a year, equivalent to the emissions of over 94 million passenger cars [13]. Thus, air dyeing, a resource-efficient, and eco-friendly dyeing method could be the way to achieve significant environmental preservation in the future. It is considered one of the most eco-friendly substitutes for conventional dyeing and coloration techniques available today.

4.2.1 History of Air Dyeing

Man first applied dye to cloth using water 4,000 years ago [14]. Water pollution began in its early stages. Since then, more chemicals have been used to dye and treat textiles, resulting in continuous and worsening water contamination [13]. Identical to the jet dyeing method, air dyeing technology circulates the material while dyeing it using air rather than water. Wilhelm Christ created the idea for air dyeing, and THEN Maschinen GmbH, a German manufacturer, created the first machine prototype in 1978. Later, Colorep, a sustainable technology company based in California, registered the patent in both Europe and the US [15, 16]. During the ITMA (International Textile Machinery Association exhibition) 1983 in Milan, the first air dyeing machine was introduced to textile entrepreneurs. Since then, the technology has advanced in response to industrial demands, significantly influencing environmentally friendly dyeing processes. The top air dyeing machines in the world right now are the THEN Airflow series of dyeing machinery from THEN, Germany (Fig. 4.4). In 2010, the THEN Maschinen GmbH officially merged with other machine brands, assuming the new name Fong's Europe GmbH [17]. The manufacturer unveiled the latest THEN Airjetwin and THEN airflow synergy machines at ITMA 2023 in Milan, utilizing them to offer superior, economical, and environmentally sustainable dyeing processes. A couple of American companies named ColorZen® and AirDye® have introduced the waterless air dyeing technique over the past few years. AirDye® utilizes 95% less water, 86% less energy, and 84% less emissions of greenhouse gases and Colorzen® technology can utilize up to 95% fewer chemicals, 90% less water, and 75% less energy [18–20]. Various air dyeing machines are now being produced under various brand names by Fong's Europe, and a few other manufacturers are listed in Table 4.1.

Table 4.1 Leading manufacturer of air dyeing machine with their model's name

Producer and origin	Models name
THEN, Germany	• Airflow AFS • Airflow AFE • Airflow AFT • Airflow AFA • Airflow Lotus • Airflow Synergy • Airjetwin
Thies, Germany	• Luft-roto • Luft-roto plus • Luft-roto plus SII • Air stream
Tong Wu, Taiwan	• EPO-500 HTHP Airflow Dyeing Machine
Cleantech, China	• HTHP Air Flow
Ningbo Dyetec, China	• New Age Airflow Dyeing Machine
Flainox, Italy	• High Temperature (HT) Machine with Air Injection
Laip, Italy	• Airjet 2000 • Air Water Flow 800
MCS, Italy	• Airjet HT • Airjet 2HT • First Vento • Universal Dyeing
Fong's, Hong Kong	• Airsoft High Temperature
Dilmenler (DMS), Turkey	• DMS 03 HT Synchron Airjet
Hisaka, Japan	• Airflow CUT-AJ
ATYC, Spain	• Airtint EVO
Alliance, France	• Riviera • Ventura

Adapted with permission from Ref. [7] and updated, Copyright © 2011, Woodhead Publishing Limited

4.2.2 *Principle of Air Dyeing*

The textile substrate is dyed in a rope format using a closed tube system using the air dyeing technique, a jet dyeing type. Since a stream of air circulates the fabric during the process, air dyeing is simpler, producing less effluent, and is comparatively faster than traditional jet dyeing, even though the operational mechanism is the same [7]. Figure 4.1 illustrates the working principle of the air dyeing machine. Here, dyeing liquor is first prepared in the mixing tank by mixing dyes and auxiliary chemicals. Then, it is injected into the annular cavity of the nozzle via a dyeing liquor pipe and atomized before being applied to the fabric. A motorized reel simply lifts the fabric rope without giving any driving motion to the fabric. The fabric is then pulled by high-pressure air once it has been lifted. This indicates that cloth transportation does not require the use of an aqueous medium or a dye bath. This greatly lowers the amount of chemicals and water used [21]. At the front part of the nozzle, high-pressure air produced by the blower is jetted onto the fabric, forcing it forward and causing the fabric to be loose enough. The fabric then enters the rear portion of the nozzle, where atomized dyeing liquor is jetted, penetrates the loose fabric, and, once dyed, is forced out of the nozzle [22]. During spraying, the atomized dyeing solution and moving fabric quickly come in complete contact within the shortest possible time and speed up the dye's diffusion rate to accomplish level dyeing. Therefore, coloration only occurs in the nozzle. Following the nozzle zone, the fabric automatically stretches and creates a balloon form with airflow aid to avoid creases and streaks. It is then plated onto a perforated PTFE rack. Immediately, all the air and leftover dyeing solution fall into the filter pit underneath the dyeing vessel, from where it is recirculated up to the nozzle via the main pump and heat exchanger. After dyeing, a spray-type washer is used for rinsing instead of a traditional dyeing water bath. The bottom valve of the machine is fully open during rinsing, which prevents wash-off water from coming into touch with the dyed fabric—something that usually happens with traditional dyeing. This technique drains hot bath rinse water after dyeing at an elevated temperature of 130 °C. As a result, it is possible to achieve significant water savings, a quicker rinse time, and optimal heat recovery [23]. Lower liquor ratios (1:2 to 1:5) and fabric speeds of 250–1,000 m/min can be achieved with this dyeing technique without compromising the quality of the fabric [1].

4.2.3 Nozzles in Air Dyeing

The nozzle is one of the essential elements of the air dyeing machine since, as we already know, the colorant application process occurs inside it throughout the process. An airflow dyeing machine's nozzle (Fig. 4.2a) comprises components such as an injector, air pipe, dye liquor entry pipe, shell, nozzle body, and annular mixing room. The dyeing liquor, a mixture of dyes and chemicals, is supplied from the machine's mixing tank with the help of a dyeing liquor entry pipe. The injector atomizes and injects it into the annular mixing room. In parallel, a blower forces airstream with high pressure into the annular mixing room, instantly combining it with the atomized dyeing liquor. The fabric is propelled forward by high-pressure air in the nozzle area while the mixture of air and dyeing liquor is simultaneously jetted at the moving objects, and dyeing is carried out throughout the process [22]. However, there is a drawback with this nozzle system: when two separate flows and pressures of dyeing liquor and air are coupled, significant turbulence is produced, which reduces efficiency and causes kinetic energy to be lost. Because of this, the fabric cannot be fully immersed in the dyeing liquor, which can easily lead to uneven dyeing and could cause a lengthy dyeing time. This situation can be resolved by altering the air and dyeing liquid flow rates, using high-quantity dyeing liquor, and

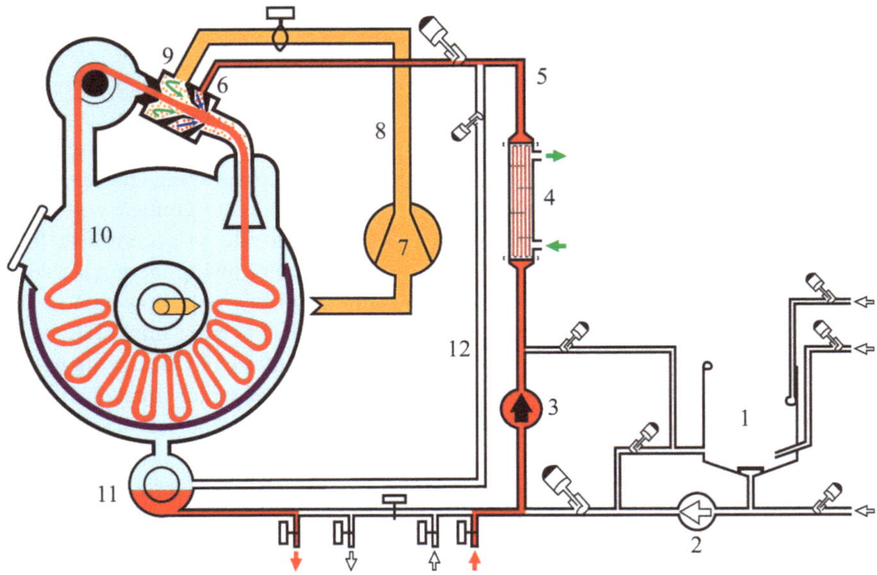

Fig. 4.1 Schematic diagram of air dyeing principle. (1) mixing tank, (2) mixing tank pump, (3) main pump, (4) heat exchanger, (5) dye liquor pipe, (6) dyeing nozzle, (7) blower, (8) air pipe, (9) air nozzle, (10) fabric, (11) collector, (12) return pipe for excess liquor. (Adapted in a modified form with permission from Ref. [23], Copyright© 2004, Springer-Verlag Berlin Heidelberg)

raising the air pressure, which may result in high energy consumption and high load of effluent.

Therefore, a new nozzle system is developed. An extra isolation board is installed in the annular mixing room (Fig. 4.2b, c) to separate high-pressure air and dyeing liquor. Therefore, the traditional nozzle injector is also removed from this nozzle. Via an airflow pipe, high-pressure air is forced into an annular cavity of air. The annular cavity maintains precise air pressure. In a similar manner, the coloring solution pushes into its annular cavity. Before being jetted onto the fabric, the coloring solution is atomized for the tiny passageway. In this nozzle system, air and dyeing fluid are not mixed and jetted onto the fabric through two separate passageways from separate annular cavities. In both Fig. 4.2b and c, the solid arrow indicates the airflow, while the dashed arrow represents the dyeing liquor flow. Before the dyeing liquor is sprayed on the fabric, high-pressure air is jetted first on it, pushing the fabric forward and causing it to become loose enough [22]. Then, dyeing liquor is jetted, enters the loose fabric, and moves the fabric forward. Compared to a traditional nozzle system, this novel nozzle offers improved level dyeing, a faster dyeing time, less energy usage, and eliminates unnecessary turbulence. Both square and round nozzles are available. The square nozzle performs best with heavier fabrics and ropes, while the round nozzle is most effective with woven fabric, synthetic fiber-simulated knitted textiles, and microfibers [24].

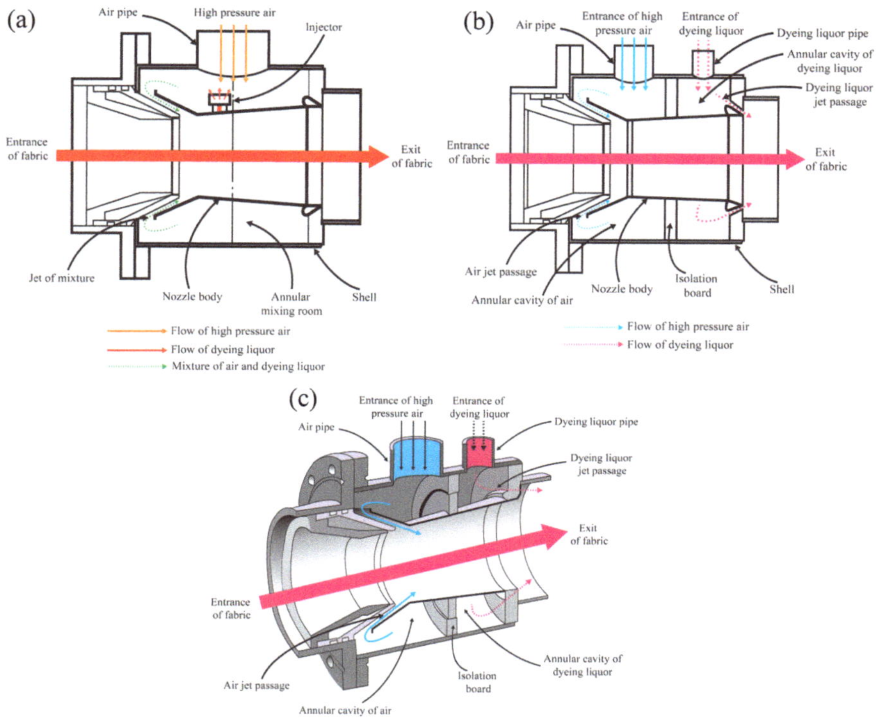

Fig. 4.2 **a** Schematic of a traditional nozzle in an air dyeing machine, **b** schematic of the newly developed novel nozzle in air dyeing machine, **c** 3D model of the nozzle. (Adapted in a modified form from Ref. [22])

4.2.4 Comparison of Overflow, Soft-Flow, and Airflow Machine Principles

In overflow dyeing, a larger rope of fabric is intended to hang at the exit side as opposed to the inlet side and is hung from the upper side winch roller (Fig. 4.3a). With the help of gravitational force larger length of materials is pulled downward and dyeing is performed mainly in the dye bath liquor [25]. The key difference between overflow and airflow dyeing is that in the latter one fabric is moved by high-pressure air passing through the nozzle at a high speed, whereas in an overflow dyeing process, fabric is moved by the gravitational force of the liquor overflow.

In Soft-flow dyeing, the fabric is circulated throughout the machine with the help of dyeing liquor in the nozzle (Fig. 4.3b). A large volume of dyeing liquor is allowed to come into much more contact with the fabric at the intensive nozzle zone, resulting in level dyeing at low fabric speed relative to high liquor velocity [27]. In the drain and fill operation, neither the fabric circulation nor the liquor is stopped. This dyeing method differs from air dyeing primarily in that the low working pressure of

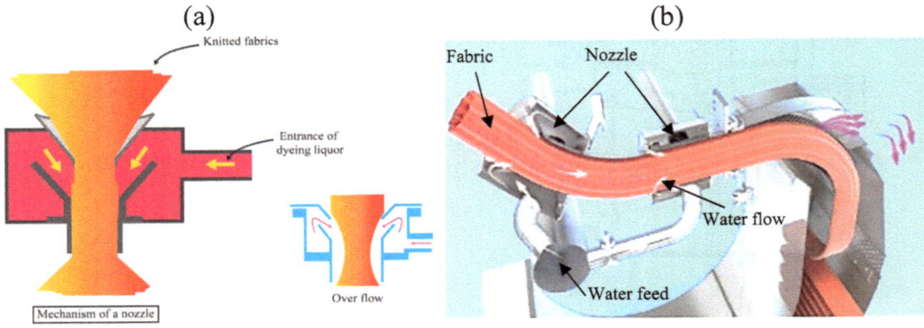

Fig. 4.3 **a** Nozzle of overflow dyeing machine (Adapted in a modified form from Ref. [26]), **b** Nozzle of soft-flow dyeing machine

the nozzle from 0.07 to 0.3 bar is made possible by a unique pump design that allows for a high-volume liquid flow.

In Air dyeing, the fabric is mostly moved within the machine by air at 3 bar operating pressure. The major difference with a jet dyeing system is that air dyeing utilizes high-pressure air, while jet dyeing employs a water jet to circulate the fabric. Additionally, in air dyeing, coloration is performed in the nozzle area (Fig. 4.1). After dyeing, the fabric passes into the storage chamber, with a very small quantity of dyeing liquor underneath the processed materials, which prevents fabric contact with the dye bath liquor. As a result, less water, chemicals, and energy are consumed. In addition, while a traditional hydraulic dyeing machine typically runs at a liquor ratio of around 1:10, an air dyeing machine may operate at a ratio much lower than 1:5 [28].

4.2.5 Developments in Air Dyeing Machines

Many more inventions and technological advancements pertaining to air dyeing machines have been made since Wilhelm Christ's initial patent in 1978. These are the latest rope dyeing technology on the market right now and are usable in atmosphere and high-temperature variants. Airflow dyeing technology is an advancement over soft flow dyeing systems, which move the cloth with dyeing liquor combined with air in a mist form and allow for the use of liquor ratios as low as 1:3 [29, 30]. A couple of American companies named ColorZen® and AirDye® have introduced the waterless air dyeing technique over the past few years.

Using the AirDye® technique, fabrics are dyed and printed rather than impregnated in the conventional dye bath. Specifically designed paper colors are applied to polyester fabric using this method. Compared with the traditional method, AirDye® utilizes 95% less water, 86% less energy, and 84% less emissions of greenhouse gases, depending on the fabric [31, 32]. The release of less hazardous waste, reduction in water usage, and many other benefits of air dye have been outlined in numerous scientific publications and magazines [8, 13, 33, 34]. Unfortunately, cotton and other natural fibers cannot be used for the air-dye process,this technology is limited to synthetic materials such as polyester. This method is similar to transfer printing of polyester fiber with disperse dyes, except instead of using water, air is utilized to allow the dyes to permeate into the fiber [35]. AirDye® is presently used by a number of textile processors in North America, Asia, and Europe [19].

ColorZen®, a revolutionary method for coloring cotton fabric, was introduced in 2012. This approach modifies and increases the receptiveness of the cotton structure. The environmentally friendly and efficient dyeing of cotton is made possible by this innovative method.

The leading producers and models of air dyeing machines are listed in Table 4.1, some of which are covered in the sections that follow [7].

Airtint EVO high temperature is introduced by Argelich Termes Y Cia. S.A., Spain has a much lower liquor ratio (1:3) that works very well with delicate materials like polynosics, modals, cupro-viscose, and lyocell. It has the ability to both defibrillate and fibrillate lyocells (Tencel). Airflow AFS and AFT are two air dyeing machines that were presented by THEN Machines and Apparatus, GmbH. With the Dye Net control and System TLON control for automatic dyeing process regulation, the new Airflow AFT is easier to operate. Thies GmbH's Air-Stream machine was another air dyeing system that made its public debut at ITMA. This machine is well-suited for processing lyocell, polynosic, and cupro fibers, uses Teflon-lined fabric chambers, liquor ratios starting at 3:1, and controllable tumbling action [36]. Recently, an air dyeing machine with a series of nozzles was designed and produced by Cleantech Solutions International, Inc., China. According to founder and CEO Mr. Jianhua Wu of Cleantech, the air dyeing machine atomizes the water under high pressure. 60–70% less water, 30% less energy, and 50% less chemicals than traditional models can be utilized by this machine [35]. This business is also investigating the waterless dyeing apparatus, which is presently in the early stages of trial and development. A detailed overview of a few latest air dyeing machines is given below.

4.2.5.1 THEN Airflow Synergy, Germany

One of the most advanced and favorite air dyeing machines on the market right now is the THEN Airflow Synergy. In terms of loading capacity, there are three varieties available: SYNERGY 250 kg/port, SYN G2 300 kg/port, and SYN 8, 250 or 300 kg/port [37]. Its lower liquor ratio is 1:2 for synthetic fibers and 1:3–4 for natural fibers which is the lowest on the market and can be maintained even when operating underloaded. 40% less energy and, a 25% reduction in the total processing time can be achieved by this machine. It can be used to dye woven and knit fabric in the range of weight of 50–800 g/m^2. THEN Airflow Synergy machines can dye fabric with up to 20% elastane content, which is often achieved with long-tube conventional machines. The latest THEN airflow high-temperature fabric dyeing machine is depicted in Fig. 4.4, together with part details.

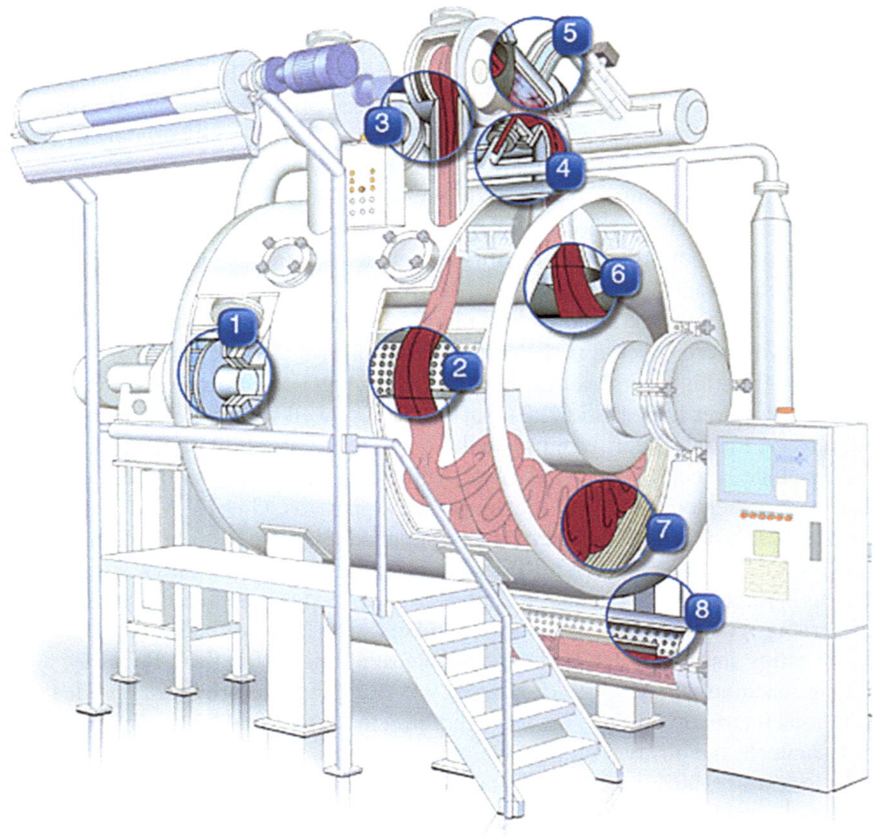

Fig. 4.4 The THEN airflow synergy dyeing machine. (1) blower, (2) aerodynamic drive, (3) fabric pipe, (4) sprayer type rinsing, (5) dye jetting area of atomized dye liquor (6) storage chamber, (7) PTFE coated perforated separator, (8) self-cleaning filter (Adapted from the Ref. [37])

4.2.5.2 THEN Airflow Lotus, Germany

The first long-tube machine in history, LOTUS, runs on the aerodynamics concept. It has surpassed the conventional bounds of jet piece dyeing, allowing for significantly increased productivity and economics when processing even the most delicate materials previously limited to beam dyeing machines. THEN Airflow Lotus machines can dye fabric with up to 45% elastane content no matter whether the material is cotton, synthetic, or blends with their finest yarn counts. This machine uses moisture-saturated air flow instead of dye liquid to move pieces of fabric with a low liquor ratio of 1:2 [35]. In addition to satisfying the highest quality standards and generating significant savings of water, and energy by up to 40%, and a reduction in dyeing time by up to 25%, it also provides flexibility that spurs creative textile solutions, marking a major advancement in green innovation [38].

4.2.5.3 THEN Airjetwin, Germany

THEN Airjetwin is the most advanced machine among all air dyeing machines manufactured by Fong's Europe. The manufacturer unveiled the newest THEN Airjetwin machines at ITMA 2023 in Milan, utilizing them to offer superior, economical, and environmentally sustainable dyeing processes. There are three series and twelve distinct capacities available for AIRJETWIN: 250 kg batch capacity, 300 kg batch capacity, and medium batch capacity. With considerable savings of up to 50% in salt, 35% in water, 30% in process time, and 20% in dyestuffs, it can be used to dye any kind of fabric. This machine facilitates efficient cleaning with an overhead cleaning system that guarantees no residual color remains. It has an integrated intelligent rinsing system that measures the color index of discharge water using specialized software, an optical device, and other features. When the color index exceeds a set point, the rinse automatically stops and saves excess water, energy, and discharge [39].

4.2.5.4 Luft-Roto Plus SII of Thies, Germany

Luft-roto series air jet was created in collaboration with DyStar that combines innovations in liquor movement, liquor penetration, and process technology with exceptional characteristics of efficient material transportation and adaptability. This machine not only facilitates a low liquor ratio of 1:3 to save large amounts of water but also minimizes the processing time and overall energy consumption [7]. All varieties of fibers and blends can be processed with this extremely versatile machine.

4.2.6 Special Features and Components of THEN Synergy Air Dyeing Machine

4.2.6.1 THEN-Time Saver

The innovative THEN-AIRFLOW® "Power Rinse" technique, when combined with the "Direct Rinse" procedure, can save a significant amount of time. An optional hot water tank can provide enough preheated water at 60 °C for rapid rinsing that saves water, energy, and extra time. "Volmedos" use a control circuit that includes a control valve and a flow monitoring device for accurate dosing of chemicals and colors [40], (Fig. 4.5a).

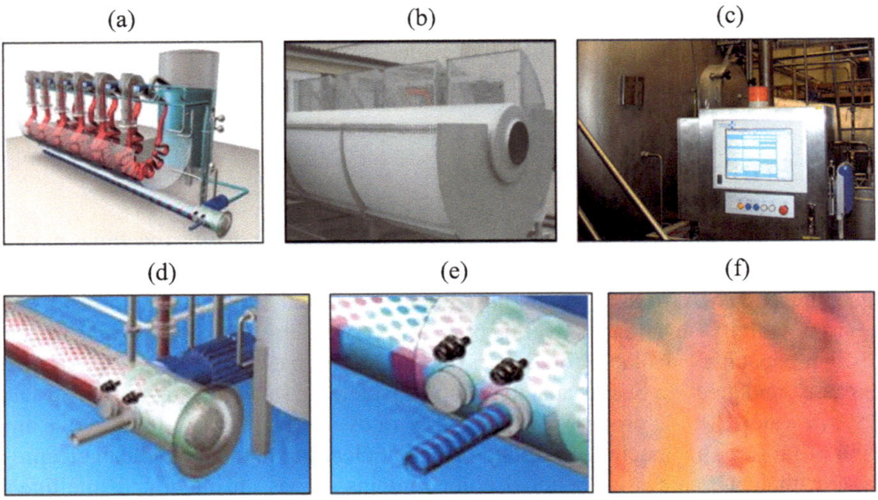

Fig. 4.5 **a** THEN-Time saver, **b** PTFE storage chamber, **c** THEN-DYNET®, **d** Self-cleaning filter, **e** THEN hot drainer, **f** Spectra dyeing effect. (Adapted from Ref. [40])

4.2.6.2 THEN Storage Chamber

Storage chamber also called dye bath chamber where long length of fabric exists in plaited form into the dyeing liquor before being pulled by winch roller. PTFE (Poly Tetra Fluoro Ethylene) rods are installed in the storage chamber. PTFE, well known by its brand name Teflon, provides excellent non-adherence properties, effectively protecting the fabric's surface. Extra dyeing liquor is quickly poured into the collection pit underneath the vessel through the perforated chamber [40], (Fig. 4.5b).

4.2.6.3 Then-Dynet®

A Windows® PC with a touch screen and a main control unit (MCU) make up the intelligent THEN-DYNET® control unit. It facilitates network functions to receive data from the control station and can be used to operate the entire process of dyeing in the machine. Real-time batch processing is displayed. In addition to the machine's set and actual values, errors and operator-initiated modifications are also recorded [40], (Fig. 4.5c).

4.2.6.4 THEN Self-Cleaning Filter

The filter with self-cleaning technology continuously maintains itself clean, in contrast to traditional filters that require manual cleaning. It significantly shortens the time needed before batch loading and guarantees a lint-free substrate. After completing the dyeing process, the lint is automatically cleaned away [40], (Fig. 4.5d).

4.2.6.5 THEN Hot Drainer

This technology enables the draining of liquor over 95 °C, although the fabric is in running condition. After every process, this is feasible and saves 15–30 min. Oligomer contamination in polyester fabric can be effectively eliminated [40], (Fig. 4.5e).

4.2.6.6 Spectra Dyeing

SPECTRA module system in air dyeing machine enables the development of a wide range of effects, such as multicolor, rainbow, and batik shading. The use of SPECTRA nozzles achieves rapid dye application. The SPECTRA technique involves stopping the flow of liquor once the fixing conditions are met. The designs seem to be unique [40], (Fig. 4.5f).

4.2.7 Advantages of Air Dyeing

- The major benefit of air dyeing is that it only needs a small amount of water to dissolve the dyestuff and chemicals while the fabric is being moved by a stream of air. As a result, dyeing at a low liquor ratio of 1:2 to 1:4 is possible that are almost one-third of the conventional jet dyeing system [23]. Table 4.2 represents the liquor ratio of different dyeing machine systems.
- Moreover, there is a continuous rinsing technology using spray-type washer. A hot bath drain is possible after dyeing at 130 °C without further contact of the dyed fabric with the rinsed water. This process decreases the overall dyeing time by around 25%, saves water, and optimal heat recovery is obtained.

Table 4.2 Liquor ratios of various fabric dyeing machines

Name of dyeing machine	Material to liquor ratio (M:L)
Air dyeing machine	1:2–1:4
Soft-flow dyeing machine	1:4–1:8
Overflow dyeing machine	1:7–1:15
Winch dyeing machine	1:20–1:40
Jigger dyeing machine	1:4–1:10

Adapted from Ref. [1], and updated

- In comparison to hydraulic dyeing machines, the air dyeing method saves up to 60% water and 50% heating energy because of the lower liquor ratios and continuous rinse procedure. Additionally, it would be possible to save up to 10% on dyestuff, 35% on salt, and 40% on chemicals [1].
- Especially in the case of reactive dyeing, a lower liquor ratio facilitates low hydrolysis of dyestuff, thereby improving the yield of dye in the dyebath and higher exhaustion.
- The consumption of dyes, chemicals, and energy in an air dyeing system compared with conventional dyeing is presented in Table 4.3. As it is seen, the air dyeing technique requires substantially less water, steam, dyes, and chemicals than conventional dyeing. However, the air jet machines use a comparatively 30–40% higher amount of electricity due to the generation of high-pressure air from a blower.
- A uniform and consistent dyeing is obtained in the air dyeing process due to the even flow of temperature by moisture-saturated air throughout the machine and fabric.
- Additionally, the dye bath chamber contains a low amount of dye solution, hence the fabric practically becomes lighter than it would be in traditional dyeing, and enables the fabric to run at high rates up to 700 m/min using an effective frequency inverter. As a result, there is little chance of draft or strain, which is especially useful for dyeing textiles with elastane fibers [41].
- Another benefit is that the fabric is dyed with minimal tension, and airflow assists in opening the rope in a balloon-like shape after the nozzle, minimizing crease marks and leading to a uniform dyeing result.
- Almost any type of fiber and its blend (apart from pure wool) and fabric weight ranging from 30 to 800 GSM, along with all commercially available dyes, are processable in air dyeing [13].
- It is easy to process fabrics with up to 20% elastane in air dyeing without compromising their dimensional stability, which is difficult to do in traditional jet dyeing [1].
- The air dyeing process injects color inside the fabric; therefore, long-lasting and unique coloration is obtained. For every 25,000 T-shirts sold, an industry can conserve 1,132,500 megajoules of heating energy, 157,500 gallons of water, and

Table 4.3 Comparison of consumption among air and conventional dyeing machines

Input materials	Unit	Air dyeing system	Conventional dyeing system
Water	l/kg	Polyester: 20 Cotton: 60	Polyester: 100 Cotton: 150
Steam	kg/kg	2–3	4–6
Dyes	g/kg	4–60	5–80
Chemicals	g/kg	7–35	12–60
Electricity	kWh/kg	0.38–0.46	0.27–0.33

Adapted from Ref. [23], and updated

57,500 kg of emissions from greenhouse gases by using air dyeing technology [32]. It leads to 84% less global warming.
- Because of the low liquor ratio, less water is contaminated during the color application procedure. In case of air dyeing, air is used instead of water, hence no effluent or harmful waste is emitted. As a result, it minimizes the issue of water scarcity and the high expense of treating effluent [13]. Therefore, this technique is one of the most sustainable alternatives to conventional dyeing techniques.
- However, it is always challenging to adopt new technology because of the high initial investment cost. This approach has a one-third higher initial cost than traditional dyeing, but it has a comparatively faster payback period because of the large savings in input raw materials [42].

4.2.8 The Economic Aspects of Air Dyeing

Air dyeing technology is the most economical of all jet dyeing methods because it operates on a low liquor ratio, enables a continuous rinsing system, and uses less water, energy, colorants, and chemicals. As air dyeing uses less water, this process will produce less waste water overall. This will result in significant savings on wastewater treatment costs. Due to an intense heating gradient of 9 °C/min, the total temperature raising time is shortened, saving 25–35% of cycle time and extra costs. Higher productivity is possible with airflow dyeing machines, which also significantly lower the material-to-liquid ratio to 1:2, saving high consumption of water [43]. For instance, dyeing a 100% cotton single jersey fabric in a dark color takes 4.5 h overall, which is 35% less time than using a conventional jet dyeing process. As a result, compared to other traditional hydraulic machines, the daily production of air dyeing machine will increase by 20–30%. Despite having a one-third higher machine price, and a higher electricity consumption than all other exhaust dyeing, air dyeing is yet inexpensive due to high savings in input material [42]. Therefore, it is now high time to replace traditional dyeing equipment with air dyeing equipment. Table 4.4 presents a detailed comparison of costs between air dyeing and soft-flow dyeing. The results state that for the same amount of dyeing production of 1,560,000 kg per year, air dyeing technology consumes 22.35% less cost, despite the high machine price and electricity cost of the airflow dyeing machine.

4.2.9 Quality Aspects in Air Dyeing

Compared to water, air is a gentler media that dyes textiles gently while making the cotton fabric's surface less hairy. Furthermore, the fabric expands automatically and forms balloons as it leaves the nozzle, ensuring smooth, high-quality materials free of wrinkles and streaks, and dyeing only occurs inside the nozzle, which contributes to the incredibly uniform coloration. The air dyeing process injects color inside

Table 4.4 Comparison of the costs of soft-flow and airflow dyeing machines

Calculation criteria	Soft-flow dyeing	Air-flow dyeing	Assumptions
Per day production in kg (A)	5,000	5,000	
Per year production in kg B = A × 26 × 12	1,560,000	1,560,000	Assuming 26 working days per month
Per day no. of batch dyeing	3	3.75	25% more production per day for air dyeing
Capacity per batch in kg	1,667	1,334	
Per day no. of dyeing machine required (D)	4	3	Assume that 500 kg machine with 475 kg loading capacity for air dyeing and 450 kg for soft-flow dyeing
Machine price in US$ (E)	180,000	240,000	One-third higher price for air dyeing machine
Total cost of machine in US$ (F = D × E)	720,000	720,000	
Depreciation % per year	10%	10%	
Depreciation cost per year (US$) (G = F × 10%)	72,000	72,000	
Per month salary of one operator in US$ (H)	150	150	
Total salary of all operator per year, I = {H × (D × 2) × 12}	14,400	10,800	Assume that one machine is managed by two operators
Per year total water cost (US$) (J = Consumption in kiloliter × US$ 0.1 × B)	21,840	9,360	Assume that soft-flow consumes 140 L of water per kg of fabric, while airflow consumes 60 L of water, and the per-kiloliter water price is 0.1 US$
Per year chemicals cost (US$) (K = chemical consumption in kg × US$ 0.5 × B)	35,100	19,500	Assume that for soft-flow m/c, chemical consumption per kg of fabric is 0.045 kg, while consumption in airflow is 0.025 kg, and the per kg of chemical price is 0.5 US$
Per year total steam cost (US$) (L = Consumption in kg × US$ 0.0075 × B)	52,650	29,250	Assume that for soft-flow m/c, steam consumption per kg of fabric is 4.5 kg, while consumption in airflow is 2.5 kg, and the per kg of steam price is 0.0075 US$

(continued)

Table 4.4 (continued)

Calculation criteria	Soft-flow dyeing	Air-flow dyeing	Assumptions
Per year electricity cost (US$) (M = Electricity consumption in kWh × US$ 0.055 × B)	27,456	32,604	Assume that soft-flow consumes 0.32 kWh of electricity per kg of fabric, while airflow consumes 0.38 kWh of electricity, and the per kWh of electricity price is US$ 0.055
Per year total cost (US$) = (G + I + J + K + L + M)	223,446	173,514	
Comparison of cost in %		22.35% lower cost	Airflow dyeing saves 22.35% cost compared to soft-flow dyeing

the fabric; therefore, long-lasting and unique coloration is obtained. In the case of conventional dyeing machines, dyeing occurs mainly in dye bath liquor, which results in many dyeing defects. The air dyeing machine features a perforated PTFE (Poly Tetra Fluoro Ethylene) chamber that effectively separates extra liquor from the dyed fabric into a collection pit. Additionally, the chamber's low coefficient of friction with the fabric results in significantly less damage to the material's surface. During the air dyeing process, up to 1% of goods are damaged only, whereas conventionally dyed fabric is damaged by 10% [32]. Furthermore, issues including shade variation, dyeing defects, hand feel, and fastness properties are also improved [44]. Air dyeing machines can dye fabric with high elastane content without any crease marks and defects. Up to 20 and 45% of the elastane content of fabrics can be dyed by THEN-AIRFLOW SYNERGY round-vessel and THEN-AIRFLOW LOTUS long-tube machines, respectively which is nearly impossible to achieve with a conventional dyeing machine. Another benefit is that superior heat recovery in materials is made possible by the hot bath rinsing capability.

4.2.10 Environmental Aspects of Air Dyeing

Air dyeing techniques create less environmental pollution, such as air, water, and noise pollution, by using less water, energy, colorants, and chemicals. This process uses air as the medium for transporting fabric rather than water and releases almost zero effluent and hazardous waste into the environment [44]. In this system, less polluted wastewater can be easily treated and recycled for reuse in dyeing, which minimizes the water scarcity issues in the world. Air dyeing technology reduces greenhouse gas emissions and results in 84% less warming of the planet, contributing to a lower carbon footprint. Therefore, this technique can be considered one of the sustainable alternatives to conventional dyeing techniques.

4.2.11 Applications of Air-Dyed Fabric

Air-dyed fabric finds diverse applications in several industries, including the apparel industry, print industry, medical sector, home textiles, and many more [13].

- **Apparel industry**: All types of fabric especially cellulosic and synthetic, specialty fabrics like tents, awnings, umbrellas, flags, uniforms, civic items, aprons, gowns, medical wear, sportswear, outerwear, swimwear, active-wear, sleep-wear, junior apparel, infant apparel, intimate and lingerie apparel, etc. Fashionable items like cosmetic bags, handbags, shopping bags, accessories, etc.
- **Print industry**: Laminate for hard surfaces, thin paper, thermos films, vinyl transfer, and melamine are some examples of industrial print applications.
- **Medical sector**: Aprons, hospital gowns, and cubicle curtains are application areas.
- **Home textiles**: Chair covering, sofa cover, pillows and mattress covering, shower curtain, tabletop, wall covering, bedsheet, blinds and curtains, ceiling tiles, carpet all these can be application areas.
- **Promotional items**: Reusable bags, branded products, tote bags, signage and retail display, and various types of banners.

4.2.12 Exciting New Options in Air Dyeing

- **Handling delicate fabrics**: When processing delicate knitwear and woven fabrics made of synthetic and cellulosic fibers with a high proportion of elastane, air dyeing is a particularly good option. Sports, leisure, swimwear, lingerie, and the automobile industries all employ these types of textiles. The fabric is simply moved by air due to the newly developed dome blower. High everyday production rates are made possible by quick cycle time, a low liquor ratio, large loading capacities of up to 200 kilos for average items, and lightweight fabrics up to 1,200 m [13].
- **Polyamide dyeing**: Polyamide makes most fabrics for swimwear, sportswear lingerie, and automotive uses. The air dyeing equipment handles fabrics composed entirely of polyamide with identical outcomes to those previously discussed.
- **Numerous air-dyed patterns**: When compared to conventional heat transfer printing, air dyeing machines can use extremely low pressure to transfer inks onto fabric and cause them to enter deeply into the fibers, producing a smooth hand feel and more even coloring [45]. Air-dyed fabrics are said to feel exactly the same as those dyed conventionally. They can be dyed in the following ways: one color on each side, both sides with print design, both sides with solid color, solid color on one side, and print design on the other [12, 33, 46].
  - **One color on each side**: With the aid of air dyeing, two distinct colors can be dyed on opposing sides of a fabric with a small amount of dyeing liquor.

- **Print to print**: Using air dyeing, the fabric can have one side printed with one design and the other with a different design.
- **Dye squared**: Using air dye technology, fabric can be dyed on both sides in the same solid color with a small amount of dyeing liquor.
- **Dye to print**: With air dyeing, the fabric can be dyed a solid color on one side and printed on the other side.

4.3 Dope Dyeing

The process of coloring by incorporation of colorants into the spinning system prior to extrusion into filaments or fibers is known as dope dyeing. This is also known as mass coloration, solution coloration, or spun dyeing. Dope dyeing is an environmentally friendly modern alternative approach to the traditional wet dyeing procedures. It is considered a green technique that has been more widely used in recent years which allows the simultaneous creation of fiber and dyeing [47, 11]. In a world where industrial wastewater disposal is subject to greater environmental constraints and water resources are becoming scarcer, this dyeing technology offers a unique chance to color fibers with minimal water [48]. It is the preferred approach for adding color to polymers that are otherwise challenging to dye, such as polypropylene, and it produces products with excellent attributes including even shade, uniform coloration, higher dye uptake rate, shorter production cycle, superior colorfastness, lower cost, and less pollution, etc. [11, 47, 48, 49, 50]. More significantly, the method eliminates the need for downstream items to be dyed and drastically lowers the amount of water, energy, and chemicals used by the textile industry as a whole.

4.3.1 Methods of Dope Dyeing

There are two methods involved in dope dyeing: continuous dope dyeing and discontinuous dope dyeing.

4.3.1.1 Continuous Method

Both powdered and masterbatch forms of the colorant are introduced to the polymer. The colorant can be introduced straight to the molten polymer, or it can be continually distributed with the aid of a dozer and combined with chips prior to the melter (screw type extruder or static melter). It is also possible to add the colorant in liquid or paste form. In both scenarios, a secondary extruder assists in continually injecting the polymer into a molten polymer throughout a continuous operation. The spinning process route in the continuous method is shown in Fig. 4.6.

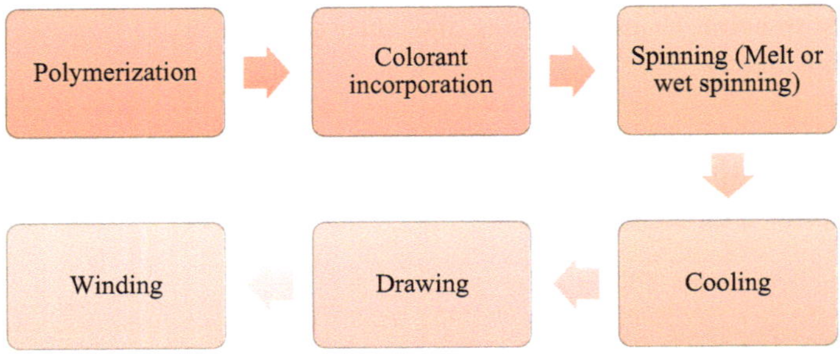

Fig. 4.6 Spinning process route in continuous method

4.3.1.2 Discontinuous Method

In this method, after leaving the poly-condensation reactor, the polymer cools and turns into chips. These granules are kept in dried form and provided to the melters as needed. Smaller volumes can be adjusted to this process more readily. The spinning process route in the discontinuous method is shown in Fig. 4.7.

4.3.2 Colorant Incorporation Techniques in Dope Dyeing

Dope-dyeing is based on the basic idea of incorporating colorants into the polymer melt before extrusion. The mixture is melted and homogenized to ensure the uniform distribution of the colorants throughout the melt. After that, spinnerets are used to extrude the colored polymer melt into fibers with the appropriate diameter [51]. Depending on the specific fiber production process, there are six main methods of incorporating colorants into manufactured fibers [52].

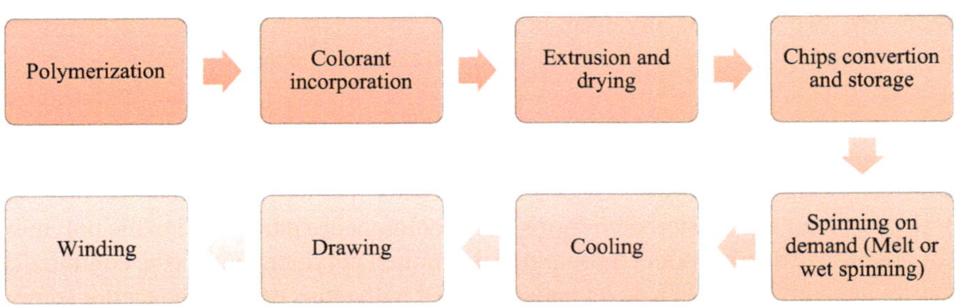

Fig. 4.7 Spinning process route in discontinuous method

4.3.2.1 Addition of Colorants During Polymerization

The process of adding color to polyester melts using carbon black (1.5–2.5%) is carried out at high vacuum (less than 0.133 kPa) and higher temperatures (260–290 °C), mainly for the manufacturing of jet-black colored polyesters to be used in automotive applications. It is estimated that this method costs roughly 5% of the conventional dyeing process of the polyester fibers using dispersed dyestuff [53]. A second option is to create the black fiber in two steps. At the first step, the carbon black content is significantly lowered to equal or less than 0.5%, and a deep grey-colored fiber is thus produced without any significant issues with agglomeration and choking of the spinneret hole. In 2nd step, by overdyeing the grey fiber using traditional dyeing techniques, it is converted into a jet-black fiber. Colored nylon, acrylic, and polypropylene fiber can also be produced by this technique.

4.3.2.2 Chip Dyeing Method

In the chip dyeing method, the polymer chips are colored first with polymer-soluble dyes and then are melt-spun. This method is mostly suitable for nylon 6 [52]. In nylon 6 production, it is viable to dye the chips simultaneously with metal complex and acid dyes during the extraction of the monomer and the following drying process. Polyester is not commercially dope dyed in this method because it is very difficult to remove moisture from dyed polyester chips entirely which causes hydrolysis of polyester resulting in a loss of strength when it is melted and spun [54].

4.3.2.3 Chip Coating Method

In the chips coating method, the process of coating chips involves rotating dry polymer chips with colorant, either by themselves or with the help of an adhesive like silicone oil, magnesium stearate, or wax. The coated and dried chips are melted and spun to produce colorful fibers or filaments [54].

4.3.2.4 Masterbatch Addition Method

With the masterbatch addition method, a significantly lower melting carrier polymer is used to create colored chips or granules with a high percentage (10–50%) of a colorant. Three primary components make up color masterbatches: pigment, dispersing agent, and carrier. In addition to these components other additives, such as antistatic agents, antimicrobial agents, nucleating agents, stabilizers, wetting agents, etc. may be employed [55]. Both inorganic and organic pigments can be used. Many of the inorganic pigments are not very susceptible to heat degradation, however, color changes can happen with organic aromatic pigments at temperatures as low as 280–300 °C (535–570 °F), which is significantly lower than those used to spin polyester or nylon. The carrier polymer is used to reduce the risk of thermal degradation of pigments. The pigments are initially mixed with the carrier polymer melting at a considerably lower temperature, like polypropylene, which has a melting point of approximately 167 °C (332 °F), and dispersing agents [56]. The carrier polymer must be compatible with both the pigment and the dispersing agents. To create colorful fiber, masterbatches of colored chips are mixed with white chips mainly in two processes. In the first process, colored fiber is created by blending the colored masterbatches of chips with white chips in the appropriate ratio and spun together (Fig. 4.8a). Rapid melting of the carrier polymer results in homogenization of the polymer bulk and pigment particles. The use of molten polypropylene as a carrier polymer for nylon is technically limited, however, it can be used for polyester [56]. In the second process, dried white chips and dried masterbatch chips are fed into a mixing chamber via separate metering stations. To spin, this homogenous mixture is sent directly to the extruder (Fig. 4.8b). The method produces a highly consistent shade. This technique works especially well with newly introduced compact (short) spinning plants [57].

The advantages and disadvantages of the masterbatch addition method are as follows [59]:

Advantages:

- Reduction in material costs compared to those of ready-made mixtures dyed in mass.
- Ability to obtain the desired properties of the finished product.
- Reduction in or elimination of the inhalation of toxic dust substances.
- Easy to control.

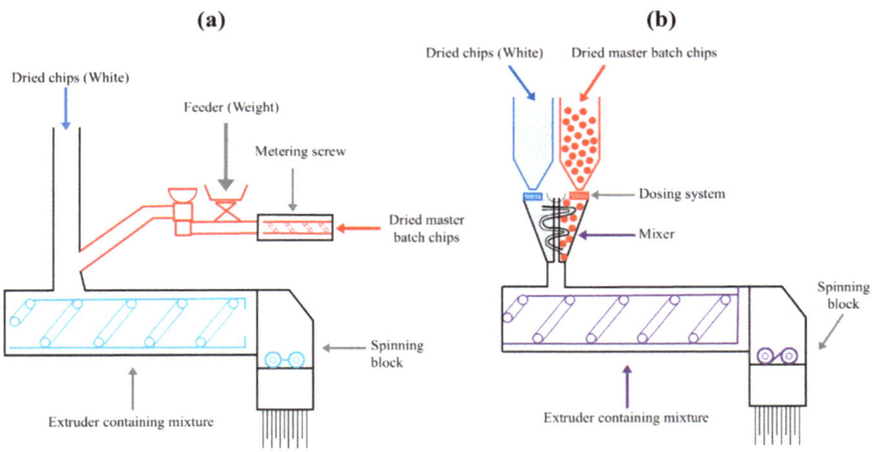

Fig. 4.8 **a** Melting of colored masterbatch and white chips together in extruder, **b** melting of colored masterbatch and white chips in separate melter chamber (Adapted in a modified form from Ref. [58])

Disadvantage:

- The only disadvantage is the possibility of incomplete mixing of the ingredients of the composition.

4.3.2.5 Injection Method

The injection method is the newest method used globally for dope dyeing. In this method, the powdered polymer soluble dyestuff is injected into the stream of polymer chips in the extruder right before spinning. The same dyestuffs can be mixed with the virgin polymer before injection to produce the desired tints [54].

4.3.2.6 Melt in Melt Metering System

In continuous spinning processes, melt in the melt metering system is carried out in a static or dynamic operational mixture. In this method, white polymer chips and polymer-soluble colorants are melted in a separate melting chamber. After that, the melted stream of colorant and the melted stream of polymer exiting from the polycondensation reactor are combined (Fig. 4.9). Fibers are then spun from the melt. This process ensures that only brief piping segments require cleaning during color change by placing the mixing point directly next to the spinning line. Compared to other ways, this enables faster and less expensive working [52, 54, 60].

4.3.3 Spinning Techniques in Dope Dyeing

Various techniques have been employed to transform the solid polymer into a liquid state. In addition, extrusion methods also vary. It has been found that spin or dope dyeing are more environmentally friendly methods of producing colored fibers either using melt or wet spinning. In melt spinning, dyes or pigments are introduced as an intermediary step, in the form of a paste, powder, or solid, into the molten polymer prior to the beginning of fiber production. The colorant is evenly distributed throughout the fiber after spinning [57]. Polyester, polyamide, and polypropylene fiber can be spun using this method.

In the wet-spinning process, the dissolved polymer, also called "dope", is extruded into an antisolvent-containing coagulation solution, thereafter the polymer solidifies into the fiber. This is a slow process, and the fiber doesn't entirely develop until it dries completely. Due to the presence of residual polymer dissolution solvent in the polymer matrix, this intermediate form also known as "never-dried" fiber, has a more "open" polymer network. It is possible to color the fibers during this stage by adding color to the treatment baths since this allows the colorants to more easily permeate the polymer network [61]. Polyvinyl alcohol, polyvinyl chloride, and acrylic fiber can be spun using this method.

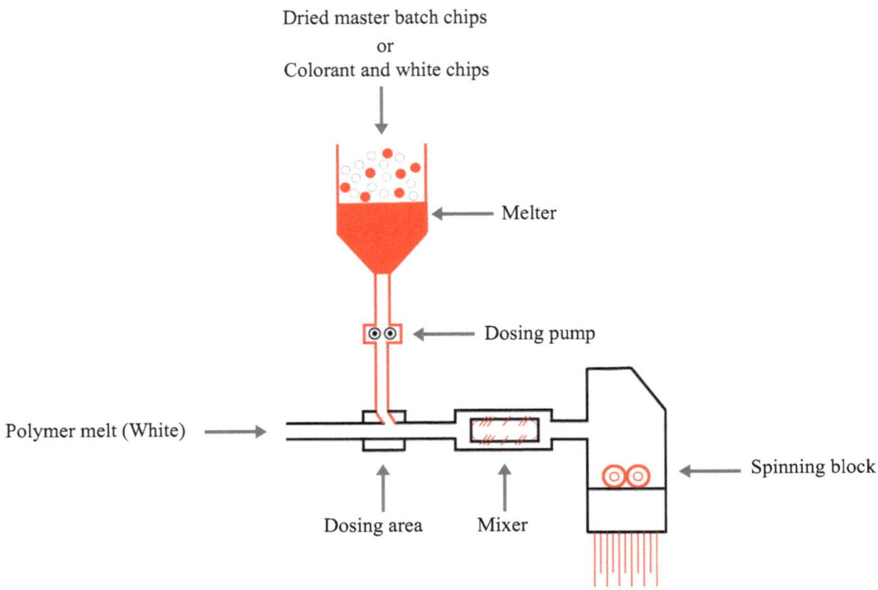

Fig. 4.9 Melt in melt metering system of dope dyeing. (Adapted in a modified form from Ref. [58])

4.3.4 *Factors Need to Consider in Dope Dyeing*

A crucial factor in the dope dyeing procedure is to ensure that the polymer-colorant in the dope mixture is physio-chemically compatible [62]. In this regard, regenerated cellulosic fibers face particular challenges since their production techniques entail treating the cellulose with highly powered reducing and/or oxidizing chemicals, which might reverse colorant stability or have an impact on the resulting fiber's tensile strength. The spun-dyeing of regenerated cellulosic materials has been made feasible despite these apparent obstacles and several strategies have been developed to attain it [63]. The choice of colorants is critical and it depends on the following:

(a) The limiting concentration: The lowest dilution at which a colorant is thermally stable at a specific temperature is called its limiting concentration. For example, when the amount of fluorescent pigment in the polymer matrix increased from 2 to 4%, the thermal stability of the dope-dyed alginate fiber using fluorescent pigment was shown to be affected. As the ratio of colorant to dissolved polymer can alter the microstructure and crystalline alignment of the fiber during extrusion, the limiting concentration also influences the tensile strength of the fibers to some extent. Furthermore, the limiting concentration of the regenerated cellulose can also be significantly impacted by the acidity or alkalinity of the spinning dope [50]. For this reason, the dope stoichiometry needs to be carefully calculated.

(b) Particle size and its distribution: A pigment's specific surface area and resulting viscosity decrease with increasing particle size. A pigment with a high specific surface area will need more "wetting out" in the dope-dyeing of regenerated cellulose to achieve the best possible dispersion. When physical dispersion forces and pigment wetting reach their peak, a narrow band distribution with a minimum of over and undersized particles will disperse more easily in a thermoplastic system, but this is ineffective for regenerated fibers; therefore, the particles must be properly distributed and dispersed before being added to the spinning dope [50].

(c) Color dispersibility: To provide fiber functionality for daily use, optimal colorant dispersion is crucial because this may ultimately impact the fiber's crystalline characteristics.

(d) The light fastness: The achievement of specific colorants light fastness depends on the optimal dispersion of mill base, pigment, and other water-insoluble colorants. It has been found that processing parameters can also influence the light fastness characteristics of dope-dyed regenerated fibers; however, this effect can be reduced to some degree if the dope-dyeing process is properly monitored.

4.3.5 Characteristics Features of Dope-Dyed Fibers

Dope-dyed fibers are characterized by their exceptional color fastness and light fastness, strong water resistance, and vibrant color with luster. In addition to having the exceptional quality of regular fiber, spun-dyed fibers also have lower color variations between batches and less environmental contamination [64, 65]. The color is uniformly distributed since the dye is incorporated during the fiber creation stage, making the finished product more homogeneous and aesthetically pleasant [57]. Besides, spun dyeing is a popular method and cost-effective option for coloring synthetic polymers, such as polypropylene, that are otherwise challenging to dye [48].

4.3.6 Mass Coloration or Dope Dyeing of Synthetic Fibers

Polyester, nylon, acrylic, polyurethane, and polypropylene are among the synthetic fibers that can be produced by dope dyeing. However, according to the fiber to be processed, the dyeing techniques and parameters may be changed. The following are the processing steps involved in mass coloration when producing synthetic fibers [66]:

- Melting and formation of dope: The dope dyeing process starts with the melting and formation of dope, where "dope" or viscous solution is created by the melting of the fiber-forming polymer. The required dye colorants and other additives that give the finished fiber certain qualities may be included in this dope.

- Mixing and homogenization of dope: The dope is carefully prepared to get the maximum dye dispersion and color constancy. This involves adding chemicals like stabilizers and modifiers as well as selecting and adding dye colorants precisely. A uniform dye distribution throughout the fiber can only be guaranteed by the complete mixing and homogeneity of the dope.
- Spinning and extrusion: The polymer solution injected with dye is prepared for spinning once the dope preparation is finished. Then continuous filament is formed by the extrusion of dope via spinneret. A consistent and long-lasting coloring is produced when the dye molecules are embedded in the fiber matrix during this extrusion process.
- Post-processing: To solidify, the colored filaments are allowed to cool immediately after the extrusion. The brilliant color of the dye is preserved during this cooling process, which also hinders any subsequent fading or bleeding. After solidifying, the fibers are prepared for further processing, including texturizing, weaving, and stretching.

4.3.6.1 Dope Dyeing of Polyester Fibers

Polyester fiber faces several challenges in traditional dyeing systems, mostly because of its high glass transition temperature and lack of reactive groups. Their hydrophobicity restricts the dyestuffs to distribute across the fiber. Previously, dyeing was carried out with carrier chemicals, but this had a negative impact on the environment. Nowadays, high-temperature and high-pressure methods are being used to dye polyester. Therefore, alternative techniques have been explored for coloring polyester fabric to overcome these challenges. Dope dyeing, often referred to as mass coloration, is one such technique that involves combining a molten polymer of polyester with a color masterbatch before the start of the melt-spinning process to create polyester filament. This unique method improves the product's environmental sustainability and saves water resources greatly. The method involves three steps: the production of a solid polyester masterbatch (granules) followed by quality checking and finally spinning of the fiber [57].

Masterbatch Preparation

Typical stages in the process of producing solid masterbatch (granules) are outlined below.

- Preparation of pre-mix: The carrier polymer and colorant (pigment or polymer soluble dye) are weighed precisely in a separate way. Wetting agents are added to pigment-based concentrates during production. Certain additives, such as UVA (Ultraviolet A for UV blocking), HALS (Hindered amine light stabilizers), or optical brighteners, can also be added for specialized end uses.

- Mixing of ingredients: To create a uniform mixture, each ingredient is added successively to the high-speed mixer. The speed of the mixture is increased gradually from low to high and the time required to prepare the mixture depends on the set temperature.
- Extrusion: The hopper of a twin-screw extruder receives the pre-mixed batch after it has cooled to room temperature. Extruders containing a closed barrel with splined shafts holding intermeshing, co-rotating screws are mostly used for masterbatch to provide highly flexible conveyance, compression mixing, shearing, and heating. The solid polymer is transferred into a molten mass by the heat from the barrel and mechanical shear from the screw. This mass is converted into a polymer strand by pressing it through the die head of the extruder [57].
- Pelletizing: After exiting the extruder and cooling in a water bath, the polymer strands are cut into granules by a pelletizer.
- Packing: To eliminate dust, after being sieved, the granules are either packaged or transported to the spinning department.

Choosing an appropriate carrier polymer is necessary for the production of masterbatches for dope dyeing. For polyester, Polybutylene Terephthalate (PBT) resin serves as the foundation for most premium concentrates. PET, or polyethylene terephthalate, can also be used for standard qualities. Despite being expensive, PBT is the recommended carrier polymer for polyester because of its reduced melting point, lower sensitivity to hydrolytic degradation, and greater thermal stability. Both pigments and dyes that dissolve in polymers are used as colorants for masterbatches. Concentrations ranging from 25 to 40% are typically used to produce dye-based masterbatches. When it comes to pigments, these concentrations are usually not more than 30%. Depending on the concentration of the master batch and the need for specific shades, the typical dose of the masterbatch to polymer is between 1 and 5%.

Quality Checking of Masterbatch

The quality of color concentrations is defined by a number of characteristics. The following characterization techniques are crucial and can serve as a deciding factor.

- Filter pressure value (FPV): The FPV is determined to measure the homogeneity of distributed colorants in the master batch. Test material is pushed through a test screen by extrusion. The pressure rise is calculated and expressed in bar/g.
- Shade matching: Using a pilot spinning machine, a masterbatch is spun into filaments to achieve shade matching. A spectrophotometer is then used to compare the shade to the standard shade.
- Fastness properties: To make sure there is no color leakage during downstream processing, knitted fabric samples are evaluated for service attributes including color fastness to light, washing, or sublimation.

- Thermal stability: Thermogravimetric analysis (TGA) is used to determine the thermal stability of masterbatch based on weight loss. Test parameter: 30 min at 290 °C.
- Granulometry: Granule weight as chips/g and distribution of shape and size are measured.

Spinning of the Yarn/ Fiber

Masterbatches and polymer are mixed just prior to the melt-spinning procedure, or before the fiber is formed. A gravimetric K-Tron feeder feeds the masterbatch into the extruder mouth after it has been dried at 100–120 °C (Fig. 4.10a). Alternatively, a side-stream extruder may be fitted to the main extruder (Fig. 4.10b). In this system, the master batch is melted in a separate extruder before being fed into the main extruder's molten virgin polymer stream. Static mixing components in the main extruder or a dynamic mixer at the outlet of the main extruder are placed to homogenize the mixture [57]. Currently, spinning pumps can also be found with built-in dynamic mixers. These pumps can take the place of traditional spinning pumps to raise the level of homogenization.

Advantages of Dope Dyeing in Polyester Synthesis

Polyester dope dyeing is an economical coloring technique used in large-scale industries like the automotive sector. The following are some benefits of polyester mass dyeing over conventional dyeing techniques [67].

- Excellent fastness grading
- High production rates that save a lot of money on energy, utilities, and labor services
- No pollution of water
- Reduced cost of coloring
- Capacity to provide texture to dyed yarns without compromising the shade

Application of Dope-Dyed Polyester

Dope-dyed polyester fibers are used in many different industries, such as home furnishings, automobiles, construction, and clothing. They are employed in the production of geotextiles, carpets, apparel, upholstery, and a variety of other goods.

4.3.6.2 Dope Dyeing of Polyamide (PA) Fibers

The polyamide (PA) chips, or granules, are dyed at boiling temperature at a liquor ratio of 1:1 to 1:4. Dyeing is carried out in a slightly acidic media to accelerate the dye

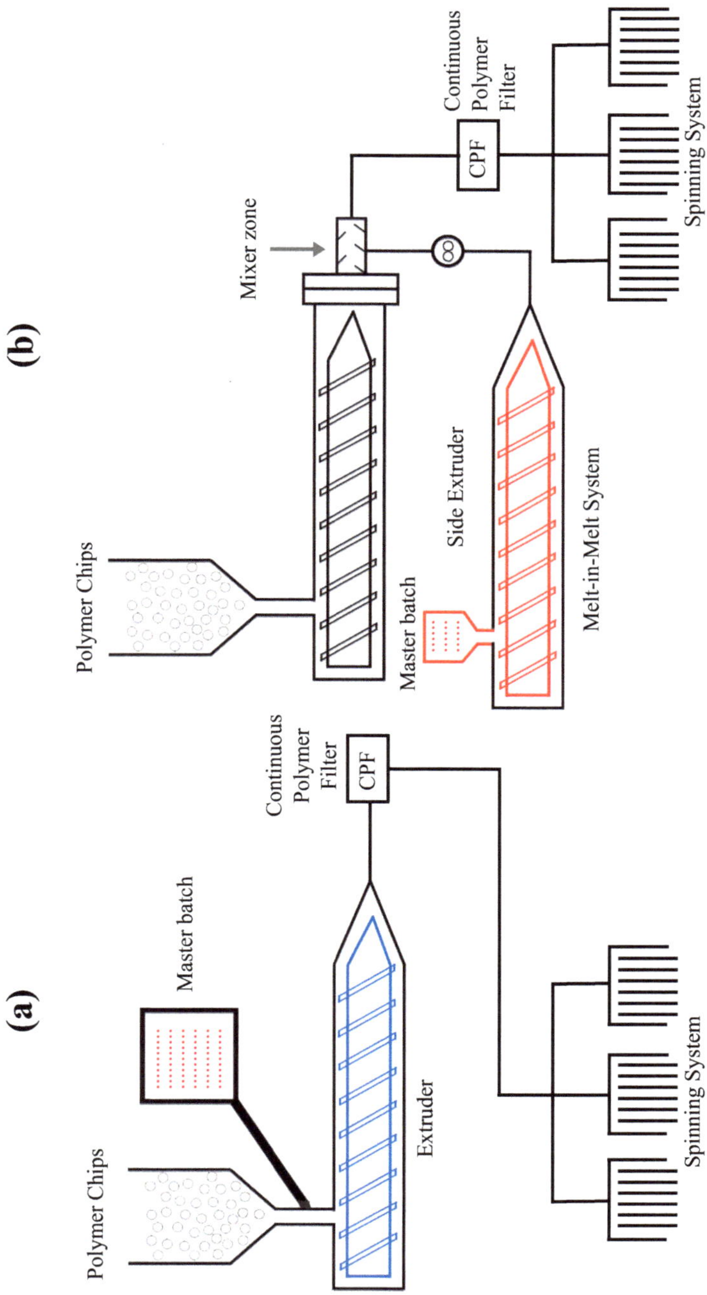

Fig. 4.10 **a** Single extruder system, **b** separate extruder system (Adapted in a modified form from Ref. [57])

exhaustion process. Meng et al. used 1,4-dihydroxyanthraquinone di glycidyl ether as a novel melting-reactive dye to color nylon, and thermal studies revealed excellent thermal stability and blending compatibility, as well as improved light fastness [68]. There are several methods for mass coloration of nylon which are described in a short follows:

Chip Dyeing Process of PA

For nylon 6, the chip dyeing procedure is often employed, whereas, for nylon 6,6, it is less common. To produce nylon 6, aqueous monomer extraction is needed. It is feasible to combine nylon chips with metal complex azo dyes and acid dyes during the monomer extraction process and the drying step that follows.

Masterbatch Addition Process of PA

This approach starts by producing colored nylon chips with 10–15% dispersed acid dyes using a polymer like nylon 6. Then, just before spinning, fresh white nylon chips are added to this masterbatch in the extruder.

Melt in Melt Metering System of PA

Here, an additional melter is used to melt nylon. Following the polycondensation, the colorant stream is combined with the melted polymer stream. Compared to other approaches, this one is less expensive.

Addition of Colorants During Polymerization of PA

Prior to polymerization into the fiber-forming polymer, colorants are directly added to the nylon monomer. After that, the polymer is melted and the colored fiber filament is extruded using a spinneret. Very few colorants are stable at high temperatures when the molten polymers are present, and nylon has a strong reducing effect on colorants.

4.3.6.3 Dope Dyeing of Polypropylene (PP) Fibers

For the woven or nonwoven market, PP textiles are important. They are difficult to dye by conventional dyeing methods since they have no reactive functional groups in their chemical structure and have high crystallinity. These factors make mass coloration technology necessary for dyeing PP.

Mass Coloration with Pigments

Before extrusion and spinning, pigments or dyes are dispersed and homogenized in PP [69]. Since melted PP cannot be colored directly by dyes or pigments, a masterbatch with a concentrated color dispersion may be effective. There are two ways to add color dispersions: (1) injecting the melt of the masterbatch concentrate into the PP melt after it has been melted and homogenized in a different extruder, or (2) adding weight dozers directly to the extrusion feed. Certain pigments, like as phthalocyanine pigments, expedite the crystallization of the PP stream beneath the spinneret. Some substances, like TiO_2 and carbon black, have less of an impact on the kinetics of crystallization [70].

Mass Coloration with Metallic Additives

An investigation has been done on the effects of colored organic and inorganic metallic additives, such as CI Yellow 37 and CI Pigment Red 108, on the tensile strength, abrasion resistance, and mass coloration of molten PP and improved results were found using this technique [71].

In a recent study, by preparing samples of PP with masterbatch and PP with liquid color concentrates (LCC), respectively, Haastrup et al., compared the properties of PP mixed with these two substances. It was found that the tensile strength, viscosity, and crystal structure obtained when using LCC instead of Masterbatch were similar [72].

4.3.6.4 Recent Approaches on Dope Dyeing of Other Synthetic Fibers

X. Wang et al. produced melt-spun red ultrahigh molecular weight polyethylene (UHMWPE) fibers by dope dyeing using a modified Fe_2O_3 pigment and a titanate coupling agent. According to their findings, the titanate coupling agent enhances the Fe_2O_3 pigment's dispersibility and offers good compatibility with the UHMWPE/PO blend during the melt-blending procedure. Using different Fe_2O_3 pigments does not significantly affect the color fastness of the colored fiber, which has great color strength and fastness [73].

Fadil et al. investigated the colorimetric characteristics and dyeing behavior of electro-spun nanofibers mixed with Remazol Yellow FG reactive dye using the dope-dyeing method. Using the electrospinning method, powdered dye was added to the pre-formed PVA solutions to create colored nanofibers. FTIR was used to detect the presence of Remazol Yellow FG dye in the dyed PVA nanofibers. An examination using scanning electron microscopy (SEM) verified that the dyed PVA nanofibers' fiber diameter had increased. According to the TGA analysis, the PVA's thermal stability increased when the dye was added to electro-spun nanofibers [74].

According to the study of Purwar et al., ratanjot (*Arnebia nobilis*) extract was used as a coloring component for dope-dyeing polyacrylonitrile filaments. The colored filaments were spun on a wet spinning machine and evaluated for their mechanical, morphological, fastness, and antibacterial properties. The findings indicate that as the content of ratanjot extract increases, it consequently increases the tenacity of dope-dyed filaments. Excellent wash-fastness and negligible antibacterial activity against tested bacteria were demonstrated by the dope-dyed filament [75].

Wan et al. used disperse orange 31 as the dye β-cyclodextrin as the additive to prepare a series of dope-dyed polyurethanes. Since β-cyclodextrin has a truncated cone shape that may form inclusions with disperse dyes and hydrogen bonds with polyurethane, it was employed as a bridge to connect the polyurethane and dye. The outcomes showed that the dope-dyed polyurethanes exhibited improved mechanical and thermal properties as well as good dye fastness, which can be attributed to hydrogen bonding between the polyurethane and β-cyclodextrin [10].

4.3.7 Dope Dyeing of Regenerated Cellulose Fibers

Regenerated cellulosic fibers such as viscose and lyocell are obtained from wood pulp by dissolution and modification of the cellulose via physical and chemical processes. Since the fibers are spun from solutions so addition of compatible colorants can yield colored fibers directly after the spinning.

4.3.7.1 Dope Dyeing of Viscose

- Dyeing with vat dyes: Mass coloration of viscose is achievable in various techniques. The most common way involves the addition of vat dyes to spinning solutions, where the vat dye is chemically reduced to its leuco form either by employing the chemical reagents that exist in the dope or by incorporating reducing agents that are finally oxidized to return it to the parent form. However, many vat dyes remain unreduced under the conditions of the spinning dope and thus cause visible specks.
- Adding reduced vat dyes directly to the spinning dope is another approach. Still, it may lack suitable viscosities for filament spinning as well as affect the uniform distribution of dye in the substrate due to premature oxidation [51].
- Dyeing with natural indigo: To create a certain size of indigo-dispersed color paste, natural indigo is added to deionized water, well mixed, and then ground in a ball mill at a speed of 100 rpm for four hours. Using a dynamic mixer and a static mixer, the color paste and viscose solution are well combined before spinning. After the homogenous mixture is extruded via a spinneret into a coagulation bath, it solidifies and is wound into two 30-hole 120D viscose filaments. On a cylindrical machine, a 55.62 g/m^2 weft plain stitch cloth is knitted. The relative standard deviation of the surface color depth value on the fabric, which is no more than 2.39%, shows that the fabric exhibits outstanding dyeing uniformity. The spun-dyed fabrics with natural indigo dye had exceptional colorfastness. It was possible

to achieve grade 5 colorfastness under artificial light, grade 3–5 colorfastness when washing with detergent, grade 4–5 colorfastness when rubbing, and grade 4–5 colorfastness under high temperatures [76].

- Dyeing with Sulphur dyes: Spun-dyeing of viscose using sulfur dyes is comparatively less explored. Earlier, researchers reported the suspension of sulfur dye intermediates in the spinning dopes before adding the dye itself, followed by the spinning. Another approach to obtain the colored filaments was to prepare a mixture of waste sulfur-dyed cotton textiles and fresh cellulose biopolymer before conducting the xanthation process and spinning. However, the latter process suffered from gaining much recognition because of the toxicity of xanthation and the production of defective fibers [51].
- Dyeing with dispersed colorants: The dispersion of finely milled organic or inorganic pigments can be added to the spinning dopes for the coloration of viscose [51, 76]. In contrast to just the surface coloration in conventional dyeing, spun dyeing does coloration throughout the entire body of the fiber, hardly causing any loss of pigments [67]. But the application of this colorant is not free from limitations also. Manufacturing properly sized pigments is slow and imposes risks of recrystallization of the colored particles. The shade and strength of the substrate also suffer a lot.
- Dyeing with dissolved colorants: The use of selected dyes, or dye derivatives, capable of dissolving either in polar water-miscible solvents or directly in the spinning dope is limited due to their instability in the strong alkaline dope and the successive strong acidic conditions imparted during regeneration. Moreover, using water-soluble dyestuffs results in poor water fastness of the formed products [51].
- Other suggested methods involve using waste cotton textiles that have been reactive dyed, combining them with fresh cellulose, xanthating the combination, and then spinning colored filaments from it. It is suggested that naphthol dye grounders be dissolved or mixed in the alkaline spinning dope, in an approach that may be classified as a quasi-mass coloration technique. Even after the fibers are woven into yarns or fabrics, the color can still be created by treating them with coupling components [51].

4.3.7.2 Dope Dyeing of Lyocell

Lyocell fibers have attracted significant attention in comparison to other synthetic cellulose fibers because of their outstanding tensile strength, resistance to wrinkles, unique gloss, and handling. Furthermore, recyclable N-methyl morpholine N-oxide (NMMO) is used as a non-toxic solvent, which makes the manufacturing process eco-friendly. Adding color to lyocell fibers is a significant approach to boost their decorative value and expand their range of uses. It has the potential to be very successful due to its advantages, which include uniform coloration, low cost, optimum colorant utilization, superior colorfastness, and short dyeing time [77]. Before spinning lyocell fiber, a cellulose solution should be mixed with specific inorganic pigments that have minimal concentrations of heavy metals that do not significantly result in a severe decomposition in the temperature of the spinning mass. According to the latest techniques, the wood pulp is first colored using a vat dye and then mixed with fresh wood pulp, which is used to spun lyocell fibers in the end. One disadvantage of this procedure is that lighter-colored fibers are produced since the undyed pulp is mixed with the dyed pulp. Figure 4.11 depicts the entire process of producing dope-dyed lyocell fiber [48].

Studies on the dope-dyeing of lyocell with pigment colorants have demonstrated that the color characteristics of the resulting fiber can be strongly influenced by the size and distribution of the particles in the dispersion. It has been suggested that a bright shade with great tensile strength is produced by small particles and their uniform distribution inside the polymer matrix. This claim is supported by the fact that the dispersion of carbon black with small particles in the lyocell spinning solution eliminates the impact of irregularities, microstructure of lyocell fibers, and surface morphology thereby guaranteeing a problem-free spinning process [48, 50, 51].

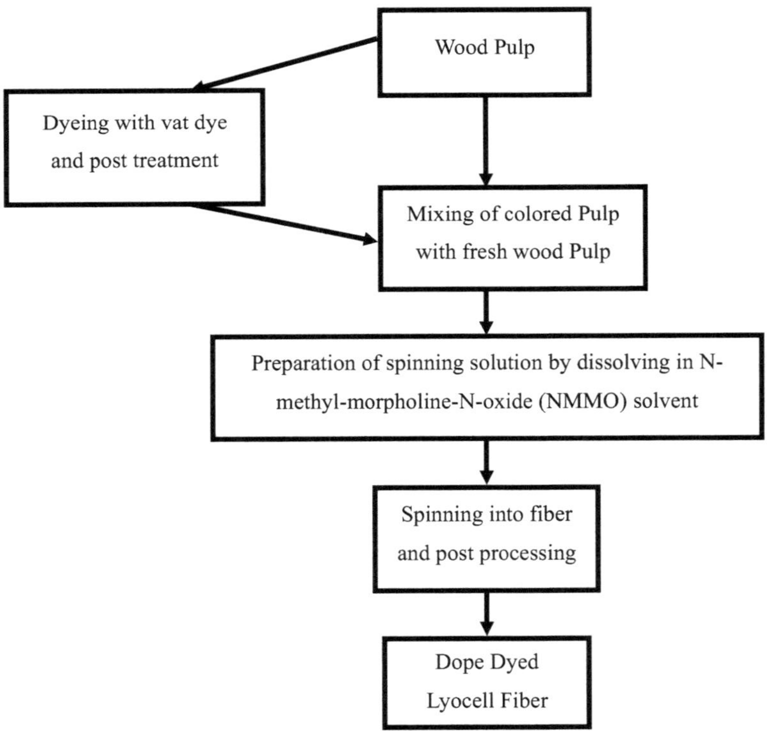

Fig. 4.11 The manufacturing process of dope-dyed lyocell fiber. (Adapted in a modified form with permission from Ref. [48], Copyright © 2006, Elsevier Ltd.)

4.3.8 Advantages of Dope Dyeing Over Conventional Dyeing

Dope dyeing provides regenerated and synthetic substrates with excellent characteristics, such as uniform coloration, and superior colorfastness both with natural and synthetic colorants [48, 51, 75, 78, 79]. Additionally, it offers many benefits over conventional dyeing, including continuous spinning, reduced cost, shorter production time, increased dye uptake rate, reduced pollution, and more. Unlike a two-step traditional fiber dyeing, dope dyeing integrates the dyestuffs in coloration during the spinning composition before extrusion into filaments, therefore, saving a significant amount of water, chemicals, and energy [57, 80]. For instance, Jiangsu Hilong Home Textiles Co., Ltd. the first IKEA supplier to produce dope-dyed synthetic textiles was reported to reduce water consumption by 80%, dyestuff consumption by more than 20%, the use of other chemical agents like alkali by 80% and electricity usage by 7% compared to traditional dyeing [81, 82]. This is the ideal technique for polymers, such as polypropylene, that are challenging to color in other forms.

Since spun-dyed yarns are highly resistant to ultraviolet (UV) fade and shade changes, fabrics composed of those are well suited for draperies, automotive fabrics, and other applications where long sunlight exposure is anticipated [83].

4.3.9 Challenges of Dope Dyeing

Although there are many benefits to dope dyeing, there are also some limitations and challenges. Ensuring the physio-chemical stability of the spinning polymer and colorant in the dope mixture is mandatory to achieve the benefits of mass coloration without affecting the mechanical properties of the fiber. In this regard, strong reducing and/or oxidizing agents, often used in the manufacturing of regenerated cellulosic may affect the stability of colorants [51]. Dope dyeing requires intensive cleaning of extruders, transfer lines, and other units before changing shade which is a complicated and time-consuming process [54]. Custom colors are impossible to produce without very large lot quantities, usually 4500 kg minimum [83]. Production of colored fiber blends is also restricted in dope dyeing [57]. Manufacturers may have less design flexibility if there is a limited selection of dye colors and shades. Furthermore, the initial cost of adding dyes to the fiber-forming solution can be higher. Long-term benefits like improved colorfastness and less water usage, however, frequently exceed these challenges.

4.3.10 Application of Dope Dyeing

Application of dope dyeing is not restricted to regenerated cellulose; rather, it has emerged as the most popular and favored technique for coloring the majority of synthetic materials due to its outstanding performance, cost-effectiveness, and ability to eliminate the waste of valuable freshwater [57, 84, 85, 86]. Therefore, it has been widely used in various fields, including automotive textiles, sports uniforms, military uniforms, and other specialized professional uniforms [47, 54]. Numerous patents related to dope dyeing have also been granted globally. The textile sector in China is actively promoting dope-dyeing technology, and as of the twelfth five-year period, dope-dyed fiber makes up 5%–10% of total fiber globally [11].

4.4 Future Trends of Air Dyeing and Dope Dyeing

It is more likely that machinery developments than process optimization will drive any meaningful advancements in textile air-dyeing technologies. The air dyeing process was promoted by the idea of Wilhelm Christ, and based on this, a prototype machine was created in 1978. Later, during ITMA 1983 in Milan, this technology first appeared to consumers. The textile dyeing industry was expected to undergo a substantial change with the introduction of air dyeing. But it's uncertain that the latest technology will be able to completely replace conventional technologies in a single day. Air-dyeing technology is still in the process of developing. One of its main advantages is the environmentally friendly nature of air dyeing, which uses less material inventory and creates smaller carbon footprints. The air dyeing technique uses up to 60% less water, 50% less energy, and 40% less chemicals, resulting in minimal textile waste. Furthermore, compared to those who use conventional dyeing techniques, manufacturers who rely on air dyeing systems find it easier to quickly adjust to shifts in worldwide trends. All of these factors encourage the increasing popularity of air dyeing in the world's textile industry. However, greater incorporation of air dyeing into industrial manufacturing is necessary for air dyeing to continue progressing. To compete with traditional dyeing machinery, machine manufacturers must also create advanced technological production dyeing machines. Nevertheless, there are a few challenges that are currently restricting the general application of this growing technology, such as the high initial cost of machine purchase and setup, maintenance issues, the operators skills, and high requirements for the operation environment. People are gradually learning more and more about the advantages of this innovative technology. When the relevant technology is further developed, all of the barriers to the application of air dyeing will be eliminated. For the textile sector to adopt this new technology, the ergonomics and features of the air dyeing machine need to be further developed.

A new Airflow Synergy G2 and Airjetwin dyeing machine from THEN, a German company, has been released recently. These machine's altered fabric storage chamber design allows for the running of ropes with higher loads, resulting in a further notable 20% reduction in water and energy consumption, with advantages for both financial and environmental reasons. The new ECO-6-Super rope dyeing machine by Fong has similar features as well. THEN is also working on creating the Airflow machine, which will be utilized to color both extremely elastic and delicate or sensitive materials, like knitted goods and textiles with up to 50% elastane. The first of these machines is called Airflow Lotus, and it is presently undergoing testing. However, more and more lightweight fabrics with high elastane contents in the finest yarn counts and gauges are anticipated in the present and future trends. Therefore, the researcher's future focus in textile dyeing might be establishing low-cost, simple air-dyeing techniques that use less water and produce almost no pollution for both natural and synthetic fabrics.

Dope dyeing is a very affordable and environmentally friendly method of dyeing fibers. However, given the recent push for eco-friendly productions in major textile-producing countries like China, India, Italy, and the USA, more research is required to determine whether dope dyeing is the next best thing to use to clean up the environmental mess left by conventional dyeing methods [50]. Dope dyeing is a promising first step toward the fundamental changes needed to clean up the dyeing industry. Textiles that are synthetic or regenerated can be dyed using this method. Dope dyeing produces fiber with superior color fastness while using up to 90% fewer chemicals, 60% less water, and less CO_2 than traditional methods. Doped dyed fibers do not bleed or fade when used or exposed to the sun. The dope dyeing method allows synthetic polymers to be dyed with natural colorants that have good color fastness properties [75]. For blue jeans, Lenzing created and introduced the TENCEL™ Modal with Indigo Color technology. Mills have also utilized TENCEL™ Modal Black for black denim. For a while now, the Pakistan-based denim manufacturer Soorty has been employing a TENCEL™ Modal assortment that has been dope-dyed. Recently, the company combined black and indigo fibers to produce deeper blue tones. This new approach to product development offers yet another illustration of the potential of these materials [87, 88]. LENZING™ ECOVERO™ black viscose is a new product from Lenzing. The LENZING™ ECOVERO™ fiber family now offers manufacturers and brands a new, even lower-impact viscose option with the introduction of this first dope-dyed fiber [89]. Nowadays, dope dyeing technology is used to color microfibers and nanofibers [74].

4.5 Conclusion

The traditional textile wet processing industry uses a lot of energy, water, and chemicals, which makes it an expensive and unsustainable operation. Because of these difficulties as well as the economic and ecological constraints placed on the textile industry, there is an immense desire to dye materials in an environmentally responsible and sustainable manner. This chapter has demonstrated completely how the water, chemical, and energy requirements for the textile dyeing process are decreased by environmentally friendly and water-saving air dyeing and dope dyeing technologies. Air dyeing technology offers a huge possibility to localize production in areas of the world that lack water resources because it eliminates the need for water at the point of color application. The air dyeing machine uses an air stream and dye liquor to transport the material at a higher speed and colors are sprayed in a jet form on the fabric from the nozzle system. Air dyeing is a machine-based technology rather than process-based. Therefore, any advancement in machinery will promote further progress in this technology. In recent years, several advanced air dyeing machines have been on the market that are discussed in this chapter. Air is a gentle medium that produces superior quality-colored fabrics with uniform color, less surface hairiness, minimal damage to the goods, long-lasting fabric, and the ability to dye fabrics with a high elastane content. Many industries, including print, home textiles, medical, garments, and many more, use air-dyed fabric for a variety of purposes. On the other hand, dope dyeing is without a doubt the more innovative method of coloring synthetic and regenerated cellulose fiber because it completely removes the need for water, energy, and chemicals, saving the product's water footprint. We will therefore see an increasing amount of dope-dyed fiber in the future due to its inherent benefits and favorable characteristics. Moreover, dope-dyed fibers excellent light and water fastness may allow for their application in woven and nonwoven technologies, including outdoor textiles, medical textiles, and the automotive industry. These dyeing techniques have a considerably low effluent load because of their high water conservation and minimal chemical add-on requirements, making them both financially and environmentally sound and sustainable options that are becoming more appropriate substitutes for traditional textile dyeing methods. These dyeing methods are ready for use, easy to use, and time- and money-saving. For this reason, it makes sense to apply these dyeing techniques widely to gain benefits in terms of both the economy and the environment.

Declaration The authors declare that they have no known competing financial interests or personal relationships that could have appeared to influence the work reported in this chapter. Generative AI and AI-assisted technologies have been used in the writing process to improve the readability and language of the work only. After using this tool, the authors reviewed and edited the content as needed and take full responsibility for the content of the publication.

References

1. Abate MT, Tadesse MG (2021) Airflow, foam, and supercritical carbon dioxide dyeing technologies. In: Innovative and emerging technologies for textile dyeing and finishing. Wiley, pp 137–164. Available at: https://doi.org/10.1002/9781119710288.ch5
2. Ahmed NSE, El-Shishtawy RM (2010) The use of new technologies in coloration of textile fibers. J Mater Sci 45(5):1143–1153. Available at: https://doi.org/10.1007/s10853-009-4111-6
3. Hasanbeigi A, Price L (2015) A technical review of emerging technologies for energy and water efficiency and pollution reduction in the textile industry. J Cleaner Prod 95:30–44. Available at: https://doi.org/10.1016/j.jclepro.2015.02.079
4. Lara L, Cabral I, Cunha J (2022) Ecological approaches to textile dyeing: a review. Sustainability 14(14):8353. Available at: https://doi.org/10.3390/su14148353
5. Rani L, Arya N (2019) Dyeing and finishing of fabrics with low or without water. Pharma Innov J 8(5):1159–1165

6. Senthil Kumar P, Yaashikaa PR (2018) Sustainable dyeing techniques. In: Muthu SS (ed) Sustainable innovations in textile chemical processes. Springer Singapore, Singapore, pp 1–29. Available at: https://doi.org/10.1007/978-981-10-8491-1_1
7. Nair GP (2011) Methods and machinery for the dyeing process. In: Clark M (ed) Handbook of textile and industrial dyeing. Elsevier (Woodhead Publishing Series in Textiles), pp 245–300. Available at: https://doi.org/10.1533/9780857093974.1.245
8. New Cloth Market (2009) AirDye® technology: coloring textiles without the use of water, Fibre2Fashion. Available at: https://www.fibre2fashion.com/industry-article/4381/airdye-technology-coloring-textiles-without-the-use-of-water. Accessed: 21 April 2024
9. Khan D, Shaukat SS (2006) The fruits of Pakistan: diversity, distribution, trends of production and use. Int J Biol Biotechnol 3(3):463–499
10. Wan T, Chen D, Bai X (2016) Preparation and relative properties of dope-dyed polyurethane modified by β-cyclodextrin. Dyes Pigments 129:18–23. Available at: https://doi.org/10.1016/j.dyepig.2016.02.011
11. Zhang YJ (2016) Research on the properties of dope-dyed bamboo pulp staple fibers. IOP Conf Ser Mater Sci Eng 137(1):012022. Available at: https://doi.org/10.1088/1757-899X/137/1/012022
12. Murugesh Babu K, Anekonda SM (2019) Air-flow dyeing—an eco-friendly water preserving fabric dyeing technology. Textile School. Available at: https://www.textileschool.com/5234/air-flow-dyeing-an-eco-friendly-water-preserving-fabric-dyeing-technology/. Accessed: 21 April 2024
13. Dhanabalan MV, Sukanya M, Lokesh MKV (2015) Air-dyeing technology-a review. Textile Today. Available at: https://www.textiletoday.com.bd/air-dyeing-technology-a-review. Accessed: 22 April 2024
14. Haque M (2016) Air dyeing technology. LinkedIn. Available at: https://www.linkedin.com/pulse/air-dyeing-technology-mynoul-haque. Accessed: 17 June 2024
15. Christ W, Von Der Eltz HU, Reuther A (1982) Method of treating textile material in jet dyeing machines, EU Patent EP0078022 assigned to Hoechst Aktiengesellscraft, p 282
16. Christ W, Von Der Eltz HU, Reuther A (1984) Process for treating textile material in jet dyeing machines. US Patent 4483032, assigned to Hoechst Aktiengesellscraft 19
17. Fong's Europe (2023a) FONG'S history, FONGS Europe GmbH. Available at: https://www.fongs.eu/company/history/fongs-history/. Accessed: 22 April 2024
18. Heida L (2014) Can waterless dyeing processes clean up the clothing industry? Yale E360. Available at: https://e360.yale.edu/features/can_waterless_dyeing_processes_clean_up_clothing_industry_pollution. Accessed: 22 April 2024
19. Mahmud I, Kaiser S (2020) Recent progress in waterless textile dyeing. J Text Sci Eng 10(6):1–3. Available at: https://doi.org/10.37421/jtese.2020.10.421
20. Maiti S et al. (2022) Textile coloration technologies. In: Emerging technologies for textile coloration. CRC Press, Boca Raton, pp 1–12. Available at: https://doi.org/10.1201/9781003140467-1
21. Murugesh Babu K (2022) Overview on significant approaches to waterless dyeing technology. In: Emerging technologies for textile coloration. CRC Press, Boca Raton, pp 171–188. Available at: https://doi.org/10.1201/9781003140467-10
22. Siu CL, Xiaohui C, Longhan X (2016) Increased ultra-low liquor ratio dyeing machine with separated flow of air and dyeing liquor. Int J Simul Syst Sci Technol 17(23):4.1–4.6. Available at: https://doi.org/10.5023/IJSSST.a.17.23.04
23. Lacasse K, Baumann W (2004) Environmental considerations for textile processes and chemicals. In: Textile chemicals. Springer Berlin Heidelberg, Berlin, Heidelberg, pp 484–647. Available at: https://doi.org/10.1007/978-3-642-18898-5_7
24. Fyi-tester (2022) Air flow dyeing machine overview, FYI Tester. Available at: https://fyitester.com/air-flow-dyeing-machine/. Accessed: 22 April 2024
25. Geocities.ws (2005b) Overflow dyeing machine, geocities. Available at: https://www.geocities.ws/dyes_pigments/overflow-dyeing-machine.html. Accessed 22 April 2024

26. Shishir MMH (2014) Jet dyeing machine, slideshare. Available at: https://www.slideshare.net/sheshir/jet-dyeing-machine. Accessed: 12 May 2024
27. Ferus-Comelo M (2006) Analysis of the factors influencing dye uptake on jet dyeing equipment. Coloration Technol 122(5):289–297. Available at: https://doi.org/10.1111/j.1478-4408.2006.00040.x
28. Geocities.ws (2005a) Airflow dyeing machine, geocities. Available at: https://www.geocities.ws/dyes_pigments/airflow-machine.html. Accessed 22 April 2024
29. Samanta KK et al (2019) Water consumption in textile processing and sustainable approaches for its conservation. In: Muthu SS (ed) Water in textiles and fashion. Elsevier, pp 41–59. Available at: https://doi.org/10.1016/B978-0-08-102633-5.00003-8
30. Saxena S, Raja ASM, Arputharaj A (2017) Challenges in sustainable wet processing of textiles. In: Muthu SS (ed) Textiles and clothing sustainability: sustainable textile chemical processes. Springer Singapore, Singapore, pp 43–79. Available at: https://doi.org/10.1007/978-981-10-2185-5_2
31. AIRDYE® (2011) DEBS corporation. Available at: https://debscorp.com/textiles/airdye/. Accessed: 24 April 2024
32. Kant R (2012) Textile dyeing industry an environmental hazard. Nat Sci 4(1):22–26. Available at: https://doi.org/10.4236/ns.2012.41004
33. Kumar D et al (2010) Dyeing without water. J Text Assoc 71(4):171–174
34. Bomgardner MM (2018) These new textile dyeing methods could make fashion more sustainable. Chemical and Engineering News. Available at: https://cen.acs.org/business/consumer-products/new-textile-dyeing-methods-make/96/i29. Accessed: 22 April 2024
35. Ammayappan L, Jose S, Arputha Raj A (2016) Sustainable production processes in textile dyeing. In: Environmental footprints and eco-design of products and processes, pp 185–216. Available at: https://doi.org/10.1007/978-981-10-0111-6_8
36. Moser LS, Wilson OE (2000) 6. Trends in preparation, dyeing, and printing. Text Progress 30(1–2):68–83. Available at: https://doi.org/10.1080/00405160008688885
37. Fong's Europe (2023b) THEN Airflow® synergy, FONGS Europe GmbH. Available at: https://www.fongs.eu/machines/then-airflow-synergy/. Accessed: 22 April 2024
38. Ate Group (2024) LOTUS, A.T.E. India. Available at: https://www.ategroup.com/textile-engineering/product-family/product-description/lotus/. Accessed: 24 April 2024
39. Fong's Europe (2023c) THEN Airjetwin, FONGS Europe GmbH. Available at: https://www.fongs.eu/machines/then-airjetwin/. Accessed: 23 April 2024
40. Fong's Europe, Fong, S (2020) THEN Airflow® Synergy. Available at: https://www.fongs.eu/site/assets/files/1055/then_synergy_2023.pdf
41. Lakshmanan SO, Ali W, Banerjee S (2022) Sustainable approaches to coloration of textile materials. In: Emerging technologies for textile coloration. CRC Press, Boca Raton, pp 13–36. Available at: https://doi.org/10.1201/9781003140467-2
42. Hasanbeigi A (2010) Energy-efficiency improvement opportunities for the textile industry. Berkeley, CA (United States). Available at: https://doi.org/10.2172/991751
43. Gopalakrishnan M, Punitha V, Saravanan D (2019) Water conservation in textile wet processing. In: Muthu SS (ed) Water in textiles and fashion. Elsevier, pp 135–153. Available at: https://doi.org/10.1016/B978-0-08-102633-5.00008-7
44. Nayak R et al (2020) Sustainable technologies and processes adapted by fashion brands. In: Nayak R (ed) Sustainable technologies for fashion and textiles. Elsevier (Woodhead Publishing Series in Textiles), pp 233–248. Available at: https://doi.org/10.1016/B978-0-08-102867-4.00011-6
45. Khalil E et al (2023) Advanced technology in textile dyeing. In: Mizanur Rahman M, Mashud M, Mostafizur Rahman M (eds) Advanced technology in textiles: fibre to apparel. Springer Nature Singapore, Singapore, pp 97–138. Available at: https://doi.org/10.1007/978-981-99-2142-3_4
46. Group, D.I. (2019) AirDye sustainable printing and dyeing. DEBS Corporation. Available at: https://debscorp.com/lang_eng/wp-content/uploads/2019/08/airdye_.pdf. Accessed: 24 April 2024

47. Tehran MA, Azimi B, Mojtahedi MRM (2011) Investigating the effect of false twist texturing process on the color coordinates variation of spun-dyed polyester filament yarns. J Eng Fibers Fabrics 6(3):54–62. Available at: https://doi.org/10.1177/155892501100600307
48. Manian AP, Ruef H, Bechtold T (2007) Spun-dyed lyocell. Dyes Pigments 74(3):519–524. Available at: https://doi.org/10.1016/j.dyepig.2006.03.015
49. Dobbelstein H (2004) Soiling behavior and processability of solution-dyed PA 6 BCF yarns. Chem Fibers Int 54:127–131
50. Tawiah B, Asinyo BK (2016) Advances in spun-dyeing of regenerated cellulose fibers. BEST: Int J Manage Inform Technol Eng 4(2):65–80. Available at: http://ir.knust.edu.gh:8080/handle/123456789/13723
51. Manian AP, Ruef H, Bechtold T (2006) Mass coloration of regenerated cellulosics-a review. Lenzinger Berichte 85:87–90. Available at: https://www.researchgate.net/publication/235948182
52. Holme I (2016) Coloration of technical textiles. In: Horrocks AR, Anand SC (eds) Handbook of technical textiles, 2nd edn. Elsevier (Woodhead Publishing Series in Textiles), pp 231–284. Available at: https://doi.org/10.1016/B978-1-78242-458-1.00009-1
53. Juanga-Labayen JP, Labayen IV, Yuan Q (2022) A review on textile recycling practices and challenges. Textiles 2(1):174–188. Available at: https://doi.org/10.3390/textiles2010010
54. Gurudatt K et al (2005) Dope-dyed polyester fibers from recycled PET wastes for use in molded automotive carpets. J Indus Text 34(3):167–179. Available at: https://doi.org/10.1177/1528083705049053
55. Hudecová A et al (2023) Colour masterbatches and their use in polylactic acid fibres dyeing. Acta Chimica Slovaca 16(1):62–72. Available at: https://doi.org/10.2478/acs-2023-0006
56. Aspland JR (1993) Chapter 14: pigments as textile colorants: pigmenting or pigmentation. Text Chem Colorist 25(10):31–37
57. Kumar Srivastava J, Ramesh DR (2023) Mass coloration of polyester fiber-emerging opportunities for green sustainable dyeing technology. Eur Chem Bull 12(13):1182–1191. Available at: https://doi.org/10.53555/ecb/2023.12.Si13.208
58. Amit kumar (2015) Mass colaration, slideshare. Available at: https://www.slideshare.net/Amitsirohi2/mass-colaration. Accessed 12 May 2024
59. Kozłowska M et al (2023) Polypropylene color masterbatches containing layered double hydroxide modified with quinacridone and phthalocyanine pigments—rheological, thermal and application properties. Materials 16(18):6243. Available at: https://doi.org/10.3390/ma16186243
60. Rakshit A (2000) Reliance industries introduces Recron dope-dyed black polyester fiber. Int Fiber J 15:72–74
61. Houghton JA et al (2024) Coloration in flow: the potential of in situ coloration of casein fibers to mitigate environmental impact of traditional dyeing methods. ACS Sustain Chem Eng 12(6):2130–2134. Available at: https://doi.org/10.1021/acssuschemeng.3c07437
62. Wang P et al (2015) Properties of alginate fiber spun-dyed with fluorescent pigment dispersion. Carbohydr Polym 118:143–149. Available at: https://doi.org/10.1016/j.carbpol.2014.11.028
63. Zhou Y et al (2003) Influence of styrene–maleic anhydride copolymers on the stability of quinacridone red pigment suspensions. J Dispersion Sci Technol 24(5):731–737. Available at: https://doi.org/10.1081/DIS-120023821
64. Fu S et al. (2013) Properties of lyocell spinning solution with the addition of carbon black/latex composite. Colloids Surf A: Physicochem Eng Aspects 428:1–8. Available at: https://doi.org/10.1016/j.colsurfa.2013.03.033
65. Manian AP et al (2008) The influence of alkali pretreatments in lyocell resin finishing—fiber structure. Carbohydrate Polym 71(4):664–671. Available at: https://doi.org/10.1016/j.carbpol.2007.07.020
66. Suzhou Topot Textiles Technology Co., L (2023) Exploring the process of dope dyeing in synthetic fiber production. topotexs.com. Available at: https://www.topotexs.com/News_details/1739526487221940224.html. Accessed: 11 May 2024

67. Terinte N et al (2014) Environmental assessment of coloured fabrics and opportunities for value creation: spin-dyeing versus conventional dyeing of modal fabrics. J Cleaner Prod 72:127–138. Available at: https://doi.org/10.1016/j.jclepro.2014.02.002
68. Meng Q et al (2001) Melting-reactive dyes for mass coloration of synthetic fibres (II)—1,4-dihydroxyanthraquinone diglycidyl ether in nylon. Dyes Pigments 50(2):127–131. Available at: https://doi.org/10.1016/S0143-7208(01)00039-0
69. Marcinčin A (1999) Dyeing of polypropylene fibers. In: Karger-Kocsis J (ed) Polypropylene: an A-Z reference. Springer Netherlands, Dordrecht, pp 172–177. Available at: https://doi.org/10.1007/978-94-011-4421-6_24
70. Gawish S, Ramadan A, Mosleh S (2018) Improvement polypropylene dyeing by modification methods. Egypt J Chem 62(1). Available at: https://doi.org/10.21608/ejchem.2018.4943.1448
71. Halbe AV (2007) Various approaches for dyeing of polypropylene, Fibre2Fashion. Available at: https://www.fibre2fashion.com/industry-article/2415/various-approaches-for-dyeing-of-polypropylene. Accessed: 11 May 2024
72. Haastrup S et al. (2016) Comparison of the performance of masterbatch and liquid color concentrates for mass coloration of polypropylene. Color Res Appl 41(5):484–492. Available at: https://doi.org/10.1002/col.21987
73. Wang X et al (2021) Preparation and properties of a modified Fe_2O_3 pigment for dope dyeing ultrahigh molecular weight polyethylene (UHMWPE) fibers by melt spinning. Fibers Polym 22(12):3343–3350. Available at: https://doi.org/10.1007/s12221-021-0962-5
74. Fadil F et al (2020) Dope-dyeing of polyvinyl alcohol (PVA) nanofibres with Remazol Yellow FG. Polymers 12(12):3043. Available at: https://doi.org/10.3390/polym12123043
75. Purwar R et al (2016) Dope dyeing of polyacrylonitrile filaments with Ratanjot'. Indian J Fibre Text Res 41:84–88
76. Zheng J et al (2021) Study on structure and properties of natural indigo spun-dyed viscose fiber. e-Polymers 21(1):327–335. Available at: https://doi.org/10.1515/epoly-2021-0036
77. Zhang L et al (2017) Dope dyeing of lyocell fiber with NMMO-based carbon black dispersion. Carbohydr Polym 174:32–38. Available at: https://doi.org/10.1016/j.carbpol.2017.06.020
78. Daneshvar E et al. (2016) Investigating the characteristics of two different methods in nanofiber yarn coloration. J Text Inst 107(7):833–841. Available at: https://doi.org/10.1080/00405000.2015.1061788
79. Mamun Kabir SM, Koh J (2017) Alkaline weight reduction and dyeing properties of black dope-dyed poly(ethylene terephthalate) microfibre fabrics. Coloration Technol 133(3):209–217. Available at: https://doi.org/10.1111/cote.12269
80. Balakrishnan NK et al (2022) Effect of colorants and process parameters on the properties of dope-dyed polylactic acid multifilament yarns. Polymers 14(22):5021. Available at: https://doi.org/10.3390/polym14225021
81. Ikea Facts Textile Dyeing (2016) Inter IKEA systems B.V. 2015. Available at: https://preview.thenewsmarket.com/Previews/IKEA/DocumentAssets/436953.pdf. Accessed 17 April 2024
82. Seidel A (2018) Closing the loop: exploring IKEA's transition to the circular economy. Pomona College. Available at: http://scholarship.claremont.edu/pomona_theses/192
83. Mahapatra NN (2019) Textile dyeing. Woodhead Publishing India Pvt Ltd
84. Kamide K, Nishiyama K (2001) Cuprammonium processes. In: Woodings C (ed) Regenerated cellulose fibres. Elsevier (Woodhead Publishing Series in Textiles), pp 88–155. Available at: https://doi.org/10.1533/9781855737587.88
85. Sui K et al (2012) Biocomposite fiber of calcium alginate/multi-walled carbon nanotubes with enhanced adsorption properties for ionic dyes. Carbohydr Polym 90(1):399–406. Available at: https://doi.org/10.1016/j.carbpol.2012.05.057
86. Wilkes AG (2001) The viscose process. In: Woodings C (ed) Regenerated cellulose fibres. Elsevier (Woodhead Publishing Series in Textiles), pp 37–61. Available at: https://doi.org/10.1533/9781855737587.37
87. Manian AP et al (2022) Dope dyeing of regenerated cellulose fibres with leucoindigo as base for circularity of denim. Polymers 14(23):5280. Available at: https://doi.org/10.3390/polym14235280

88. Nadir Khan N (2024) Inside soorty's dope-dyed TENCEL™ modal developments, carved in blue. Available at: https://carvedinblue.tencel.com/inside-soortys-dope-dyed-tencel-modal-developments/. Accessed: 6 May 2024
89. Sj Studio (2023) LENZING™ ECOVERO™ Black Viscose Offers Lower Carbon Impact. Sourcing journal. Available at: https://sourcingjournal.com/sustainability/sustainability-materials/lenzing-ecovero-black-viscose-lower-carbon-footprint-durable-color-refibra-461112/. Accessed: 6 May 2024

Chapter 5
Electrochemical Dyeing in Textile Industry: A Sustainable Approach

Sayantika Debnath and Sankar Roy Maulik

Abstract The textile industry, a significant contributor to environmental degradation, is undergoing a paradigm shift towards sustainable practices. Among the innovative solutions, electrochemical dyeing has emerged as a promising method that addresses concerns related to environmental well-being with traditional processes. Electrochemical dyeing works on the principles of electrochemistry to achieve colour on textile substrates. While electrochemical dyeing presents a promising avenue towards sustainability, it is challenging. The initial setup costs for electrochemical dyeing systems can be higher than traditional methods, requiring careful economic evaluations for widespread adoption. The technical complexity of electrochemical processes may pose challenges for small-scale manufacturers, necessitating specialised knowledge and equipment. Additionally, scaling up electrochemical dyeing operations presents logistical and technical hurdles that must be addressed for the technique to become economically viable on a larger scale. Despite these challenges, ongoing research and development efforts are focused on optimizing electrochemical parameters, exploring advanced electrode materials, and finding innovative solutions to facilitate the wider adoption of this eco-friendly dyeing method. The sustainability impact of electrochemical dyeing extends beyond environmental benefits. As consumer awareness of eco-friendly practices grows, there is also an ever-growing market demand for sustainable textiles. Electrochemical dyeing aligns with these consumer preferences, positioning itself as a competitive advantage for textile manufacturers aiming to differentiate themselves in a market increasingly focused on sustainability.

Keywords Eco-friendly · Electrochemical dyeing · Recyclable · Textiles

S. Debnath · S. Roy Maulik (✉)
Department of Silpa-Sadana, Visva-Bharati (A Central University), Sriniketan, India
e-mail: sankar.raymaulik@visva-bharati.ac.in

M. Shahid et al. (eds.), *Advancements in Textile Coloration*,
https://doi.org/10.1007/978-981-96-5091-0_5

5.1 Introduction

Sustainability is now a critical factor in the textile and apparel industry. The dyeing and printing sector of these industries is primarily known for its hazardous impacts on the environment [1]. Wet processing and dyeing are responsible for consuming huge volumes of water. These are later released as harmful chemicals into water-bodies, leading to scarcity of fresh water, ecological damage, and health issues. They also contribute to air pollution [2]. The adoption of responsible alternatives, viz., the utilisation of eco-friendly dyes and other auxiliaries and the implementation of effluent treatment systems, can reduce its ecological footprint to a large extent [3]. Consumers nowadays have an ever-growing demand for ethically sourced products following a sustainable supply chain [4]. Consumers' concerns about pollution and the safety of humans have led the textile industry to adopt several sustainable and eco-friendly manufacturing techniques.

The United Nations (UN) adopted seventeen Sustainable Development Goals (SDGs) in 2015 as a part of the 2030 agenda for sustainable development across various fields for global development [5]. All 196 members of the UN signed 'The Paris Agreement' in the same year to reduce the emissions caused by greenhouse gases [6]. The textile industry is directly challenged by some of the SDGs laid by the UN, like Goal 9, 'Industry, Innovation, and Infrastructure', and Goal 12, 'Responsible Consumption and Production', which urge the textile industry to efficiently manage chemicals and waste across their life cycle while drastically decreasing their emission into the biosphere, which is essential for mitigating adverse effects on humans as well as the environment [7].

Adopting waterless dyeing, using clean and green chemistry, and implementing closed-loop systems for recycling wastewater can help the textile and apparel industry minimise its ecological footprint to some extent [1]. This chapter aims to demonstrate one promising advancement that can be an effective tool in textiles for mitigating pollution and conserving time and resources. Traditionally, electrochemical methods are utilised to recover metals and synthesise compounds [8]. Recent research has, however, suggested a plethora of new alternative applications. Among the advancements, a few have been suggested as solutions for various issues in the textile industry [9].

Electrochemical procedures for dyeing have minimal consumption of chemicals. Therefore, it is being considered as an alternative in the dyeing sector. Unlike the traditional methods, sulphur, vat or indigo dyes, when processed through electrochemical methods, often do not require large amounts of harmful reducing agents commonly used. The reduced usage of chemicals is one of the reasons that electrochemical dyeing techniques benefit economic and environmental concerns [10].

In the Indian context, cellulosic fibres are the most commonly used natural textile fibre [11]. Vat dyes are popular for their excellent all-round colourfastness properties when applied to cellulosic fibres. Despite their popularity, the insolubility in water complicates the dyeing process [12]. The complexity of the dyeing mechanism lies with the reduction and subsequent solubilization of insoluble dye molecules under strongly alkaline conditions into sodium salts of leuco vat, which then become easily soluble in water and have an affinity for cellulosic fibres. This process of vatting is carried out with a suitable reducing agent having sufficient negative redox potential in alkaline conditions [13].

Several reducing agents responsible for reduction have been reported and among them sodium dithionite ($Na_2S_2O_4$), sulphinic acid derivatives and hydroxyl carbonyl compounds are popular [14]. Each of these reducing agents has disadvantages yet $Na_2S_2O_4$ is quite popular in dyeing with the vat class of dyestuff. In non-electrochemical processes, the reducing agent cannot be recovered from the dye bath making the whole system non-recyclable. Toxic compounds based on sulphur like sulphates, sulphites, sulphides, dithionite, and thiosulphates are generated in the dye bath and later discarded as effluents in the natural water bodies, resulting in freshwater contamination [15].

Figures 5.1 and 5.2 depict the mechanism of dyeing cellulosic fiber with anthraquinone and indigoid classes of vat dyes.

Sulphur dye, a water-insoluble dye, has good to moderate light and washing fastness, but poor chlorine bleaching fastness, is mainly preferred for application on low-cost cellulosic textiles [16]. The dyeing mechanism of sulphur dyes shown in Fig. 5.3 is similar to that of vat dyes. The dye becomes substantive to cellulosic fibres only after reducing dye molecules in their leuco form, followed by solubilization under alkaline conditions. Chemically, sulphur dye consists of sulphur linkages in its chromophore. The sulphur compounds mainly exist in sulphide, disulphide, and polysulphide forms in heterocyclic ring-like structures [16]. These sulphur linkages

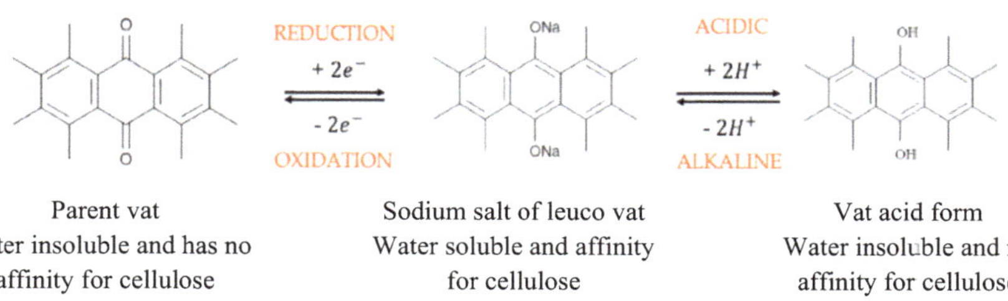

Fig. 5.1 Principle of dyeing cellulosic fibre with vat dyes

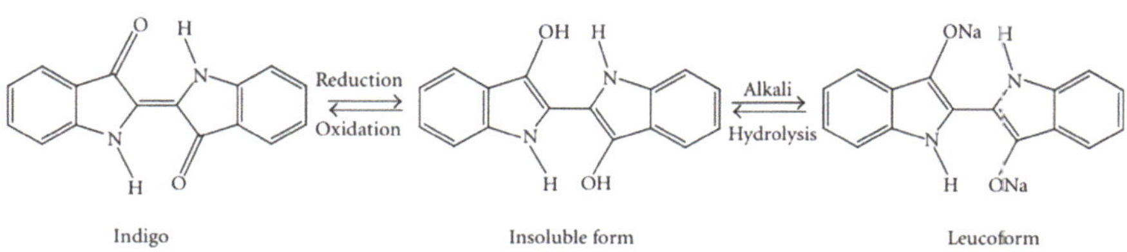

Fig. 5.2 Reaction of indigo dye to leuco-form

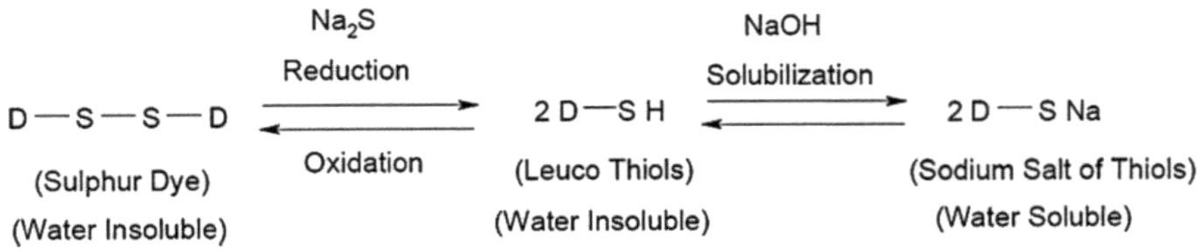

Fig. 5.3 Reaction of sulphur dye to leuco-form

are reduced to their leuco thiols and later thiolates during the reduction process to make the dye molecules water-soluble and substantive to cellulosic fibres [17, 18]. The most commonly preferred reducing agent for sulphur dyes is sodium sulphide (Na_2S), which is applied in alkaline conditions, similar to vat dyes. The use of other agents like $Na_2S_2O_4$ has also been reported but with severe disadvantages, such as the emission of toxic gases like H_2S and the contamination of air and water with sulphur-based effluents released after reduction. Other sulphur-based compounds reported include thiourea and hydroxyalkyl sulphinate. These compounds have low equivalent mass and small sulphur content. As a result, the effluents contain much lower sulphur-based harmful compounds. Despite the low sulphur content, pollution cannot be completely avoided [18].

Reactive dyes, with their excellent fastness properties, brilliancy and large number of shades, etc. are the most widely preferred dyes in the textile industries for the colouration of cellulosic fibres. However, the cellulosic fibres cannot be dyed with reactive dye unless large amounts of inorganic salts like sodium sulphate (Na_2SO_4) or sodium chloride (NaCl) are used as an exhausting agent [19]. These inorganic salts cannot be recycled or destroyed, and industries are compelled to release them as effluents that cause severe environmental damage [20]. Though recent developments suggest that dyeing with reactive dyes can be performed with minimal or no salt, these technologies still need to be put on large-scale commercialization. Electrochemical dyeing, on the other hand, is eco-friendly as it is free of heavy salt usage, and it has the benefits of easy control, minor fibre damage, and reduced resource consumption [21].

5.2 Reducing Agents: The Drivers of the Process

5.2.1 Conventional Reducing Agents for Vat Dyes

Several reduction methods like the traditional method of fermentation, reduction through ferrous sulphate ($FeSO_4$)-lime, zinc-lime method, thiourea dioxide method, etc. are suggested in the literature [22]. Among the reducing agents, $Na_2S_2O_4$ or sodium dithionite remains the universally accepted agent for the reduction of vat and indigo dyes as other agents were not commercially successful. The temperature

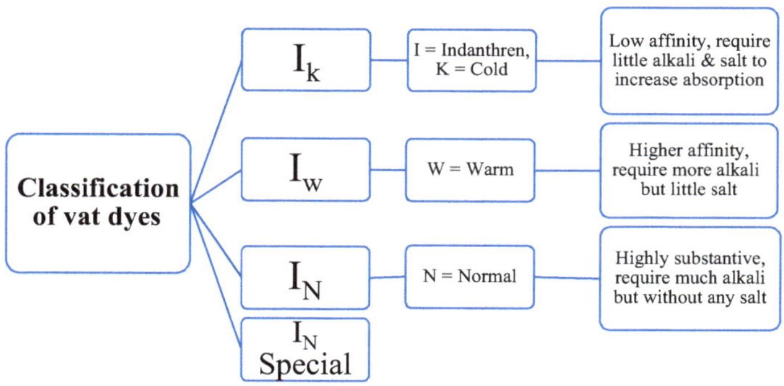

Fig. 5.4 Classification of vat dyes based on the ease of application

of reduction ranges from 30 to 60°C and makes it a lot more favourable compared to other reducing agents [23]. The amount of reducing agent required for specific classes of vat dye depends on the compactness of the structure of the dye molecule and the total number of the >C = O groups that can be reduced in that dye molecule. Based on this criterion for ease of application, vat dyes are classified into four classes [24]. Figure 5.4 depicts various classes of vat dyes based on their application. Due to the low affinity of the IK class, these dyes are not preferred for dyeing cellulosic fibres. The remaining three classes show higher affinity and substantivity for cellulosic fibres, and reduction and dyeing occur at temperatures ranging from 50 to 60°C. Though dyeing with each class of vat dye is similar, thoroughly regulating optimal reaction conditions and adjusting temperature, quantity of reducing agents, and NaOH helps achieve successful dyeing. The unstable nature of $Na_2S_2O_4$ in alkaline conditions is responsible for adding higher amounts than the calculated stoichiometric ratio during the vatting process to increase efficacy [23, 24].

5.2.2 *Environmentally Friendly Reduction Process for Vat Dyes*

Using $Na_2S_2O_4$ or sodium dithionite threatens the environment due to effluent discharge and pollution. Recent research investigated some eco-friendly alternatives to tackle pollution-related problems, as discussed below.

5.2.2.1 Complexes of Iron (II)

In the past, the 'Copperas method' was popular for reducing vat dyes. Iron (II) salts were utilised in this method [25]. The reaction of calcium hydroxide with ferrous sulphate resulted in heavy ferrous hydroxide ($Fe(OH)_2$) sedimentation due to low solubility and precipitation of $CaSO_4$ in the dye bath. Several researchers have reported that $Fe(OH)_2$ is a strong reducing agent in alkaline conditions, and its reducing effect gets even stronger with the rise of pH. However, weak solubility results in precipitation under alkaline solutions. This indicates the low reduction potential of $Fe(OH)_2$. It is so low that even at temperatures around 60°C, indigo reduction is impossible. The formation of a coordination complex between a suitable ligand and $Fe(OH)_2$ prevents precipitation, preserving its true potential. According to Werner's Coordination Theory, $Fe(OH)_2$ can combine with compounds containing weaker ligands, such as tartaric acid, gluconic acid, or citric acid, to form a complex. The formation of this complex makes $Fe(OH)_2$ water-soluble and can perform as a reducing agent with its true potential [24].

Semet et al. reported that Fe (II) salt and gluconic acid can be utilized in a molar ratio of 1:1 at 60°C, when the gluconic acid is neutralized during the treatment of effluents, making the process eco-friendly and sustainable [26]. Iron (III) hydroxide $Fe(OH)_3$, a flocculant, is insoluble and is produced from $Fe(OH)_2$ in alkaline conditions during the dyeing process. Scientists investigated the ways of maintaining the stability of $Fe(OH)_2$ in solution using various combinations of ligands. This gave rise to the 'single-ligand system' and the 'two-ligand system'. Dyeing cotton with indigo follows the single ligand system where ferrous salts are complexed with suitable molar ratios of weak acids like citric acid, tartaric acid, or triethanolamine reacted with sodium hydroxide (NaOH), which shows successful dyeing. This method is unsuitable for dyeing cotton with other classes of vat dyes; therefore, the two-ligand system was suggested. In this method, the ferrous salts are complexed with citric acid and triethanolamine or tartaric acid and triethanolamine. Later, numerous studies were reported on the suitability of the two-ligand system over the single-ligand method due to the increased solubility of ferrous salts [26, 27].

5.2.2.2 Hydroxyacetone

A sophisticated dyeing method has been developed utilizing hydroxyacetone as the reducing agent. It shows a reduction potential of about −810 mV. The pollutants released from this reducing agent can go through biological degradation. This reduces the chemical oxygen demand of the effluents. The effluents released are free from harmful sulphur residues. The only drawback in the case of hydroxyacetone is its high cost, which makes it less popular with the mass-producing industries [24].

5.2.3 *Sulphide-Based Conventional Reducing Agents for Sulphur Dyes*

The conventionally used sulphur dyes (90% approx.) are still reduced through compounds based on sulphide. Compounds like sodium polysulphide, available in an aqueous state or compounds found in a solid or liquid state like sodium sulphide (Na_2S), and sodium hydrosulphide (NaHS) can reduce sulphur dyes. However, these compounds pose a serious threat to the environment by releasing harmful residues and emission of toxic gases like hydrogen sulphide (H_2S). But there have been efforts to find alternatives for reducing the impact on the environment [24, 28].

5.2.4 *Environment-Friendly Agents for the Reduction of Sulphur Dyes*

5.2.4.1 Glucose

Since ancient times, it has been widely known that sulphur dyes can be reduced with glucose and it is the preferred reducing agent for reducing sulphur black [29] and indocarbon during single-stage printing. Glucose cannot produce satisfactory results in open exhaust dyeing processes such as jiggers and winches. Acceptable outcomes can only be achieved when the temperature is above 90°C. Adding a stabilising agent and dyeing in highly alkaline conditions results in a significant improvement. The reduction potential that glucose and sodium hydroxide (NaOH) provide, typically adequate for reducing sulphur dyes, is from -550 to -600 mV [24, 29].

5.2.4.2 Mercaptoethanol

β-Mercaptoethanol based reducing agent can replace sodium sulphide. When applied with NaOH, it could be used to apply sulphur dyes through the exhaust dyeing or one-bath pad-steam process. The results are generally satisfactory, but the yields are marginally lower than those of the sulphide system. The significant benefit of mercaptoethanol is that it does not release obnoxious odours or harmful effluents during and after application due to the absence of sulphur compounds. The drawbacks include the inability to dissolve sulphur dye residues that remain insoluble in water and its high cost, making it less commercially viable [24].

5.3 Introduction to Electrochemical Dyeing

Daruwalla first investigated the electrolysis of the dye bath to reduce vat dyes. The vat dye was reduced by applying direct voltage without using large amounts of $Na_2S_2O_4$ and a powerful reduction of sulphoxylate ions ${SO_2}^{2-}$ species was generated. Its redox potential was much higher than that of sodium dithionite, which indicated a 30% reduction in sodium dithionite consumption [24, 30]. This behaviour of $Na_2S_2O_4$ was due to the decomposition of hydrosulphite, which generated a free SO_2 radical. The simple equation is demonstrated as follows:

$$S_2O_4^{2-} \rightarrow 2SO_2^-$$

The drawback of this study was its non-recyclable nature. The dye bath could not be regenerated or recycled, and the unstable dye residue was discarded as harmful effluent [24, 30, 31]. Grotthuss first reported that it is possible to reduce vat and indigo dyes to their leuco forms by passing an electric current where electrons from the current break down the insoluble dye to its leuco state. Roessler et al. (2003) reported a similar concept, which was more environmentally friendly and efficient than the earlier research, was applied in the electrochemical dyeing technique, where the liquid in the electrochemical process could be recycled [15, 18].

5.4 Classification of Electrochemical Dyeing Processes

The main aim of electrochemical dyeing is to present itself as an eco-friendly alternative and replace conventionally preferred sulphur-based chemicals. The electrochemical dyeing procedure can be broadly classified into three classes: direct reduction technique, indirect reduction technique, and electrocatalytic hydrogenation technique.

5.4.1 Direct Electrochemical Reduction

Direct reduction of dyestuff is carried out by facilitating physical interaction between the surface of the specialized electrode and the dye molecules. Electrochemical reduction happens when insoluble dye molecules come into contact with the cathode surface. A negative potential is created at the electrode surface, which lets electrons move and reduces the dye [32]. The initiation of the process requires a reducing agent. The dye solution is maintained at an alkaline pH to ensure its solubility and maintain the overall efficacy of the reduction and dyeing procedure [18, 32].

Initiating the direct reduction process requires producing a small amount of leuco dye through the conventional method introduced into the reduction system [15, 32].

Fig. 5.5 Mechanism of direct electrochemical reduction of vat dye

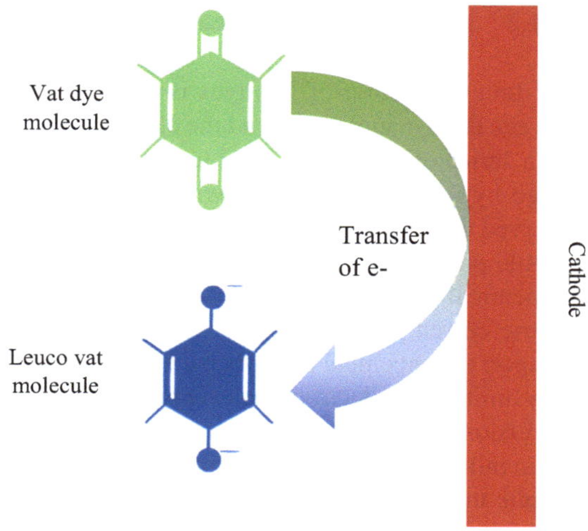

This approach is successful for sulphur dyes. It was also observed that the dye concentration needed to achieve a specific colour was comparatively higher compared to a conventional reduction process. This disparity can be attributed to the limited interface between the cathode and dye particles [15, 18, 32]. Figure 5.5 illustrates the direct electrochemical reduction of vat dye.

5.4.1.1 Direct Electrochemical Reduction in Sulphur Dyeing

Due to its lower reduction potential, the sulphur dye can be reduced easily compared to vat or indigo dyes. The electrochemical properties of sulphur suggest that previously reduced sulphur dye can effectively act as a self-medicating agent for oxidised sulphur dye. So, the pre-reduced leuco sulphur dye undergoes oxidation during the dyeing process due to oxygen within the dye bath. The oxygen produced is subsequently recycled through a direct self-mediated cathodic reduction mechanism. The cathodic reduction generates a leuco form of sulphur dye, which serves as the in-situ reducing agent and functions as a dyestuff within the dyeing process [18]. The redox potential is crucial in traditional chemical reduction [33], and the depth of colour intensifies proportionally with higher reduction potential. Similarly, the colour values and shade of the dyed material can be altered by applying varying ranges of current densities administered throughout the process of electrochemical dyeing. The intensity of the current significantly impacts the structure of leuco-sulphur dyes in electrochemical dyeing; given its dual function as a reducing agent, it results in maximum production of colour. Therefore, various ecological and economic benefits can be expected from this method free from sulphide-based reduction [18, 28, 33].

5.4.1.2 Direct Electrochemical Reduction in Vat Dyeing

The direct reduction of vat dyes is facilitated through the direct physical contact between the dye molecules and the electrode surface. Researchers used voltammetry and a series of electrochemical tests to find a new way to make the leuco form of indigo by reducing it directly with electricity in an aqueous medium. It has been observed that the indigo radical anion can undergo quick reduction, which results in a stable product. Diffusional transport mechanisms slow down the rate of the electrochemical reaction. At more negative potentials than 1.1 V (vs. Ag/AgCl 3 M), electric current density increases due to hydrogen liberation at the cathode. Mass transport enhancement towards the electrode could lead to radical generation, potentially limiting the electrochemical process. The reduction of the indigo radical follows a diffusion-controlled 1H/1e electrochemical pathway, where pH plays a significant role in influencing the peak potential, having minimal impact on the reduction current. Despite this, the radical concentration remains low, limiting the current and rate of reactions [34].

5.4.1.3 Other Processes Based on Direct Mechanism

Direct Electrochemical Reduction of Dye Pigment via the Radical Process

This is a unique type of direct electrochemical reduction process. In this case, the reduction system does not have a mediator or require a reducing agent. The reaction is sustained by generating a radical anion by comproportionation reaction between the dye molecules and its leuco form. The radical at the next stage is reduced through the electrochemical reduction process and acts as an intermediate species. Initiation of this process requires the production of a leuco form of dye through the conventional method, and then the reaction is self-sustained [35].

In the year 2000, Morte patented this reduction technology. Rosseller and his team conducted more research to study the reduction mechanism and reported that the electrochemical reaction rate was slowed down due to the diffusion transport of radical anion. The density of electric current was increased to improve the rate of reaction, resulting in rapidly decreasing efficiency of current [15, 34].

The reduction of the dye molecules takes place on the cathode surface; therefore, to promote better reduction quality, the surface area of the cathode must be large. The radical process successfully reduces sulphur dye. However, due to limited interaction between the dye particles and the cathode, the dye concentration required to achieve the desired shade becomes significantly higher. Since the reduction procedure was deemed rate-limiting, further research focused on increasing the concentration of radicals and enhancing the flow at the cathode region, several drawbacks proved that this technology might not be industrially feasible [34, 36].

Fig. 5.6 Electrode with precoated catalyst layer

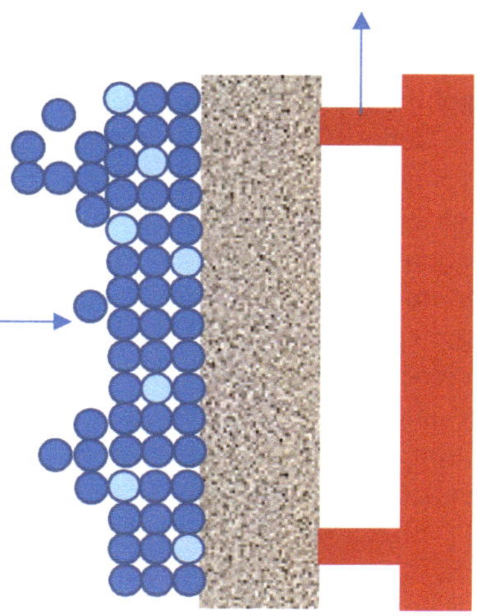

Direct Electrochemical Reduction on Graphite Electrode

The lack of sufficient contact between the electrode of the electrochemical cell and the insoluble dye particle is considered the primary drawback in successfully implementing direct electrochemical reduction [36]. Figure 5.6 shows the concept of a precoated layer cell, which was proposed to mitigate the above-mentioned drawback. It was reported that the indigo dye particles could be brought into contact with the electrode based on the concept of filtration. It is made by a conductive filter fabric support and a layer of Raney-nickel polarised by a cathode that is formed by precoating filtration. The drawback in this case was the large drop in pressure and the clogging risk in the reactor, which led to its discontinuation [30, 33]. Studies were conducted to improve this technology's feasibility by applying granular graphite, acting as a material with a large surface area in a fixed and fluidised bed reactor. Roessler and Crettenand conducted experiments on fixed-bed cathodes containing graphite granules. They are low-cost and have a very high and stable surface area, making graphite a suitable electrode material. Unlike other methods of direct electrochemical reduction, here, in the dye solution, a direct transfer of electrons is observed between the dye particles and the graphite cathode, making a leuco form of dye [30, 37]. Bechtold and Brunner, in 2005 [33], studied the surface chemical activity of graphite to boost the reduction rate. The functionalities of oxygen-containing groups like quinone and hydroquinone heavily influence its chemical activity.

The generated quinoid functional groups on treatment with acidified hydrogen peroxide (H_2O_2) have the potential to double the rate of reduction in graphite. This method is known as pre-anodisation, which is effective in enhancing the working of graphite-based electrodes. Several studies were conducted to improve the efficiency of graphite electrodes [18, 33]. Studies have also been reported on the application of iron deposition on the surface of graphite cathode during the electrochemical vat dyeing process [38]. Additives like ammonium thiocyanate and gelatine added through electrodeposition can enhance the iron deposits to improve dyeing. The implementation of additives was backed by an analysis of cyclic voltammograms, which proved the efficacy of the process.

Yang et al. conducted a study utilising carbon felt as the cathode to perform the direct electrochemical reduction of vat dye. The electrolytic tests and cyclic voltammetric analysis confirm that carbon felt can be considered for direct electrochemical reduction, and the washing fastness and the obtained shade were comparable with conventional dyed samples, proving the efficacy of the technology [39].

5.4.2 *Indirect Mthod or Mediator-Enhanced Electrochemical Reduction*

The indirect method of electrochemical reduction is performed between separated surfaces; here, the direct interaction between the dyestuff and the cathode is not necessary, as in the case of the direct electrochemical dyeing process. A mediator or electron carrier is commonly implemented through the setup, and the reduction can occur at separate places (Fig. 5.7). It has been reported that a soluble redox mediator can significantly increase the reaction rate [18, 24, 31, 32]. Mediators play a crucial role as reducing agents in this reduction method, undergoing alternating cycles of reduction and oxidation. After the reduction process, the mediator undergoes oxidation; it is then reduced at the cathode, ensuring its availability for the cycle of reduction of dye to continue [40].

A reversible redox system is employed to generate a constant reduction potential throughout the reaction. On completion of the dyeing process, any residual unexhausted dye present is precipitated by air oxidation and it is eliminated through filtration, enabling the mediator-infused dye liquor to be recycled for the next dyeing process. This particular characteristic of a closed cycle makes the system beneficial in terms of cost-effectiveness and eco-friendliness.

5.4.2.1 Mediator Systems

Redox mediators are known chemicals possessing electrochemical properties, which enable electron transfer between the dye molecules and the surface of the cathode. The mediator acts as a shuttle carrying electrons. A variety of redox mediators are

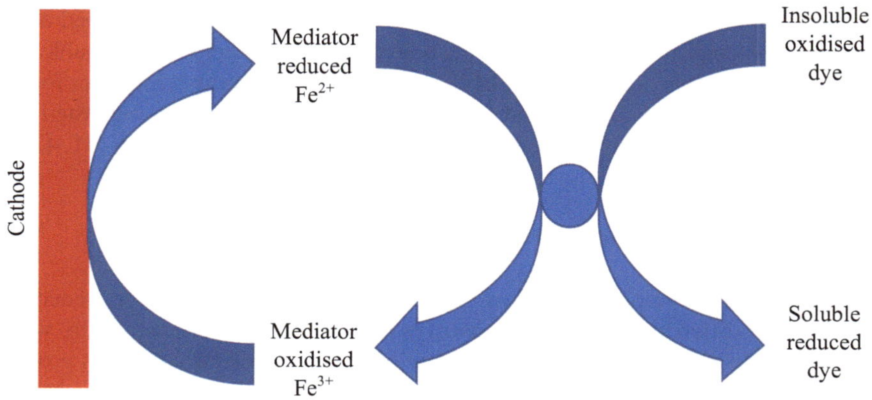

Fig. 5.7 Indirect electrochemical dye reduction

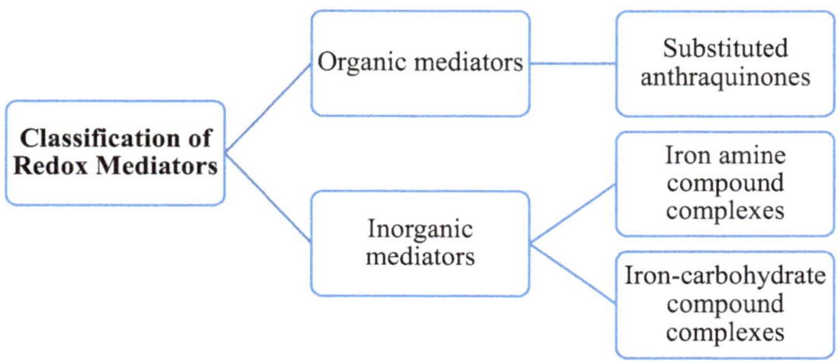

Fig. 5.8 Classification of redox mediators

functioning as reversible redox systems to facilitate the indirect electrochemical reduction process [18]. Figure 5.8 describes the classification of redox mediators.

5.4.2.2 Indirect Electrochemical Dyeing Process

In 1994, Bechtold [40] and his team suggested an innovative procedure regarding electrochemical reduction through an indirect method in an electrochemical cell containing triethanolamine (TEA), ferric sulphate and alkali. A cathode made of copper was the source of electrons along with a reference electrode (Ag/AgCl/3M KCl) to monitor the system's reduction potential. The ferric ions are first converted to ferrous ions, then the dyestuff is added. TEA stabilises the ferrous ions and acts as a reducing agent for the dye. After the dye is reduced, it is shifted to another trough for further dyeing process. The striking advantage of this method is its independence from temperature changes. There is direct control over the reduction potential of the system, without any impact of temperature on it, making it more flexible and accurate than the conventional methods. This redox-mediator system is sustainable and commercially viable due to its recyclability. The anthraquinone as a mediator for the reduction of dispersed dyestuff and the cyclic voltammetric analysis suggested that the mediators successfully conducted indirect electrochemical reduction [41, 42].

5.4.2.3 Electrocatalytic Hydrogenation

Electrocatalytic hydrogenation and traditional electrochemical dye reduction methods are different as the former involves in-situ hydrogen production by water electrolysis, which chemically reduces dyes on the electrode surface. This process avoids the high pressure, temperatures, and storage challenges of chemical hydrogenation by directly generating hydrogen on the catalyst electrode surface. The hydrogen evolution reaction competes with dye reduction. Though gaseous hydrogen formation still limits current efficiency, metal catalysts (e.g., Raney-nickel, platinum black) improve efficiency by facilitating hydrogen adsorption and bond splitting [32]. Due to availability, cost and alkaline media stability, various metal catalysts have been tested for electrocatalytic hydrogenation, with Raney-nickel chosen for industrial research [43].

Platinum-black electrodes have poor durability, and large electrode surface areas are required for efficient industrial reduction rates, but they are highly active. Though enhanced electrode materials, like precoat-layer cells and noble metals on graphite granules, improved hydrogenation, it faced challenges with cost, long-term stability, and mechanical strength [44]. The reduction of indigo shows high conversion with low current efficiency and has been effective with this method. Attempts were made to reinforce Raney-nickel electrodes using binders, like RVC carbon and lanthanum phosphate to improve stability, to improve stability, but catalytic activity was reduced. Electro-deposition methods for creating Raney-nickel electrodes also proved short-lived [15]. Although environmentally and economically advantageous by eliminating reducing agents, electrocatalytic hydrogenation requires further research to improve electrode durability and efficiency for large-scale applications.

5.5 Merits of the Electrochemical Dyeing Process

Figure 5.9 depicts a graphical representation of traditional dyeing against the electrochemical dyeing process.

Hazardous chemicals like sodium dithionite and sodium sulphide are used in traditional reduction methods for vat and sulphur dyes, which contaminate the air and sewage systems, creating disposal challenges. Significant reduction of chemical load in wastewater, enhancement of ecological profile, and long-term sustainability result from electrochemical reduction, which uses electric potential for dye reduction. This method reduces sewage concentrations from current levels of 640–1600 to 200 mg/

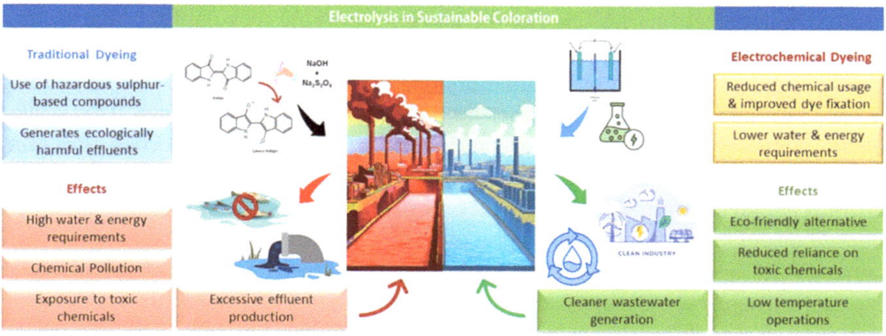

Fig. 5.9 Traditional dyeing and electrochemical dyeing method

L. Chemical costs and effluent load can be reduced by 50–75% by using controls like redox potential, mediator concentration, charge flow, cell current measurement and electrochemical reduction [45, 46].

Dyeing with the electrochemical method has shown several economic benefits for the industry. Firstly, due to the recyclability of the process, several cycles of dyeing can be completed. This allows less water consumption, and the effluent load is greatly minimized, saving tonnes of water, chemicals, auxiliaries and overall energy. It has been reported that the cost of chemicals can be lowered to 80% along with the 85% reduced cost observed in water recycling systems in the industry. Secondly, the maintenance cost of the electrochemical device parts, such as electrodes and electrochemical cells, is low, which can, in turn, be cost-effective [46].

5.6 Conclusion

In the present article, numerous environmentally safe alternatives have been reported to mitigate harmful effluent discharge in freshwater bodies by the conventional sulphur-based reducing agents and ensure that the reactants being utilised are safe and do not pose a threat to the environment. This chapter discussed several eco-friendly reducing agents and their working mechanisms in the context of electrochemical reduction. The electrochemical system, in particular, utilises electric current instead of hazardous sulphur-based compounds, reducing the effluent discharge burden and its treatment. Various electrochemical reduction processes are discussed, but the major classification shows two principal ways, namely, direct and indirect methods of electrochemical reduction, and observed that the direct method is simple and easy, but the effectiveness is not good due to poor contact between the dye and the cathode. The modified multi-cathode electrolysers or Raney-nickel electrodes could enhance the efficiency of the direct method. In the case of indirect electrochemical reduction, mediator systems such as substituted anthraquinones and iron complexes are discussed. These mediator systems are responsible for improved reduction efficiency through better electron transfer. The mediator systems are recyclable, which reduces costs and saves resources. Electrolytic hydrogenation, another indirect reduction method, proved beneficial on both laboratory and industrial scales. In this process, in-situ hydrogen is generated to react with dyes on the electrode surface. This system aims to achieve high reduction rates, color values, and fastness properties, providing a promising, eco-friendly alternative to conventional toxic chemicals. Drawbacks in terms of achieving high electric current efficiency remain. This affects the establishment of current reduction rates. In practice, a slight reduction is performed through conventional agents to sustain the electrochemical reduction and entirely reduce the dye bath. Therefore, it can be concluded that electrochemical dyeing technology represents a step towards sustainable manufacturing techniques. It is possible to reduce the textile industry's carbon footprints with more refined research to achieve sustainability goals and improve industries' future.

References

1. Rahaman MdT, Pranta AD, Repon MdR et al (2024) Green production and consumption of textiles and apparel: importance, fabrication, challenges and future prospects. J Open Innov Technol Mark Complex 10:100280. https://doi.org/10.1016/j.joitmc.2024.100280
2. Samanta KK, Pandit P, Samanta P, Basak S (2019) Water consumption in textile processing and sustainable approaches for its conservation. In: Muthu SS (ed) Water in textiles and fashion. Elsevier, pp 41–59
3. Roy Choudhury AK (2014) Environmental impacts of the textile industry and its assessment through life cycle assessment. In: Muthu SS (ed) Roadmap to sustainable textiles and clothing. Springer Singapore, Singapore, pp 1–39
4. Madumali S, Weerasinghe BA, Thibbotuwawa A, Perera HN (2023) Transparency in textile & apparel supply chains: a systematic review. J South Asian Logist Transp 3:71–101. https://doi.org/10.4038/jsalt.v3i2.71
5. THE 17 GOALS | Sustainable development. https://sdgs.un.org/goals. Accessed 29 Oct 2024
6. The Paris Agreement | United Nations. https://www.un.org/en/climatechange/paris-agreement. Accessed 29 Oct 2024
7. Sehrawat A (2023) Applications of green chemistry principles in textile wet processing. J Adv Sci Res 14:1–5. https://doi.org/10.55218/JASR.202314901
8. Rai V, Liu D, Xia D et al (2021) Electrochemical approaches for the recovery of metals from electronic waste: a critical review. Recycling 6:53. https://doi.org/10.3390/recycling6030053
9. Sala M, Gutiérrez-Bouzán MC (2012) Electrochemical techniques in textile processes and wastewater treatment. Int J Photoenergy 2012:1–12. https://doi.org/10.1155/2012/629103
10. Kalapriya K, Gurumallesh Prabu H, Ramakrishnan G (2019) Electrochemical dyeing of polyester fabric using disperse dyes. Rasayan J Chem 12:319–323. https://doi.org/10.31788/RJC.2019.1215004
11. Dhir YJ (2022) Natural fibers: the sustainable alternatives for textile and non-textile applications. In: Jeon HY (ed) Natural fiber. IntechOpen
12. Aspland JR (1997) Textile dyeing and coloration. American association of textile chemists and colorists, Research Triangle Park, NC
13. Bechtold T, Burtscher E, Amann A, Bobleter O (1992) Reduction of dispersed indigo dye by indirect electrolysis. Angew Chem Int Ed Engl 31:1068–1069. https://doi.org/10.1002/anie.199210681
14. Makarov SV, Horvath AK, Silaghi-dumitrescu R, Gao Q (2016) Sodium dithionite, rongalite and thiourea oxides: chemistry and application. World Scientific. https://doi.org/10.1142/q0028

15. Roessler A, Jin X (2003) State of the art technologies and new electrochemical methods for the reduction of vat dyes. Dyes Pigments 59:223–235. https://doi.org/10.1016/S0143-7208(03)00108-6
16. Chakraborty JN (2011) Sulphur dyes. In: Clark M (ed) Handbook of textile and industrial dyeing. Elsevier, pp 466–485
17. Chakraborty JN, Chakraborty JN (2015) Fundamentals and practices in colouration of textiles. CRC Press
18. Yusuf M, Shahid M (2022) Emerging technologies for textile coloration, 1st edn. CRC Press
19. Wolela AD (2021) Effect and role of salt in cellulosic fabric dyeing. Austin Publ Group 6
20. Niu T, Wang X, Wu C et al (2020) Chemical modification of cotton fabrics by a bifunctional cationic polymer for salt-free reactive dyeing. ACS Omega 5:15409–15416. https://doi.org/10.1021/acsomega.0c01530
21. Xiong W, Yang Y, Hou Y et al (2021) Electrochemical dyeing of cotton fiber for environmental protection. In: Yang M, Lanzinha JCG, Bao X (eds) Advances in transdisciplinary engineering. IOS Press
22. Chakraborty JN, Chavan RB (2004) Dyeing of cotton with vat dyes using iron(II) salt complexes. AATCC Review 4(7):17–20
23. Shore J (1995) Cellulosics dyeing. Society of Dyers and Colourists, Bradford, West Yorkshire, England
24. Kulandainathan MA, Patil K, Muthukumaran A, Chavan RB (2007) Review of the process development aspects of electrochemical dyeing: its impact and commercial applications. Color Technol 123:143–151. https://doi.org/10.1111/j.1478-4408.2007.00082.x
25. Omender K, Chakraborty JN (2020) Eco-friendly vat dyeing of cotton using alkaline iron (ii) salt as reducing agent. TEKSTILEC 63:305–320. https://doi.org/10.14502/Tekstilec2020.64.305-320
26. Semet B, Sackingen B, Gurninger GE (1995) Iron (II) salt complexes as alternatives to hydrosulphite in vat dyeing. Melliand Engl 3:E45–E46
27. Chavan R, Chakraborty J (2000) Enhancement in indigo uptake on cotton through metal salt pretreatment. Ind J Fibre Text Res 25:130–137
28. Shukla SR, Pai RS (2004) Sulphur dyeing using non-sulphide reducing agents. Ind J Fiber Text Res 29:454–461
29. Zouhaier R, Sofiène D, Faouzi S (2014) The use of glucose as ecological reducing agent for sulphur dyes: optimization of experimental conditions. Eur Sci J 10:436–446
30. Roessler A (2003) New electrochemical methods for the reduction of vat dyes. PhD Diss. ETH Zurich
31. Božič M, Kokol V (2008) Ecological alternatives to the reduction and oxidation processes in dyeing with vat and sulphur dyes. Dyes Pigments 76:299–309. https://doi.org/10.1016/j.dyepig.2006.05.041
32. Ahmed NSE, El-Shishtawy RM (2010) The use of new technologies in coloration of textile fibers. J Mater Sci 45:1143–1153. https://doi.org/10.1007/s10853-009-4111-6
33. Bechtold T, Brunner H (2005) Electrochemical processes in textile processing. In: Nuñez M (ed) New developments in electrochemistry research. Nova Science Publishers, New York, pp 1–55
34. Roessler A, Crettenand D, Dossenbach O et al (2002) Direct electrochemical reduction of indigo. Electrochim Acta 47:1989–1995. https://doi.org/10.1016/S0013-4686(02)00028-2
35. Roessler A, Dossenbach O, Meyer U et al (2001) Direct electrochemical reduction of Indigo. CHIMIA 55:879. https://doi.org/10.2533/chimia.2001.879
36. Bond AM, Marken F, Hill E et al (1997) The electrochemical reduction of indigo dissolved in organic solvents and as a solid mechanically attached to a basal plane pyrolytic graphite electrode immersed in aqueous electrolyte solution. J Chem Soc Perkin Trans 2 1735–1742. https://doi.org/10.1039/a701003f
37. Roessler A, Crettenand D (2004) Direct electrochemical reduction of vat dyes in a fixed bed of graphite granules. Dyes Pigments 63:29–37. https://doi.org/10.1016/j.dyepig.2004.01.005

38. Kulandainathan MA, Kiruthika K, Christopher G et al (2008) Preparation of iron-deposited graphite surface for application as cathode material during electrochemical vat-dyeing process. Mater Chem Phys 112:478–484. https://doi.org/10.1016/j.matchemphys.2008.06.010
39. Yang Z, Shen W, Chen Q, Wang W (2021) Direct electrochemical reduction and dyeing properties of CI Vat Yellow 1 using carbon felt electrode. Dyes Pigments 184:108835. https://doi.org/10.1016/j.dyepig.2020.108835
40. Bechtold T, Burtscher E, Turcanu A, Bobleter O (1994) The reduction of vat dyes by indirect electrolysis. J Soc Dye Colour 110:14–19. https://doi.org/10.1111/j.1478-4408.1994.tb01586.x
41. Bechtold T, Burtscher E, Turcanu A (1999) Anthraquinones as mediators for the indirect cathodic reduction of dispersed organic dyestuffs. J Electroanal Chem 465:80–87. https://doi.org/10.1016/S0022-0728(99)00057-1
42. Kulandainathan MA, Muthukumaran A, Patil K, Chavan RB (2007) Potentiostatic studies on indirect electrochemical reduction of vat dyes. Dyes Pigments 73:47–54. https://doi.org/10.1016/j.dyepig.2005.10.007
43. Roessler A, Dossenbach O, Marte W, Rys P (2002) Electrocatalytic hydrogenation of vat dyes. Dyes Pigments 54:141–146. https://doi.org/10.1016/S0143-7208(02)00035-9
44. Roessler A, Crettenand D, Dossenbach O, Rys P (2003) Electrochemical reduction of indigo in fixed and fluidized beds of graphite granules. J Appl Electrochem 33:901–908. https://doi.org/10.1023/A:1025876114390
45. Abdelileh M, Manian AP, Rhomberg D et al (2020) Calcium-iron-D-gluconate complexes for the indirect cathodic reduction of indigo in denim dyeing: A greener alternative to non-regenerable chemicals. J Clean Prod 266:121753. https://doi.org/10.1016/j.jclepro.2020.121753
46. Holme I (2002) Recent developments in colorants for textile applications. Surf Coat Int Part B Coat Trans 85:243–264. https://doi.org/10.1007/BF02699548

Chapter 6
Renewable Residues-Derived Dyes and Mordants: Progress and Applications

Brenda Juana Flores, Luis Fernando Flores, Yaco Rozzisi, and María Yamila Chumbita

Abstract Natural dyes are widely available from various sources, including plants, minerals, animals, insects, and microorganisms. Compared to synthetic dyes, they are generally biodegradable, non-carcinogenic, and non-toxic, with reduced effluent discharge and minimized harm to ecosystems. In textile applications, natural dyes provide brilliant colors and added functional benefits, such as UV protection, antioxidant and antimicrobial activities, and lower allergenic potential. However, their use in the textile sector is constrained by the significant quantities required and their lower affinity for textile fibers, making mordants essential for enhancing dye fixation and color vibrancy. Renewable biomordants, including tannin-rich compounds and oil-based mordants, are emerging as alternatives to traditional salt-based options. Both natural dyes and biomordants can be derived from renewable residue feedstock. Nonetheless, these sustainable approaches face challenges, such as the variability and availability of raw materials, cost-effectiveness, and efficient dyeing procedures. Despite these challenges, many dyes and mordants currently available in the textile market are sourced from renewable residues within a circular and carbon economy. This chapter reviews the development and application of renewable residue-derived dyes and mordants, emphasizing their potential to advance sustainability in textile production.

B. J. Flores · Y. Rozzisi · M. Y. Chumbita
Escuela de Producción, Tecnología y Medio Ambiente, Universidad Nacional de Río Negro, Villa Regina, Rio Negro, Argentina

B. J. Flores · L. F. Flores (✉) · Y. Rozzisi · M. Y. Chumbita
Club de Biotecnología Agrícola y Ambiental, Villa Regina, Rio Negro, Argentina
e-mail: lfflores.lab@gmail.com

M. Shahid et al. (eds.), *Advancements in Textile Coloration*,
https://doi.org/10.1007/978-981-96-5091-0_6

Graphical Abstract

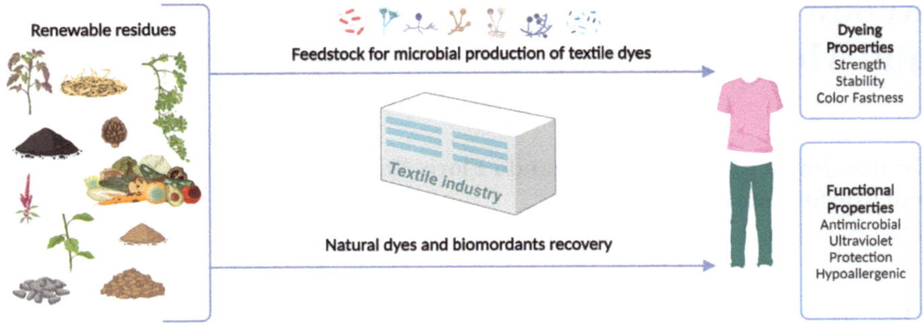

Keywords Renewable residues · Natural dyes · Biomordants · Sustainable dyeing

6.1 Synthetic Dyes and Mordants

In 1856, mauve, recognized as the first synthetic dye, was developed by Perkin. With the synthesis of additional textile dyes and their practical benefits, most textiles were dyed synthetically by 1910 [1]. In addition to being cheaper, synthetic dyes are chemically more stable over time, do not degrade under physical, chemical, or biological agents, and provide a consistent and durable color [2]. Nowadays, these dyes are primarily derived from the petrochemical industry, with diverse chemical structures and high complexity. Approximately 800,000 tons of these products are produced globally per year, 75% of which are used entirely by the textile industry [3]. Dye chemists classify synthetic dyes according to their chemical structures, grouping them into phthalocyanine, azo, formazan, and anthraquinone dyestuffs. Additionally, textile dyes are classified according to the types of fibers they dye, such as cellulose, protein, and synthetic fibers [4, 5]. Among the extensive range of dyes available in the commercial market, azo dyes represent the largest category in terms of production volume, accounting for approximately 60–70% of global organic dye production. This preference is due to the fact that they have a simple and low-cost synthesis compared to other dyes, as well as their high molar extinction coefficient, significant structural diversity, and excellent fastness for use in textiles. [6–8]. A classification of synthetic dyes according to the fabrics they dye and their chemical composition is shown in Fig. 6.1.

The fastness of the dyes is enhanced through the application of mordants [9], which can be classified into dulling mordants (ferrous and copper sulfates) and brightening mordants (potassium dichromate, alum, and stannous chloride) [10, 11]. A general classification of mordants and mordanting processes is shown in Fig. 6.2.

- *Pre-mordants*: Premordanting involves applying the mordant before the staining step, allowing the mordant to adhere to the fabric more effectively [13]. This method generates an integrated layer of dye, mordant, and textile surface. The bonding of the metal with the fabric and the dye generates a color that responds

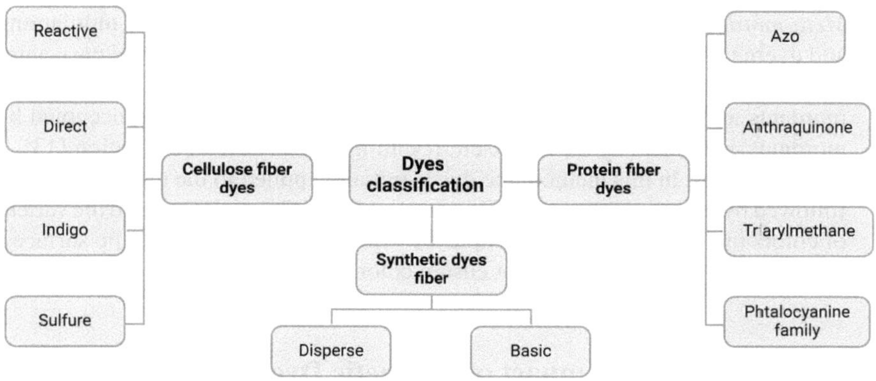

Fig. 6.1 Classification of synthetic dyes. Adapted from [5]

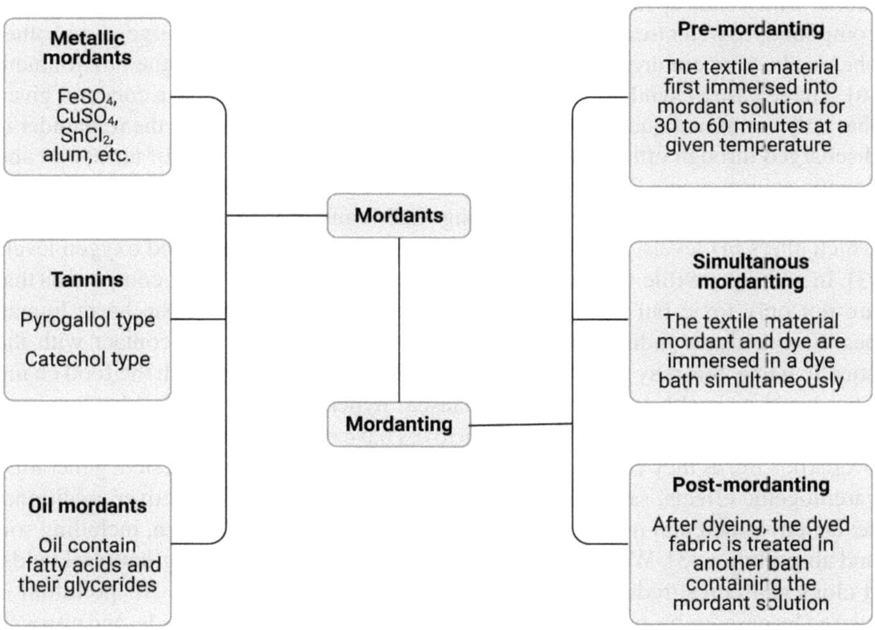

Fig. 6.2 Mordants and mordanting methods. Adapted from [12]

quickly to factors such as washing, rubbing and light. Simultaneously, chelation occurs, allowing for effective dissipation of photonic energy and optimizing the light fastness of the dyed materials [13]. Hence, due to the efficient use of resources throughout the process, this technique is considered more sustainable for mordanting.

- *Meta-mordants*: When meta-mordents are applied, the processes of mordanting and dyeing occur simultaneously. However, this practice is considered unsustainable due to the waste of resources. Furthermore, the complexity of combining mordants and dyes may lead to some regions of the material being occupied by mordants and others by the dye itself, resulting in inconsistent coloration [13].
- *Post-mordants*: In this method, the dye is initially applied to the untreated fabric, followed by the mordanting step. This practice is mainly used to expand the variety of colors by forming mordant complexes with dye molecules on textile surfaces. However, this technique does not ensure color fastness [13].

6.2 Environmental Impact of Synthetic Dyes and Mordants

During the dyeing process, between 10 and 15% of the dye remains unfixed on the fabric, which ends up in textile wastewater [3]. However, due to the resistance of these compounds to water treatment (due to their significant stability to detergents and other chemicals, temperature, and light) most of them are released into the environment [6]. The release of synthetic mordants, such as metallic salts, is also a concern, given that only a minimal quantity of these salts adheres to textiles, while the remainder is discharged through effluents, thereby contributing to the pollution of terrestrial and aquatic environments [3, 14]. Synthetic dyes and mordants pollution have alarming effects on aquatic environments, affecting water transparency and light transmission, which alters pH levels, inhibits photosynthesis and reduces dissolved oxygen levels [3]. In addition, textile wastewater also contains recalcitrant organic compounds that are not only toxic but also mutagenic and carcinogenic, which can harm human health, directly or indirectly [1, 6]. When the waste comes into contact with the aquatic fauna, this may affect humans as the pollutants move through the food chain. This leads to health issues, such as nausea, hypertension, cramps, bleeding, and ulcerations. For instance, triarylmethane dyes have even more worrying effects after overexposure, as they can penetrate the skin and irritate it when ingested, generating carcinogenic effects, sarcomas, and methemoglobinemia [2]. Moreover, additional severe environmental impacts occur throughout the production chain, including soil and air pollution [3]. When polluted textile wastewater flows into agricultural fields, it clogs soil pores, reducing productivity in the affected areas [4]. Air pollution is caused by gaseous by-products such as sulfur derivatives, formaldehyde, and nitrogen oxides, typically released into the atmosphere with insufficient or no treatment [3]. As a consequence of its negative impact on the environment, the textile industry has been recognized as one of the major sources of water pollution due to the use of multiple harmful chemicals [1]. Hence, increased attention has been paid to the environmental problems brought about by the textile industry. New standards have been established and control bodies have been created to enforce more restrictive legislation regarding waste treatment [15]. For instance, the European Union, United States, China, India, Brazil, Japan, and Australia have published articles that restrict the use of substances classified as mutagenic, carcinogenic, or toxic for reproduction [15, 16]. However, these regulations might be absent or inadequate in some developing countries [3]. Therefore, great emphasis must be placed on developing new strategies for the textile sector to achieve sustainability and a circular economy.

6.3 Natural Dyes and Mordants

Due to the growing global concern about environmental sustainability and pollution in the textile sector, there has been a renewed interest in natural dyes, which are biodegradable, non-toxic, and do not contribute to the generation of toxic effluents in wastewater [1]. Natural dyes have been used all over the world since 3500 BC [1]. The first technique consisted of crushing pigments and rubbing them on the fabric. With the passage of time, vegetable matter and crushed fruits began to be used which were boiled together with the fabric and tested for water and light resistance. Among the most recognized ancient dyes are indigo blue, which came from the leaves of *Indigofera tinctoria*, madder obtained from the roots of *Rubia tinctorium*, and yellow obtained from the stigmas of saffron flowers (*Crocus sativus* L.) [17]. Until the nineteenth century, natural dyes predominated as the main textile dyes [17]. However, as mentioned above, the arrival of synthetic dyes in 1856 [18] implied almost a complete substitution of natural dyes due to their advantageous characteristics, including a broader spectrum of colors, enhanced dyeing quality, and greater reproducibility, resulting in lower costs [19]. However, there is a reemergence of natural dyeing with the help of synthetic and natural mordants, as discussed in the following sections.

6.3.1 Natural Dyes

Natural dyes are colorants (pigments and dyes) obtained from natural sources [17, 20]. Most of them are biodegradable, environmentally friendly, and non-toxic, with the potential to promote more sustainable dyeing practices [21, 22]. For textile applications, they not only add color but also serve additional functions such as UV protection, antioxidant, and antimicrobial activities [21, 23]. Natural dyes can be classified in several ways according to their chemical structure, origin source, and application method [23, 24] as it is shown in Fig. 6.3. Classification according to their source includes dyes obtained from animals, plants, microorganisms, and minerals [25]. Based on the chemical structure, they include betalains (purple betacyanin and yellow betaxanthin), carotenoids (red, orange, and yellow), flavonoids (yellow, blue, and orange), indigoids (blue), quinonoids (yellow to red), dihydropyran-based and tannins (yellow, gray, black and brown). Finally, according to the application method, their classification includes direct dyes, acid dyes, basic dyes, mordant, vat dyes, and disperse dyes [23].

On the basis of sources	On the basis of chemical structure	On the basis of applicaiton method
• Vegetable Origin • Insect/Animal Origin • Mineral Origin • Microbial Origin	• Betalains • Carotenoids • Flavonoids • Indigoids • Quinonoids • Dihydropyran based • Tannins	• Acid dyes • Basic dyes • Direct dyes • Mordant dyes • Vat dyes • Disperse dyes

Fig. 6.3 Classification of natural dyes [26]

As mentioned, natural dyes can be classified into four categories based on their sources: plant, animal, mineral, and microbial-derived dyes.

- *Plants*: Dyes are sourced from various plant parts, including seeds, leaves, roots, stems, and fruits [20, 26]. In addition, they possess antifungal, antimicrobial, disinfectant, deodorant, and insect repellent properties [27]. Common plant dyes include yellow from saffron stigmas, red from henna leaves, and indigo from indigo leaves [27].
- *Animals*: Animal dyes are typically derived from dried insect bodies and secretions [27]. The most important dyes produced by insects are kermes, cochineal, and lac [10]. Furthermore, since ancient times, purple and violet dyes have been produced by shellfish and mollusks [26].
- *Mineral*: Among the most important minerals for their use as dyes are ocher or limonite, which provides yellow, brown, and red colors; manganese, from which dark tones are obtained; malachite which provides green tones and azurite, blue [26].
- *Microorganisms*: Dyes produced by bacteria, yeasts, fungi, and algae represent a valuable alternative source of natural pigments [28]. These natural dyes may exhibit antimicrobial, antioxidant, anticancer, insect repellent, and ultraviolet protective properties [29]. Some fungal species produce stable dyes, such as pre-anthraquinones, anthraquinone, and carboxylic acids [26]. Moreover, certain bacteria can produce a wide range of unusual colors, such as gold, silver, brown, fluorescent green, blue, white, and black [30].

With this scope, natural dyes can be distinguished from synthetic dyes through various aspects, as presented in Table 6.1.

6.3.2 *Natural Mordants*

Most natural dyes require a mordant to adhere properly to the fiber and increase their durability [31]. Traditionally, mordants for the fixation of natural dyes, were

Table 6.1 Differences between natural and synthetic dyes

Natural dyes	Synthetic dyes
• They come from natural sources • They can come from by-products and wastes • Low availability • Low dye adsorption • Poor fastness properties • Poor shade reproducibility • High processing costs • Require mordants • Environmentally friendly • Non-toxic • Biodegradable	• Produced by artificial materials derived from hydrocarbon compounds such as naphthalene and benzene • Reasonably priced • Ease of application • Wide range of colors • Reproducible shades • Good fastness properties • Toxic • Mutagenic and carcinogenic • Non-biodegradable

of chemical origin, but as stated in previous sections, they have negative effects on the environment [31]. In this context, biomordants or natural mordants are emerging alternatives to enhance natural dyes fixation into fabrics [29, 31].

A wide variety of biomaterials and plants can be used as biomordants in the dyeing process and their effectiveness is strongly influenced by their chemical compositions and structures [3]. The biomordant mechanisms for achieving adhesion include the establishment of van der Waals and π–π interactions, and hydrogen bonds, making them an ecological and biological alternative to metallic and synthetic mordants [18, 32]. However, the different variants, mordant classes and concentrations can affect the shades of the dyed fabrics, resulting in darker, brighter or more intense natural colors, which may be a favorable or undesirable result [3, 29, 31]. Hence, the differences between natural and synthetic mordants should be considered, as presented in Table 6.2.

Table 6.2 Differences between natural and synthetic mordants

Natural mordants	Synthetic mordants
They include plants, tannic acid, tartaric acid, tannins, bio-residues, and by-products	Chloride derivatives, oxides, hydroxides, and sulfates
Environmentally friendly and sustainable	Most of them are not environmentally friendly
Bonded via π–π interactions, hydrogen bonds, inclusions, or van der Waals interactions	Bonded by coordinated/covalent bonds, hydrogen bonds, and other interactions

6.4 Renewable Residues as Feedstocks for Natural Dyes and Mordants

The current production of natural dyes is mainly based on the cultivation of pigment-rich plants, which not only affects food security but also leads to higher production costs [18, 19, 25]. To reduce costs, alternative sources, such as renewable residues generated by other industrial activities, should be explored for use in the extraction or production of natural dyes [8, 18].

Residues are classified as renewable when they come from a renewable and sustainable source, and their generation does not unsustainably damage the environment or deplete natural resources [33]. These wastes, by-products, or resources out of place, generally come from biological origin and can be regenerated in a short time by human intervention or through natural processes, including forestry and agricultural waste, manure, wood, and food waste. These renewable residues represent a renewable carbon source that can support energy, economic, and environmental goals in the textile industry [33].

6.4.1 *Industrial Residues*

The food and beverage industry generates substantial amounts of natural dye-containing waste streams, which can be utilized for textile dyeing, providing a sustainable and economical resource for these operations [17, 34]. For instance, olive oil industry wastewater can be used for dyeing synthetic fibers such as polyamide [35]. Also, dyes derived from black currant skins (residues from the juice pressing process) have been used to dye casein fibers [36]. Tomato processing wastes also have significant potential for dyeing textile fabrics, such as polyamide, wool, and silk, yielding shades ranging from reddish to yellowish with good washing and lightfastness properties [37]. Moreover, wine industry-derived grape pomace is a renewable residue rich in natural dyes that can be extracted and applied to wool fabric [38].

6.4.2 *Forestry Residues*

Forestry operations produce large quantities of residues such as wood chips, bark, leaves, and branches, owing to deforestation of forests and firebreaks of species such as cedar, deodar, jackfruit, eucalyptus, and walnut. These by-products contain several natural pigments and can be used for textile dyeing [39]. For instance, Turkish red pine bark extracts have been applied to various fabrics, while mango and eucalyptus bark extracts have been used to dye wool, silk, and cotton [39]. Debarking wastes of coniferous plants, such as Siberian fir, larch, and spruce extracts have also shown a variety of beige-orange shades on cotton fibers [40]. Additionally, recent research has shown that the use of maritime pine and eucalyptus bark extracts can be used as natural textile dyes without the need for metallic mordants, due to their high polyphenol content [41].

6.4.3 *Agricultural Residues*

Agricultural residues are by-products of various agricultural operations [17]. These wastes, such as onion peels, pomegranate wastes, walnut and almond husks, are rich in natural pigments with good fastness, antimicrobial activity, and UV protection [17, 39]. Additional examples of natural dyes for textile applications, extracted from forest and agricultural wastes, are presented in Table 6.3.

6.5 Renewable-Residues-Derived Mordants

Recently, alternative plant-based mordants such as tannins and oil mordants have been investigated [18, 53]. These types of plant-based mordants are also referred to as biomordants, and they encompass a wide chemical spectrum, such as tannins, tartaric acid, tannic acid, and metal-containing plants, which may be present in multiple renewable biowaste streams [31].

6.5.1 *Tannins*

Tannins are astringent products found in a wide variety of plants, including wood, bark, fruit, leaves, pods, and roots. For mordanting applications, three distinct types of bonds are involved: ionic, hydrogen, and covalent bonds with proteins (such as silk and wool) and cellulose fibers [31, 54]. However, tannins encompass a wide variety of organic compounds that often differ widely in their chemical compositions and reactions. This diverse group is structurally divided into two different classes, according to the type of phenolic nuclei involved, and the manner in which they bind together. The first class is hydrolyzable tannins, whereas the other class is condensed tannins [13]. Hydrolyzable tannins are hydrolyzable by mineral acids, specific enzymes such as tannase, or water. Condensed tannins are not easily hydrolyzed to simpler molecules as their structure consists of a nucleus that is biosynthetically related to flavonoids [41, 55].

Table 6.3 Renewable plant residues as sources of textile dyes

Residue	Colorant compounds	Color	References
Grape pomace	Procyanidin, prodelphinidin, glucosylated procyanidin, and gallated procyanidin	Purple-bronze	[39]
Pomegranate rind	Flavan-3-ols, granatonine, or punicalagins	Yellow	[42]
Onion peel	Quecertin	Yellow and reddish-brown	[43]
Walnut shell	5 hydroxy-1, 4-napthoquinone)	Brown	[39]
Almond shell	(+)-catechin, (−)-epicatechin, kaempferol, and isohamnetin	Brown–green	[44]
Peanut shell	Flavonoid, phenols, and tannins	Brown	[45]
Groundnut/peanut skin	Catechin, vanillin, and epicatechin	Brown	[39]
Tamarind seeds	Procyanidin trimer, epicatechin, and procyanidin B2	Yellow–brown	[39]
Safflower seeds		Yellow	[46]
Banana	Luteolin and tannins	Light brown	[47]
Flower waste	Lutein and patulinin	Orange	[39]
Pumpkin skin			[48]
Mango leaves	Mangiferin	Green	[49]
Jack fruit tree bark		Yellow	[50]
Eucalyptus bark		Yellow–brown	[50]
Beetroot	Cyanins and betanins	Red–yellow	[51]
Used marigold flowers	Patultrin and patuletin	Yellow–orange	[52]

6.5.2 Oil-Based Mordants

Oil mordants are natural oils containing fatty acids and their glycerides, such as ricinolic, oleic, palmitic, and stearic [31]. Vegetable oils like castor oil and Turkey red oil (TRO) are examples of oil-based biomordants [10]. Oil-based mordants are capable of forming complexes with primary mordants, such as alum [31]. The –COOH group of fatty acids reacts with metal salts to produce –COOM, where "M" represents the metal. Alum, being water-soluble, does not bind to cotton fibers and can be easily washed out of the fabric after mordanting. Moreover, when oils are treated with concentrated sulfuric acid, the sulfonic acid group $–SO_3H$ is introduced, resulting in sulfonated oils that exhibit greater metal-binding capacity than natural oils in general [31]. Consequently, this type of biomordants provides superior color fastness and hue [10].

Biomordants, like metal mordants, can vary in effectiveness depending on their type and concentration, impacting color yield, fastness, and shade variations when used with different natural dyes. In addition, they can lead to different shades depending on the natural dye applied in conjunction with them. These effects can be attributed to the different interactions between the metal ions in the biomordants, natural dye molecules, and textile fabrics [47, 56]. A brief list of reported renewable biomordants in the literature is presented in Table 6.4.

6.6 Renewable Residues as Feedstock for Microbial Production of Textile Dyes

During the last decade, microbial pigments have emerged as environmentally friendly dyeing alternatives for the textile industry [80, 81]. The most commonly used microorganisms in this process are bacteria and yeasts, which are considered safer than filamentous fungi, as fungi can produce mycotoxins. Moreover, their scalability through fermentation-based biotechnological processes makes them suitable for the development of a sustainable textile dyes industry [82]. Furthermore, genomic and metabolic engineering strategies allow for the regulation and selective expression of pigment-production-related genes, achieving high production yields [82]. However, to effectively integrate microbial pigments for textile applications, it is essential to establish green production methods that utilize renewable by-products as substrates to support microbial growth and pigment production, such as those listed in Table 6.5.

The industrial production of microbial dyes involves a number of technical and biological considerations to ensure the success of the process [93]. Scale-up is the fundamental step to perform a thorough analysis of the dyes from a laboratory-scale fermenter [93]. This provides essential information about pigment production and allows the identification of the physical, chemical, and biological factors that influence the scaling process. Careful evaluation and adaptation of fermentation parameters and culture conditions will ensure the reproducibility and efficiency of large-scale processes for each type of microorganism and pigment [94].

The pretreatment of biomass residues is a fundamental step in microbial pigment production to increase the availability of essential nutrients for microorganisms. Pretreatment methods can be classified as physical, chemical, biological, or combined, each with its own advantages and limitations [95]. Physical methods include crushing, grinding, and sieving, which fragment the waste to increase its surface area and facilitate the action of microorganisms. Chemical methods involve the use of chemical agents such as acids, bases, or organic solvents to break down the structural components of waste. Biological or enzymatic methods employ microbial enzymes to degrade complex substrates into simpler forms that are available to the

Table 6.4 Renewable-feedstocks-derived mordants

Residue	Bio-mordant	Natural dye	References
Acacia Nilotica (babul bark)	Babul bark extract	*Nyctanthes arbortristis* (coral jasmine)	[57]
Acacia nilotica bark extract	Tannin-rich extract	*Kerria lacca* and *Rheum emodi*	[58]
Green almond shells	Gallic acid	*Phoenix dactylifera* extracts	[59]
Pomegranate rind	Tannin-rich extract	Red sandalwood (*Ptrecarpus santalinus*)	[60]
Banana sap *Palayankodan*	Tannin-rich extract	Banana stem extract	[61]
Castanea sativa Mill	Tannin-rich extract	Extracts of *Helichrysum italicum* and *Rubia peregrina* L.	[62]
Pomegranate peel extract (*Punica granatum*)	Tannin-rich extract	Eucalyptus bark extract	[63]
Lemon peel (*Citrus lemon* L.)	Naringin	Extracted from spent cofee grounds	[64]
Oak (*Quercus*)	Tannin-rich extract	Iranian madder and weld	[65]
Black cherry stem (*Prunus serótina*)	Ferrous	Hops flower	[66]
Onion skin (*Allium cepa*)	Tannin-rich extract	Banana leaf extract (*Musa paradisiaca*)	[67]
Pomegranate rind (*Punica granatum*)	Tannin-rich extract	Banana leaf extract (*Musa paradisiaca*)	[67]
Pomegranate peels	Tannin-rich extract	*Cassia obovata* extract	[68]
Pine-nut hull	Catechin	*Cassia obovata* extract	[68]
Pomegranate peels	Tannin-rich extract	*Vitis vinifera* L. leaves extract	[69]
Quebracho bark tree extract	Tannin-rich extract	Nut shell extract	[70]
Rhus coriaria extract	Tannin-rich extract	Madder root extract (*Rubia tinctorum* L.)	[71]
Eucalyptus extract	Tannin-rich extract	Madder root extract (*Rubia tinctorum* L.)	[71]
Terminalia chebula extract	Tannin-rich extract	Madder root extract (*Rubia tinctorum* L.)	[71]
Quercus castaneifolia extract	Tannin-rich extract	Madder root extract (*Rubia tinctorum* L.)	[71]
Pomegranate extract	Tannin-rich extract	Madder root extract (*Rubia tinctorum* L.)	[71]

(continued)

Table 6.4 (continued)

Residue	Bio-mordant	Natural dye	References
Sumac extract (*Rhus coriaria* L.)	Tannin-rich extract	Madder and Reseda	[72]
Banana peel (*Musa sapientum* L.)	Tannin-rich extract	Marigold flower's petals (*Tagetes erecta* L.)	[73]
Acacia mearnsii sawdust	Tannin-rich extract	*Hibiscus sabdarifa* flowers extract	[74]
Acacia mearnsii sawdust	Tannin-rich extract	*Camellia sinensis* leaf extracts	[74]
Acacia mearnsii sawdust	Tannin-rich extract	*Curcuma longa* root extract	[74]
Acacia mearnsii sawdust	Tannin-rich extract	*Allium cepa* peel	[74]
Tamarindus indica L. seed coat	Tannin-rich extract	*Turmeric* extract	[75]
Tamarindus indica L. seed coat	Tannin-rich extract	*Pomegranate rind* extract	[75]
Mango bark extract	Tannin	*Castanopsis indica* extract	[76]
Zeera (*Cuminum cyminum*) extracts and Neem *(Azadirachta indica)*	Tannin	Arjun bark (*Terminalia arjuna)*	[77]
Residues rich in tannins/fallen leaves of *Chinese tallow/Sapium sebiferum* L.	Tannin	*Sapium sebiferum* extract	[78]
Chinese tallow flora leaves, *Cinnamomum Camphora and Folium Artemisiae Argyi*	Tannin	Seeds of *Eriobotrya japonica* L. (EJSS)	[21]
Quercus aegilops extract	Tannin	*Crocus sativus* extract	[79]

microorganisms. Moreover, combining pretreatment methods may improve the efficiency of waste decomposition for subsequent microbial bioconversion into textile dyes [95].

In addition to pretreatment, detoxification of lignocellulosic hydrolysates may be necessary to remove compounds that inhibit microorganisms or negatively affect the fermentation process. These compounds include lignin degradation products, furfural, and fatty acids. Detoxification is usually performed through specific physicochemical or biological methods to neutralize or eliminate such compounds [94].

Microbial fermentation is a fundamental process in pigment production and can be performed via two different mechanisms, as described below.

- *Solid-state fermentation (SSF)*: In this fermentation technique the substrate is in a solid state with a low amount of water, such as fruit and vegetable waste, rice, rice straw, wheat bran, paper pulp, coconut fiber, and bagasse. Microorganisms

Table 6.5 Microbial dyes produced using renewable residues-based substrates

Molecule	Microorganism	Color	Waste feedstock	References
Prodigiosin	*Serratia rubidaea*	Red	Spoiled coconut	[83]
Prodigiosin	*Serratia marcescens* SS1	Red	Sago industrial waste	[84]
Prodigiosin	*Serratia marcescens* UCP 1549	Red	Waste soybean oil	[85]
Prodigiosin	*Serratia marcescens* CC17	Red	Fishery processing waste	[86]
Violacein	*Chromobacterium violaceum*	Violet	Sugarcane bagasse and solid pineapple waste	[87]
Violacein	*Chromobacterium violaceum* UTM5	Violet	Liquid pineapple waste	[88]
Violacein	*Chromobacterium violaceum*	Violet	Waste soybean meal	[89]
Monascus pigments	*Monascus purpureus* CMU001	Red	Brewer's spent grain	[90]
Monascus pigments	*Monascus purpureus* ATCC16362	Red	Date waste substrates	[91]
Monascus pigments	*Monascus purpureus* ATCC16436	Red and orange	Glycerol and corn cob	[92]
Talaromyces albobiverticillius pigments	*Talaromyces albobiverticillius*	Red	Pineapple peel waste	[8]

grow and produce metabolites on the surface or within the solids of the substrate, presenting advantages such as higher yield and productivity, as well as the possibility of using the fermented product directly as colorants without having to isolate the product. This technique is particularly suitable for microorganisms adapted to low moisture content, such as fungi [93].

- *Submerged Fermentation (SmF)*: In submerged fermentation, the substrate is immersed in an aqueous medium to support microorganism growth and pigment production. It is characterized by the rapid consumption of substrates and is commonly used for bacteria and yeast, as they require a higher moisture content during the fermentation process. Renewable residues-based substrates include discarded fruit and vegetable juices, molasses, and wastewater [95].

Both techniques have advantages and disadvantages. SSFs offer an environment conducive to microbial growth and pigment production at a low cost with simplified bioreactors that require less energy. However, they can be difficult to regulate and scale, and product purification can be difficult. On the other hand, SmF allows for rapid substrate consumption and more precise process regulation but can generate more waste and require higher energy consumption [95].

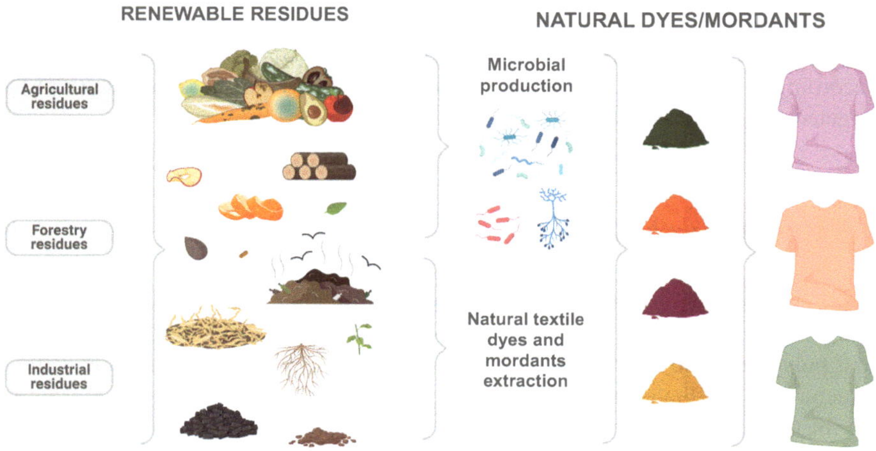

Fig. 6.4 Schematic representation of the use of renewable residues in the manufacture of dyes and mordants (Created with BioRender.com)

In large-scale production, as the bioreactor volume increases, hydrostatic pressure becomes a determining factor that influences enzyme activity, cell membrane permeability, and metabolic flux [93]. Therefore, managing this pressure and its associated effects on cell viability and process performance is essential for successful scaling-up. Chemical stress of the renewable residue also has to be considered, as it may introduce microbial contamination, inhibitors, or harmful impurities. Therefore, it is critical to ensure the sterility and quality of the waste feedstock during the initial stages of seed cultivation [93]. To address these challenges, genetically improved strains would be beneficial for scaling up processes in a more predictable and efficient manner. However, it is important to consider regulatory requirements, stability, costs, and toxicity of microbial dyes for textile applications [94]. In addition, optimizing fermentation conditions, including substrate concentration, nitrogen source, and moisture, is essential for maximizing the pigment yield and profitability of industrial applications [93]. Figure 6.4 shows the two pathways for dyes and mordants production from renewable residues, extraction, and microbial fermentation.

6.7 Extraction of Natural Dyes and Mordants from Renewable Residues

Renewable residues-substrates contain a great diversity of molecules and macromolecules such as carbohydrates, water-insoluble fibers, and proteins, which can be co-extracted alongside the desired dye or mordant. Therefore, the choice of the extraction technology must be appropriate for each renewable feedstock [3, 96]. Moreover, the moisture in the feedstock must also be considered. If it is dry, it can be chopped or ground into a powder, whereas if it is wet, it can be dried or ground into a paste prior to extraction [26]. Hence, extraction technologies should be applied according to the type of renewable residues and the physicochemical nature of the targeted dye or mordant [3].

6.7.1 Aqueous Extraction

Traditionally, aqueous extraction has been the most commonly applied technique to extract natural pigments from biowaste feedstocks. In this technique, the biomass is first pulverized and subsequently immersed in boiling water to disrupt the cellular structures to facilitate the extraction of dyes [3]. The parameters to be considered are the temperature, extraction time, pH, renewable residue/water ratio, and pigment concentration in the biomass [10]. The extraction can be repeated several times, and the resulting extract must be filtered for subsequent fabric dyeing [26]. Although the extract can be readily applied to textile fabrics, this procedure has some disadvantages, such as large volumes of water required, long extraction times, high temperatures, low yields of recovered natural dye, co-extraction of undesired compounds, and limitation to water-soluble dyes [3, 26]. In addition, both boiling and high temperatures may affect the extraction of temperature-sensitive dyes, resulting in a reduced yield [3].

6.7.2 Solvent Extraction

Organic solvent extraction is similar to aqueous extraction. Commonly used solvents are methanol, ethanol, and acetone (for water-soluble compounds), dichloromethane, hexane, and petroleum ether (for liposoluble pigments), or different mixtures of solvents, which can lead to higher extraction yields [3, 26].

The solvent selection depends on the specific type of renewable residue and the nature of the natural mordant or dye [26]. For instance, water-alcohol mixtures can facilitate the extraction of both water-soluble and water-insoluble components [3]. Time, temperature, degree of grinding of the residue, equipment, and type of solvent are the principal variables affecting the extraction efficiency [26].

An essential advantage over aqueous extraction is the low water consumption, high yields, and the possibility of performing the extraction at lower temperatures to limit the degradation of thermolabile pigments [26]. Furthermore, this technology is simple to carry out, with low investment in equipment. In addition, the extracted pigments and biomordants can be purified easily, and solvents can be removed and recovered by distillation for reuse [3, 26]. However, other components like chlorophylls and waxy materials may also be co-extracted [3]. In addition, toxic residual solvents and greenhouse gas emissions may be released [3, 26].

6.7.3 Alkaline or Acid Extraction

Extraction under acidic or alkaline conditions can increase the yield of glycoside-related dyes. Alkaline extraction is particularly effective for dyes and mordants with phenolic groups that are soluble under alkaline conditions. However, these conditions might damage the dyed material because most dyes are pH-sensitive [3]. Optimal extraction conditions involve specific NaOH concentrations, temperature, and duration.

6.7.4 Supercritical Fluid Extraction

Supercritical fluid extraction (SFE) represents a novel alternative to standard solvent extraction for extracting organic compounds from different matrices [3, 10]. Supercritical fluids are substances above their critical temperature and critical pressure, exploited due to their properties such as those of both gases and liquids. As a result, they have high diffusivity and solvent power, low viscosity, and low surface tension, which accelerates the mass transfer. This facilitates penetration of the supercritical fluid into the renewable residue structure, which leads to a fast and efficient extraction [26]. Several substances can be used as supercritical fluids, including methanol, ethane, n-butene, n-pentane, nitrous oxide, and water. However, carbon dioxide (CO_2) stands out as the most extensively used substance for several reasons, such as its low pressure and critical temperatures (73 atm and 31.2 °C), it is non-flammable and non-toxic, and its recognized safety for the environment and human health in specific conditions. In addition, CO_2 is relatively inexpensive and highly pure, and it is easy to separate from the extract [3].

The advantages of SFE include efficiency and high yields with low extraction times, the possibility of using low or ambient temperatures, and the use of solvents generally recognized as safe (GRAS). The resulting extract contains no residual solvent, and the extraction can be coupled to chromatographic techniques, such as supercritical fluid chromatography (SFC) or gas chromatography (GC) [3]. Nevertheless, the high cost of SFE, inefficient extraction of polar compounds, and possible risks for workers due to operating at pressures significantly higher than atmospheric pressure pose limitations that necessitate the implementation of preventive safety measures [3].

6.7.5 Ultrasonic-Assisted Extraction

Ultrasound-assisted extraction (UAE) requires the dye-containing matter to be dissolved in water or any other suitable solvent as a medium for the transmission of ultrasound waves. Under these conditions, ultrasound causes the development of microbubbles that oscillate and collapse, generating microjets that collide with cell surfaces, destabilize them, and allow the release of intracellular dyes or mordants [3]. Frequency, energy, and pulses are key parameters to enhance the cavitation effects, diffusion, and emulsification [26]. Although high extraction yields can be achieved, a drawback is the heat generated during ultrasonication, which may damage thermosensitive dyes. Furthermore, extraction time, required solvents, and energy consumed should be assessed to successfully apply UAE [3, 96].

6.7.6 Microwave-Assisted Extraction

In microwave-assisted extraction (MAE), dye- and mordant-containing residues are treated with solvents in the presence of microwaves [3, 96, 97]. Cell lysis is induced by the ability of different cell components to absorb microwaves [26]. The advantage lies in the lower solvent requirement and faster processing, which directly impacts energy consumption [26, 97]. Additionally, microwaves can also be employed during dyeing with the extracted dyes, potentially leading to savings in time, energy, and costs [96].

6.7.7 Enzyme and Fermentation-Based Extraction

Enzyme extraction is an alternative to the use of solvents for extracting pigments from biomass matrices, providing an environmentally friendly approach. Plant tissues can be broken down by enzymes such as cellulase, pectinase, and amylase under appropriate pH and temperature conditions. For instance, cellulase breaks down cellulose and hemicellulose, modifying the cytoplasm and loosening the cell wall, thereby increasing pigment diffusion into the extraction medium [3].

This technique is suitable for extracting pigments from renewable residues such as bark or roots from forestry or agricultural industries. An advantage is the ability to control conditions, although this approach can be expensive [26].

Alternatively, colors can be extracted from renewable residues through enzymatic hydrolysis during microbial fermentation, although this method is slower and may produce odors due to the decomposition of organic matter [26].

According to the nature of the natural dye or mordant, it is essential to standardize the extraction technique while optimizing the relevant parameters and their associated costs to enhance the commercial viability of these renewable materials for the textile industry [10]. Figure 6.5 illustrates the process of extracting natural textile dyes from biomass residues and pigment-producing microbes using both conventional and modern techniques.

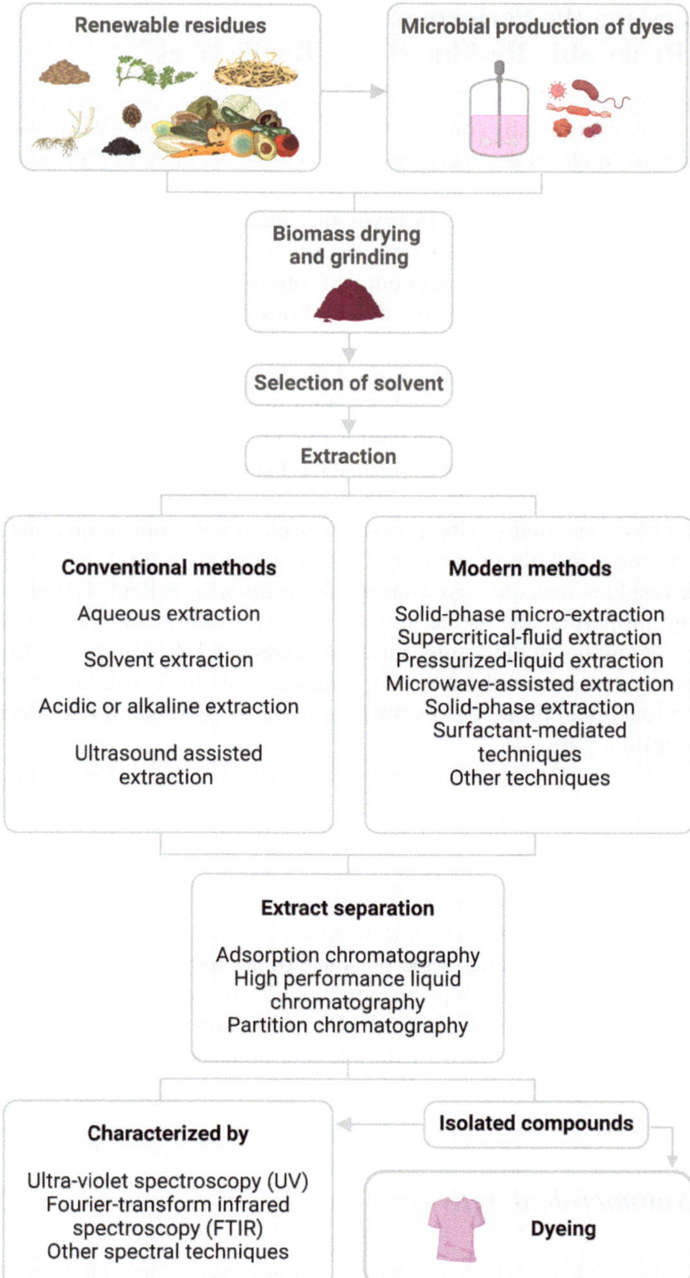

Fig. 6.5 Extraction methods of textile natural dyes from renewable residues. Adapted from [12]. (Created with BioRender.com)

6.8 Assessing the Performance of Renewable-Residue-Based Textile Dyes

The dyeing process typically consists of three stages. The first stage involves pretreatment to remove undesired impurities using detergents, enzymes, or aqueous alkaline substances. Some fabrics may be bleached with chlorine dioxide or hydrogen peroxide [4]. The second stage involves the application of dyes to the fabric to achieve color fastness and uniformity according to its intended use. To this end, chelating agents, softeners, bases, or emulsifying oils, among others, are used [5, 6]. Finally, the finishing stage treats the dyed fabrics again with auxiliary compounds to enhance their quality. These treatments can improve water resistance, provide protection against microorganisms, and offer resistance to dirt or UV light [4, 6].

Various pretreatment, dyeing, and finishing methods have been reviewed in the literature. However, in the context of natural dyeing, most natural dyes have the drawback of being fugitive, even when a mordant has been applied, which has limited their use in the textile industry [10]. Therefore, dyes derived from renewable residues must be assessed based on their dyeing properties such as color uniformity and intensity, and color fastness to light or washing [18]. In addition, natural dyes derived from renewable residues may possess antimicrobial, insect-repellent, UV-protective, or deodorizing properties due to the presence of glycosides, saponins, anthocyanins, flavonoids, or tannins in the residues used as feedstock [98]. These functional properties must also be assessed when using natural dyes for textile dyeing processes [18]. Some functional properties to consider in the assessment of dyes derived from renewable residues are.

6.8.1 *UV Protection*

Natural dyes derived from renewable residues can confer UV protection, mainly when derived from plants due to the presence of photosynthetic pigments, tannins, and flavonoids [25, 99]. The UV protection provided by natural dyes can be evaluated using spectrophotometry to determine the UV protection factor [25, 100]. Generally, protection is expected to improve with a higher concentration of dye in the fabric, and darker dyes are more effective [26].

6.8.2 *Antimicrobial Activity*

Unlike synthetic fibers, which do not decompose easily, natural fibers are more vulnerable to microbial degradation due to their biological origin and moisture content. Microbial attacks can cause pH changes or dye discoloration in fabrics and produce unpleasant odors or skin irritation [98]. Therefore, natural dyes with antimicrobial properties are of great interest. Different microbial strains, including *Klebsiella pneumoniae* and *Staphylococcus aureus*, are used for testing. For instance, antibacterial activity has been found in dyes derived from brewery waste extracts [99]. Extracts such as pomegranate peels and sweet potato leaves have shown antimicrobial activity [25, 98]. This is due to positively charged compounds like aminoguanidine and other nitrogenous compounds that can destroy negatively charged bacterial cell walls [98]. Polyphenols, steroids, tannins, and flavonoids are also antibacterial molecules present in natural pigments [25].

6.8.3 Deodorization Performance

Deodorization of fabrics is a functional advantage for garments, although the mechanisms of deodorization by natural dyes are unclear. Substances such as acetic acid and ammonia are used to evaluate the deodorizing effect of natural dyes [18].

6.8.4 Insect-Repellent Finishing

Textiles with insect-repellent properties can provide more effective protection for users. When chemical agents are used, concerns about duration and safety arise, however, natural and sustainable dye sources may represent an alternative [101].

6.8.5 Antioxidant Activity

Oxidation damages materials and human skin through a chain reaction of free radical or reactive oxygen species (ROS) production. Factors such as alcohol, toxins, polluted air, and smoking can create free radicals that affect human health. Therefore, treating textiles with antioxidants is a way to deliver antioxidants to the skin and deactivate free radicals, protecting against oxidative stress and damage [101].

These are just some functional properties of natural dyes that can add value to the textile industry [101]. They represent a promising way for customers to receive functional textiles that go beyond aesthetics, offering comfort, protection, health benefits, and hygiene [18, 98, 101].

Recently, in the context of sustainability, the assessment of environmental and sustainable performance has been proposed [7]. This involves considering the by-products used in obtaining natural dyes and mordants, their appropriate valorization, suitable dyeing techniques, and their environmental impact compared to current practices [1]. In this context, Life Cycle Assessment (LCA) should be employed to identify critical environmental points and improvement opportunities to promote circular economy principles and sustainable practices in the development of dyes and mordants based on renewable residues [7].

6.9 Challenges and Limitations of Natural Dyes and Mordants Applications in Textile Industry

Several technical challenges remain to be overcome when implementing natural textile dyes in mainstream textile processes. The cost–benefit ratio of both the extraction of compounds and their application in fabrics must be improved in order to scale up to an industrial scale [1]. Achieving reproducibility in shades is challenging because plant-derived products vary by harvest season, location, species, maturity, and extraction and processing conditions, as these dyes are typically mixtures of secondary metabolites [17, 76]. Also, the low availability of renewable residues that generate dyes and their logistics may represent additional production challenges [53]. Moreover, the natural dye industry must dye fabrics soon because it is difficult to store the extracted dyes for long periods under dry conditions [76]. However, it is complex to establish a consistent procedure with the use of renewable-residues-based natural dyes since both the dyeing process and the finishing treatments depend not only on the compound but also on the textile material, the water quality, its pH, the amount of mordant used and the dyeing time [17, 102].

Waste-derived natural dyes are not synthetically designed for application in the textile industry, therefore, they are usually not very stable and have low affinity to fabrics [1]. Fabrics such as cotton fibers, have surfaces with a high density of negative charges because of their composition, presenting hydroxyl groups and ether bonds, which presents a challenge because natural dyes often carry a negative charge due to polyphenolic compounds. This compromises the results in terms of color intensity and fastness [54]. Therefore with very few exceptions, almost all natural dyes require the use of mordants to adhere efficiently to textile materials. However, many of these compounds do not bind to the fabric during dyeing, leading to potential environmental issues with effluent discharge [17].

Additionally, naturally dyed fabrics tend to change their color when exposed to sunlight, perspiration, water, and air, so the results are highly dependent on the use of the fabrics and the washes performed [17]. Due to the low color affinity, intensity, and fastness of natural dyes, larger quantities are needed to achieve the desired color compared to synthetic dyes. This leads to longer, costlier processes and increased water consumption, complicating efforts to find greener dyes [17]. Moreover, although natural dyes provide an improvement in the protective properties of the fabric against UV rays, prolonged UV exposure reduces or even eliminates these protective characteristics [54].

Based on the above, it is evident that to achieve the application of natural dyes on an industrial scale, it is necessary to overcome a variety of technical limitations. As a result, potential solutions involve the use of improved mordants, whether metallic or biological, enzymes, fabric modifications, or modifications of the extracted natural dyes [1].

6.10 Benefits of Dyes and Mordants Production from Renewable Residues

Despite the limitations described above, the production of mordants and dyes from renewable residues offers numerous environmental, economic, social, and sustainability-related benefits.

First, the utilization of renewable residues is aligned with the circular economy, which affects sustainability. This renewable dye-and-mordant production system promotes an approach that adds value and contributes to sustainable development [36, 103]. One of the key benefits is that producing dyes and mordants from renewable residues does not compete with feed and food production, which helps contribute to food security [33]. It avoids damage to the environment through deforestation and competition with pigmented food crops that may require agrochemicals that pollute terrestrial and aquatic habitats [33]. For instance, extracting textile dyes from edible fruits or vegetables with high dye concentrations poses a risk of deforestation and affects food security [1]. Microorganisms, on the other hand, offer significant advantages, such as large-scale production capabilities and ease of genetic manipulation, without affecting biodiversity [94].

Within the environmental dimension, the utilization of renewable residues as feedstock for the production of textile dyes and mordants reduces dependence on petroleum derivatives. This results from reduced consumption of synthetic dyes derived from non-renewable resources, such as fossil fuels, which contain hazardous chemicals [10, 97]. Consequently, they have harmful impacts after their use in the textile industry, when they are discharged into the environment, even if they have been treated, contributing to soil and water pollution [5, 103]. Conversely, producing natural textile dyes reduces harmful chemical discharge, thus benefiting ecosystem health [103]. With this vision, natural dyes and dyestuffs extracted from renewable resources such as residues, by-products, and waste can have a lower environmental impact, in addition to contributing to waste management [103]. Another characteristic is that natural dyes are biodegradable, although they must be treated as effluents to prevent threats to ecosystems [26]. Furthermore, when renewable energy is incorporated into the production of waste-based dyes and mordants and in their application during dyeing, the environmental impact can also be reduced [7]. Merging dyeing and finishing processes when using functional dyes can enhance energy and water efficiency [32]. Consequently, the textile industry's environmental footprint can decrease [7]. All of these aspects are important because the textile industry manufactures essential goods globally and is the second largest industry globally, but one of the most polluting sectors [104, 105].

Social, economic, and cultural dimensions can also benefit from the production of waste-based mordants and dyes. On one hand, it can reduce funding needed for waste treatment [106]. These wastes can be valorized into value-added products for the textile industry, depending on their availability, low cost, and the design of an efficient extraction or fermentation dyeing process [3, 39]. Jobs can also be generated, especially in rural communities where by-product-content colorants may be available. In developing countries, these biorefineries can generate economic and social benefits through local production or exports [107]. In this sense, the extraction of dyes and mordants from renewable residues also diversifies income sources. It generates supply chains that can be more sustainable while reducing dependence on a single industry [100, 103]. At the cultural level, cultural heritage is conserved, and traditional knowledge is preserved and passed on to the next generation [103]. Also, natural dyes and mordants can contribute to improving the safety and health of both workers by avoiding the use of hazardous materials such as synthetic dyes and consumers because of their functional properties [23]. Moreover, natural dyes are suitable for various skin types and even in garments for newborns [26]. In addition, natural textile finishing is becoming a tangible prospect for strengthening a more sustainable and eco-friendly textile industry [32].

6.11 Companies Producing Renewable-Residues-Derived Dyes and Mordants

The projected expansion of the textile market between 2021 and 2028 is 4.4% [18, 96]. With this expansion, by 2030, the global textile industry market is forecasted to reach US$ 3 trillion in retail sales [105]. Consequently, the production of green textile materials that meet health and safety demands is on the rise [53]. This encouraged textile companies to take advantage of this market [103]. Consequently, in recent years, new products that use renewable residues as raw materials have entered the market, contributing to a circular design in the textile industry [36]. Some of the companies that have developed natural dyes and natural mordants from renewable residues are shown in Table 6.6.

6.12 Future Trends in a Greener and Sustainable Future Dyeing

With more and more customers placing a positive value on nature-friendly products, there is a growing market niche for pigments derived from renewable residues [117]. Thus, an increase in global demand for natural dyes obtained from renewable feedstocks is expected [18]. The adoption of sustainable dyes and mordants is an important step to achieving a more sustainable future in the textile sector, benefiting customers

Table 6.6 Commercially available dyes and mordants produced using renewable residues

Dyes/mordants	Renewable residue feedstock	Company/References
Diresul® Fiber-Teak	100% textile-waste-based raw material	[108]
Diresul® Fiber-Ochre	50% textile-waste-based raw material	[108]
Diresul® Fiber-Maroon	50% textile-waste-based raw material	[108]
Diresul® Fiber-Slate	50% textile-waste-based raw material	[108]
Diresul® Fiber-Graphite	50% textile-waste-based raw material	[108]
Diresul® Earth-Oak	100% almond shells	[108]
Diresul® Earth-Clay	90% beet residues	[108]
Diresul® Earth-Cotton	100% cotton plant waste	[108]
Diresul® Earth-Forest	90% saw palmetto residues	[108]
Diresul® Earth-Sand	90% bitter orange residues	[108]
Diresul® Earth-Stone	70% saw palmetto residues	[108]
Mallow®	Pomegranate peel	[109]
Nimbus®	By-product of lac processing	[109]
Bee®	Acacia catechu waste	[109]
Various	Agricultural waste	[110]
Various	Plant and food wastes	[111]
Various	Roots, shells, leaves, fruits, and offcuts wastes	[112]
Various	Textile-waste	[113]
Various	Plant waste	[114]
Various	Avocados, onions, blueberries, and red cabbage food wastes	[115]
Various engineered microorganisms	Plant-byproducts	[116]

and the environment [41]. Moreover, the demand for natural pigments also encourages the possibility of creating new business models and circular economy models based on sustainable dyes and mordants for the textile industry derived from renewable biomass. This could be accompanied by the emergence of new supply chains for raw materials and products that employ environmentally responsible practices [45, 53].

To strengthen the future application of natural pigments in the textile industry, innovative and lower cost extraction and production techniques are required [19, 118]. In this way, the high-cost limitations that restrict large-scale applications can be overcome. Although the use of renewable residues as feedstock for pigments and mordants may be available at low or no cost, their availability and transportation still need to be considered for cost-effective production [18]. Additionally, new sources of natural dyes should be researched in renewable residues that are currently overlooked in the textile industry's valorization studies [53].

In the coming years, renewable residues could be considered an appropriate source and a competitive alternative to synthetic equivalents if there is greater support in research, transfer, and industry adoption [31]. Collaboration between researchers and industry should address these technical aspects, such as the best renewable biomass sources and their influence on pigment chemistry and mordants, their lightfastness, quality, safety, toxicity, and added functional advantages [10, 31]. Optimization of the extraction in terms of quality and concentration should be accompanied by optimal dyeing conditions [41].

Regarding the microbial production of pigments from by-products, known strains should be exploited to their maximum potential, and new pigment-producing strains identified to meet the existing demand for microbial-synthesized colors. Moreover, their identification should be accompanied by toxicological tests, leading to better consumer acceptance [93]. The residues to be used as substrates should also be improved through modern pretreatment techniques, growth and bioprocessing conditions, appropriate scale-up, and industrial plant design [93].

The future promises an expansion of the textile industry, which puts pressure on policymakers and governments to impose regulations on the use of synthetic dyes, energy, and water consumption [5]. In this context, as demonstrated throughout this chapter, "biowaste-derived dyes will become a reality and can be promoted and implemented through appropriate policies [26]. As a result, further dialogue is expected between the sectors producing renewable residues and the textile industry that can valorize them through the production of mordants and dyes in the framework of a circular and sustainable bioeconomy [104].

6.13 Conclusion

Growing interest in sustainability and concerns regarding the harmful effects of synthetic textile dyes and mordants have driven the demand for more sustainable alternatives in the textile sector. In response, natural dyes and mordants derived from renewable residues have emerged as valuable alternatives due to their multiple advantages such as non-toxicity, biodegradability, and additional benefits such as antimicrobial and UV absorption capabilities. However, their effective use depends on improving their interaction with textile fibers. To achieve this, technological challenges in the production process, whether through recovery or bioconversion of renewable residues, must be addressed to enhance performance while reducing costs. As consumers increasingly choose environmentally friendly products, encouraging research and optimizing production, extraction, and dyeing techniques is essential to make these dyes both technically and economically viable. Many companies are making progress toward the wider adoption of renewable dyes and mordants in the textile industry. As a result, these materials will contribute not only to the development of aesthetically pleasing and functional textiles but also to a more sustainable future.

Acknowledgements All authors acknowledge financial support from the former Ministry of Science, Technology, and Innovationfrom Argentina for funding the project "Club de Biotecnología Agrícola y Ambiental" at Universidad Nacionalde Río Negro, Argentina (Project 103B/UNRN-40-A-1025). Access to scholarly resources provided by TheWikipedia Library of the Wikimedia Foundation is also acknowledged.

References

1. Phan K, Raes K, Van Speybroeck V, Roosen M, De Clerck K, De Meester S (2021) Non-food applications of natural dyes extracted from agro-food residues: a critical review. J Clean Prod 301:126920. https://doi.org/10.1016/j.jclepro.2021.126920
2. Ardila-Leal LD, Poutou-Piñales RA, Pedroza-Rodríguez AM, Quevedo-Hidalgo BE (2021) A brief history of colour, the environmental impact of synthetic dyes and removal by using laccases. Molecules 26:3813. https://doi.org/10.3390/molecules26133813
3. Pizzicato B, Pacifico S, Cayuela D, Mijas G, Riba-Moliner M (2023) Advancements in sustainable natural dyes for textile applications: a review. Molecules 28:5954. https://doi.org/10.3390/molecules28165954
4. Khattab TA, Abdelrahman MS, Rehan M (2020) Textile dyeing industry: environmental impacts and remediation. Environ Sci Pollut Res 27:3803–3818. https://doi.org/10.1007/s11356-019-07137-z
5. Slama HB, Chenari Bouket A, Pourhassan Z, Alenezi FN, Silini A, Cherif-Silini H, Oszako T, Luptakova L, Golińska P, Belbahri L (2021) Diversity of synthetic dyes from textile industries, discharge impacts and treatment methods. Appl Sci 11:6255. https://doi.org/10.3390/app11146255
6. Drumond Chequer FM, De Oliveira GAR, Anastacio Ferraz ER, Carvalho J, Boldrin Zanoni MV, De Oliveir DP (2013) Textile dyes: dyeing process and environmental impact. In: Gunay M (ed) Eco-friendly textile dyeing and finishing. InTech
7. Yousaf MA, Aqsa R (2023) Integrating circular economy, SBTI, digital LCA, and ESG benchmarks for sustainable textile dyeing: a critical review of industrial textile practices. Glob NEST J. https://doi.org/10.30955/gnj.005145
8. Umesh M, Suresh S, Santosh AS, Prasad S, Chinnathambi A, Al Obaid S, Jhanani GK, Shanmugam S (2023) Valorization of pineapple peel waste for fungal pigment production using *Talaromyces albobiverticillius*: insights into antibacterial, antioxidant and textile dyeing properties. Environ Res 229:115973. https://doi.org/10.1016/j.envres.2023.115973
9. Manian AP, Paul R, Bechtold T (2016) Metal mordanting in dyeing with natural colourants. Coloration Technol 132:107–113. https://doi.org/10.1111/cote.12199
10. Kumar A, Konar A (2011) Dyeing of textiles with natural dyes. In: Akcakoca Kumbasar E (ed) Natural dyes. InTech
11. Yang R, Li J, Cheng G, Inta A, Yang L (2023) Textiles dyeing with pomegranate (*Punica granatum*) peel extract using natural mordant. J Nat Fibers 20:2282056. https://doi.org/10.1080/15440478.2023.2282056
12. Yadav S, Tiwari KS, Gupta C, Tiwari MK, Khan A, Sonkar SP (2023) A brief review on natural dyes, pigments: recent advances and future perspectives. Results Chem 5:100733. https://doi.org/10.1016/j.rechem.2022.100733
13. Prabhu K, Bhute AS (2012) Plant based natural dyes and mordnats: a review. J Nat Prod Plant Resourc 2:649–664
14. Zhang Y, Jia Y, Cao J, Yu X, Zhang M (2023) Fabrication of bio-colored and functional wool using natural Pu'er tea extract. JFST 79:278–287. https://doi.org/10.2115/fiberst.2023-0025
15. Mota IGC, Neves RAMD, Nascimento SSDC, Maciel BLL, Morais AHDA, Passos TS (2023) Artificial dyes: health risks and the need for revision of international regulations. Food Rev Intl 39:1578–1593. https://doi.org/10.1080/87559129.2021.1934694

16. Hessel C, Allegre C, Maisseu M, Charbit F, Moulin P (2007) Guidelines and legislation for dye house effluents. J Environ Manage 83:171–180. https://doi.org/10.1016/j.jenvman.2006.02.012
17. Ragab M, Hassabo A, Othman H (2022) An overview of natural dyes extraction techniques for valuable utilization on textile fabrics. J Text Color Pol Sci. https://doi.org/10.21608/jtcps.2022.130253.1115
18. Rahman MM, Koh J, Hong KH (2024) Sustainable chitosan biomordant dyeing and functionalization of cotton fabrics using pomegranate rind and onion peel extracts. J Nat Fibers 21:2290856. https://doi.org/10.1080/15440478.2023.2290856
19. Bechtold T, Mussak R, Mahmud-Ali A, Ganglberger E, Geissler S (2006) Extraction of natural dyes for textile dyeing from coloured plant wastes released from the food and beverage industry. J Sci Food Agric 86:233–242. https://doi.org/10.1002/jsfa.2360
20. Ebrahim S, Mosaad M, Othman H, Hassabo A (2021) A valuable observation of eco-friendly natural dyes for valuable utilisation in the textile industry. J Text Color Pol Sci. https://doi.org/10.21608/jtcps.2021.97342.1090
21. Zhang Y, Zhou Q, Rather LJ, Li Q (2021) Agricultural waste of *Eriobotrya japonica* L. (Loquat) seeds and flora leaves as source of natural dye and bio-mordant for coloration and bio-functional finishing of wool textile. Indus Crops Prod 169:113633. https://doi.org/10.1016/j.indcrop.2021.113633
22. Ben Ticha M, Slama N, Dhouibi N, Boudokhane C, Dhaouadi H (2022) Valorization of *Juglans regia* L. bark residues as a natural colorant based on response surface methodology: a challenging approach to a sustainable dyeing process for acrylic fabrics. Sustainability 14:4134. https://doi.org/10.3390/su14074134
23. Elsahida K, Fauzi AM, Sailah I, Siregar IZ (2019) Sustainability of the use of natural dyes in the textile industry. IOP Conf Ser: Earth Environ Sci 399:012065. https://doi.org/10.1088/1755-1315/399/1/012065
24. Kapoor R, Ghosh P, Kumar M, Sengupta S, Gupta A, Kumar SS, Vijay V, Kumar V, Kumar Vijay V, Pant D (2020) Valorization of agricultural waste for biogas based circular economy in India: a research outlook. Biores Technol 304:123036. https://doi.org/10.1016/j.biortech.2020.123036
25. Fang J, Meng C, Zhang G (2022) Agricultural waste of Ipomoea batatas leaves as a source of natural dye for green coloration and bio-functional finishing for textile fabrics. Ind Crops Prod 177:114440. https://doi.org/10.1016/j.indcrop.2021.114440
26. Balachandran B, Sabumon PC (2023) Scope of natural dyes and biomordants in textile industry for cleaner production. In: Muthu SS (ed) Novel sustainable process alternatives for the textiles and fashion industry. Springer Nature Switzerland, Cham, pp 73–106
27. Verma S, Gupta G (2017) Natural dyes and its applications: a brief review. 4
28. Lagashetti AC, Dufossé L, Singh SK, Singh PN (2019) Fungal pigments and their prospects in different industries. Microorganisms 7:604. https://doi.org/10.3390/microorganisms7120604
29. Mouro C, Gomes AP, Costa RV, Moghtader F, Gouveia IC (2023) The sustainable bioactive dyeing of textiles: a novel strategy using bacterial pigments, natural antibacterial ingredients, and deep eutectic solvents. Gels 9:800. https://doi.org/10.3390/gels9100800
30. Lara L, Cabral I, Cunha J (2022) Ecological approaches to textile dyeing: a review. Sustainability 14:8353. https://doi.org/10.3390/su14148353
31. Benli H (2024) Bio-mordants: a review. Environ Sci Pollut Res 31:20714–20771. https://doi.org/10.1007/s11356-024-32174-8
32. Atakan R, Martínez-González I, Díaz-García P, Bonet-Aracil M (2023) Sustainable dyeing and functional finishing of cotton fabric by *Rosa canina* extracts. Sustainability 16:227. https://doi.org/10.3390/su16010227
33. Flores LF, Osorio Gonzalez CS, Saini R, Brar SK (2024) Renewable residues as feedstock for drop-in biofuel production. In: Brar SK, Osorio Gonzalez CS, Saini R, Soccol CR (eds) The microbiology of the drop-in biofuel production. Springer Cham
34. İşmal ÖE (2017) Greener natural dyeing pathway using a by-product of olive oil; prina and biomordants. Fibers Polym 18:773–785. https://doi.org/10.1007/s12221-017-6675-0

35. Haddar W, Baaka N, Meksi N, Elksibi I, Mhenni MF (2014) Optimization of an ecofriendly dyeing process using the wastewater of the olive oil industry as natural dyes for acrylic fibres. J Clean Prod 66:546–554. https://doi.org/10.1016/j.jclepro.2013.11.017
36. Blackburn RS, Houghton JA, Stenton M, Tidder A (2023) A dye–fibre system from food waste: dyeing casein fibres with anthocyanins. Coloration Technol cote 12718. https://doi.org/10.1111/cote.12718
37. Baaka N, Ksibi I, Mhenni MF (2016) Optimisation of the recovery of carotenoids from tomato processing wastes: application on textile dyeing and assessment of its antioxidant activity. Nat Prod Res 31:1–8. https://doi.org/10.1080/14786419.2016.1226828
38. Baaka N, Ben Ticha M, Haddar W, Hammami S, Mhenni MF (2015) Extraction of natural dye from waste wine industry: optimization survey based on a central composite design method. Fibers Polym 16:38–45. https://doi.org/10.1007/s12221-015-0038-5
39. Singhee D (2020) Review on natural dyes for textiles from wastes. In: Kumar Samanta A, Awwad NS, Majdooa Algarni H (eds) Chemistry and technology of natural and synthetic dyes and pigments. IntechOpen
40. Siberian Federal University, Krotova I, Boev V (2022) Natural dyes from coniferous plant debarking waste. Tr Ser Food Ind 1:406–420. https://doi.org/10.17516/2782-2214-0029
41. Santos J, Monteiro S, Oliveira S, Magalhães P, Magalhães FD, Martins JM, De Carvalho LH (2022) Application of forest by-products in the textile industry: dyeing with pine and eucalyptus bark extracts. In: IECF 2022. MDPI, p 30
42. Adeel S, Ali S, Bhatti IA, Zsila F (2009) Dyeing of cotton fabric using pomegranate (*Punica granatum*) aqueous extract. Asian J Chem 21
43. Mohan R, Geetha N, Jennifer DH, Sivakumar V (2020) Studies on natural dye (Pelargonidin) Extraction from onion peel and application in dyeing of leather. IJRES 7:12–16. https://doi.org/10.14445/23497157/IJRES-V7I1P103
44. Mirjalili M, Karimi L (2013) Extraction and characterization of natural dye from green walnut shells and its use in dyeing polyamide: focus on antibacterial properties. J Chem 2013:1–9. https://doi.org/10.1155/2013/375352
45. Pandey R, Patel S, Pandit P, Nachimuthu S, Jose S (2018) Colouration of textiles using roasted peanut skin- an agro processing residue. J Clean Prod 172:1319–1326. https://doi.org/10.1016/j.jclepro.2017.10.268
46. Shin Y-S, Son K-H, Yoo D-I (2008) Dyeing properties and color of silk fabrics dyed with safflower yellow dye. J Korean Soc Clothing and Textiles 32:928–934. https://doi.org/10.5850/JKSCT.2008.32.6.928
47. Yıldırım L, Erdem İşmal Ö (2019) Banana peel in dyeing of polyamide/elastane blend fabric. RJTA 23:124–133. https://doi.org/10.1108/RJTA-07-2018-0043
48. Karthikeyan MG, Vidya A (2020) Production and application of natural dye from skin of yellow pumpkin vegetable. Int J Recent Sci Res 11(3):37828–37839
49. Luo F, Lv Q, Zhao Y, Hu G, Huang G, Zhang J, Sun C, Li X, Chen K (2012) Quantification and purification of mangiferin from chinese mango (*Mangifera indica* L.) cultivars and its protective effect on human umbilical vein endothelial cells under H_2O_2-induced stress. Int J Mol Sci 13:11260–11274. https://doi.org/10.3390/ijms130911260
50. Bishal A, Ali KA, Ghosh S, Parua P, Bandyopadhyay B, Mondal S, Jana M, Datta A, Das KK, Debnath B (2023) Natural dyes. Its origin, categories and application on textile fabrics in brief. Eur Chem Bull 12(8):9780–9802
51. Sivakumar V, Anna JL, Vijayeeswarri J, Swaminathan G (2009) Ultrasound assisted enhancement in natural dye extraction from beetroot for industrial applications and natural dyeing of leather. Ultrason Sonochem 16(6):782–789. https://doi.org/10.1016/j.ultsonch.2009.03.009. Epub 2009 Mar 28. PMID: 19410496
52. Vankar PS, Shanker R, Wijayapala S (2009) Utilization of temple waste flower – Tagetes erecta for dyeing of cotton, wool and silk on industrial scale. J Textile Apparel Tech Manag 6:1–15
53. Ahmed T, Toki GFI, Mia R, Islam SR, Al Fahad MdA, Adeel S, Alpár T (2024) New trends in printing applications of natural dyes and pigments. In: Renewable dyes and pigments. Elsevier, pp 139–163

54. Pinheiro MNC, Symochko L, Castro LM (2023) Valorization of cork industry by-products as sustainable natural dyes for textiles. ACS Sustain Chem Eng 11:10555–10565. https://doi.org/10.1021/acssuschemeng.3c02576
55. Silva PMDS, Fiaschitello TR, Queiroz RSD, Freeman HS, Costa SAD, Leo P, Montemor AF, Costa SMD (2020) Natural dye from *Croton urucurana* Baill. bark: extraction, physicochemical characterization, textile dyeing and color fastness properties. Dyes Pigments 173:107953. https://doi.org/10.1016/j.dyepig.2019.107953
56. Haji A (2010) Functional dyeing of wool with natural dye extracted from *Berberis vulgaris* wood and *Rumex hymenosepolus* root as biomordant. Iran J Chem Chemical Eng-Int English Edn 29:55–60
57. Kushwaha A, Singh S, Chaudhary K (2023) Eco-friendly multifunctional dyeing of pineapple using Nyctanthes Arbortristis dye and Acacia Nilotica bio-mordant. Sustain Chem Pharm 34:101146. https://doi.org/10.1016/j.scp.2023.101146
58. Rather L, Khan M, Mohammad F (2017) Biomordanting potential of *Acacia nilotica* (Babul) in conjunction with *Kerria lacca* and *Rheum emodi* natural dyes. J Nat Fibers. https://doi.org/10.1080/15440478.2017.1414657
59. Souissi M, Guesmi A, Moussa A (2018) Valorization of natural dye extracted from date palm pits (*Phoenix dactylifera*) for dyeing of cotton fabric. Part 2: optimization of dyeing process and improvement of colorfastness with biological mordants. J Clean Prod 204:1143–1153. https://doi.org/10.1016/j.jclepro.2018.08.325
60. Barkaat S, Mehboob M, Adeel S, Rehman F, Amin N, Habib N, Hosseinnezhad M (2023) Sustainable microwave-assisted extraction of santalin from red sandal wood powder (*Ptrecarpus santalinus*) for bio-coloration of mordanted silk fabric. Separations 10:118. https://doi.org/10.3390/separations10020118
61. Mariamma TR, Jose SK (2018) Dyeability of cotton fabric with banana stem extract. Int J Chem Phys Sci 7:103. https://doi.org/10.30731/ijcps.7.2.2018.103-108
62. Sabatini F, Lonardo S, Pinelli P, Camilli F, Zipoli G, Romani A (2014) UV protective properties of cotton and flax fabrics dyed with multifunctional plant extracts. Dyes Pigments 105:89–96. https://doi.org/10.1016/j.dyepig.2014.01.027
63. Anandhan M, Prabaharan T (2018) Environmental impacts of natural dyeing process using pomegranate peel extract as a dye. Int J Appl Eng Res 13:7765–7771
64. Tehrani M, Shahmoradi F, Beni Z, Rahimi M (2023) Extracted dyes' stability as obtained from spent coffee grounds on silk fabrics using eco-friendly mordants. Environ Sci Pollut Res 30:68625–68635. https://doi.org/10.1007/s11356-023-27157-0
65. Hosseinnezhad M, Gharanjig K, Jafari R, Imani H, Razani N (2021) Cleaner colorant extraction and environmentally wool dyeing using oak as eco-friendly mordant. Environ Sci Pollut Res 28:7249–7260. https://doi.org/10.1007/s11356-020-11041-2
66. Thakker A, Sun D (2022) Innovative plant-based mordants and colourants for sustainable application on cotton fabric. J Nat Fibers 19(16):14346–14364. https://doi.org/10.1080/15440478.2022.2064391
67. Karuppuchamy A, Annapoorani G (2020) Dyeing of banana/cotton union fabric with natural dye and mordants. Int J Appl Res 5(8):24–26
68. Hasan MU, Adeel S, Batool F, Ahmad T, Tang R-C, Amin N, Khan SR (2022) Sustainable application of *Cassia obovata*–based chrysophanic acid as potential source of yellow natural colorant for textile dyeing. Environ Sci Pollut Res 29:10740–10753. https://doi.org/10.1007/s11356-021-16447-0
69. Mansour R, Gannoun S, Mhenni MF (2016) Exploring the potential uses of *Vitis vinifera* L. leaves as raw material for textile dyeing without metal mordants. Fibers Polym 17:1621–1626. https://doi.org/10.1007/s12221-016-5033-y
70. Chirila L, Popescu A, Stanculescu IR, Cutrubinis M, Cerempei A, Sandu I (2016) Gamma irradiation effects on natural dyeing performances of wool fabrics. Rev Chim 67(12):2628–2633

71. Jahangiri A, Ghoreishian SM, Akbari A, Norouzi M, Ghasemi M, Ghoreishian M, Shafiabadi E (2018) Natural dyeing of wool by madder (*Rubia tinctorum* L.) root extract using tannin-based biomordants: colorimetric, fastness and tensile assay. Fibers Polym 19:2139–2148. https://doi.org/10.1007/s12221-018-8069-3
72. Hosseinnezhad M, Gharanjig K, Belbasi S, Saadati S, Saeb M (2018) The use of sumac as a natural mordant in green production of Iranian carpet. Fibers Polym 19:1908–1912. https://doi.org/10.1007/s12221-018-7961-1
73. Phromphen P (2023) Optimization of marigold flower dye using banana peel as a biomordant. J Nat Fibers 20:2153193. https://doi.org/10.1080/15440478.2022.2153193
74. Pinheiro L, Kohan L, Duarte LO, Garavello ME de PE, Baruque-Ramos J (2019) Biomordants and new alternatives to the sustainable natural fiber dyeings. SN Appl Sci 1:1356. https://doi.org/10.1007/s42452-019-1384-5
75. Prabhu KH, Teli MD (2014) Teñido ecológico utilizando tanino de tegumento de *Tamarindus indica* L. como mordiente natural para textiles con actividad antibacteriana. J Saudi Chem Soc 18:864–872. https://doi.org/10.1016/j.jscs.2011.10.014
76. Bhandari NL, Shrestha S, Bhandari G, Parajuli N, Silwal SB, Adhikari R (2021) Extraction of dye from *Castanopsis indica* for its use in textile dyeing and medicinal purpose with natural mordant. Egypt J Chem 64:6681–6690. https://doi.org/10.21608/ejchem.2021.48519.2991
77. Habib N, Adeel S, Ali F, Amin N, Khan SR (2021) Environmental friendly sustainable application of plant-based mordants for cotton dyeing using Arjun bark-based natural colorant. Environ Sci Pollut Res 28:54041–54047. https://doi.org/10.1007/s11356-021-14536-8
78. Zhou Q, Rather LJ, Ali A, Wang W, Zhang Y, Rizwanul Haque QM, Li Q (2020) Acabado bioactivo respetuoso con el medio ambiente de textiles de lana utilizando extractos ricos en taninos de residuos de sebo chino (*Sapium sebiferum* L.) / hojas caídas. Dyes Pigments 176:108230. https://doi.org/10.1016/j.dyepig.2020.108230
79. Özomay M (2023) Sustainable and environmental dyeing with MAUT method comparative selection of the dyeing recipe. Sustainability 15:2738. https://doi.org/10.3390/su15032738
80. Elkhateeb W, Elnahas MO, Daba G (2023) Wide range applications of fungal pigments in textile dyeing. In: Satyanarayana T, Deshmukh SK (eds) Fungi and fungal products in human welfare and biotechnology. Springer Nature, Singapore, pp 289–304
81. Das A, Das S, Das N, Pandey P, Ingti B, Panchenko V, Bolshev V, Kovalev A, Pandey P (2023) Advancements and innovations in harnessing microbial processes for enhanced biogas production from waste materials. Agriculture 13:1689. https://doi.org/10.3390/agriculture13091689
82. Paillié-Jiménez ME, Stincone P, Brandelli A (2020) Natural pigments of microbial origin. Front Sustain Food Syst 4:590439. https://doi.org/10.3389/fsufs.2020.590439
83. Siva R, Subha K, Bhakta D, Ghosh AR, Babu S (2012) Characterization and enhanced production of prodigiosin from the spoiled coconut. Appl Biochem Biotechnol 166:187–196. https://doi.org/10.1007/s12010-011-9415-8
84. Abirami S, Kumar SD, Murugan A (2022) Dyeing of jute fabrics with prodigiosin produced from sago waste and their applications. J Pure Appl Microbiol 16:147–156. https://doi.org/10.22207/JPAM.16.1.03
85. Dos Santos RA, Rodríguez DM, da Silva LAR, de Almeida SM, de Campos-Takaki GM, de Lima MAB (2021) Enhanced production of prodigiosin by *Serratia marcescens* UCP 1549 using agrosubstrates in solid-state fermentation. Arch Microbiol 203:4091–4100. https://doi.org/10.1007/s00203-021-02399-z
86. Nguyen VB, Wang S-L, Nguyen AD, Phan TQ, Techato K, Pradit S (2021) Bioproduction of prodigiosin from fishery processing waste shrimp heads and evaluation of its potential bioactivities. Fishes 6:30. https://doi.org/10.3390/fishes6030030
87. Ahmad WA, Yusof NZ, Nordin N, Zakaria ZA, Rezali MF (2012) Production and characterization of violacein by locally isolated *Chromobacterium violaceum* grown in agricultural wastes. Appl Biochem Biotechnol 167:1220–1234. https://doi.org/10.1007/s12010-012-9553-7
88. Aruldass CA, Rubiyatno R, Venil CK, Ahmad WA (2015) Violet pigment production from liquid pineapple waste by *Chromobacterium violaceum* UTM5 and evaluation of its bioactivity. RSC Adv 5:51524–51536. https://doi.org/10.1039/C5RA05765E

89. Gohil N, Bhattacharjee G, Gayke M, Narode H, Alzahrani KJ, Singh V (2022) Enhanced production of violacein by *Chromobacterium violaceum* using agro-industrial waste soybean meal. J Appl Microbiol 132:1121–1133. https://doi.org/10.1111/jam.15277
90. Silbir S, Goksungur Y (2019) Natural red pigment production by *Monascus purpureus* in submerged fermentation systems using a food industry waste: Brewer's spent grain. Foods 8:161. https://doi.org/10.3390/foods8050161
91. Asghari M, Jahadi M, Hesam F, Ghasemi-Sepro N (2021) Optimization of Monascus pigment production on date waste substrates using solid state fermentation. Appl Food Biotechnol 8:247–254. https://doi.org/10.22037/afb.v8i3.34278
92. Embaby AM, Hussein MN, Hussein A (2018) Monascus orange and red pigments production by *Monascus purpureus* ATCC16436 through co-solid state fermentation of corn cob and glycerol: an eco-friendly environmental low cost approach. PLoS ONE 13:e0207755. https://doi.org/10.1371/journal.pone.0207755
93. Venil CK, Dufossé L, Renuka Devi P (2020) Bacterial pigments: sustainable compounds with market potential for pharma and food industry. Front Sustain Food Syst 4:100. https://doi.org/10.3389/fsufs.2020.00100
94. Grewal J, Wołącewicz M, Pyter W, Joshi N, Drewniak L, Pranaw K (2022) Colorful treasure from agro-industrial wastes: a sustainable chassis for microbial pigment production. Front Microbiol 13:832918. https://doi.org/10.3389/fmicb.2022.832918
95. Lopes FC, Ligabue-Braun R (2021) Agro-industrial residues: eco-friendly and inexpensive substrates for microbial pigments production. Front Sustain Food Syst 5:589414. https://doi.org/10.3389/fsufs.2021.589414
96. Toki GFI, Sharif MdN, Hossen MdA, Rahman A, Mia R, Repon MdR, Sk MS, Almutairi TM, Hossain MK (2024) Sustainable coloration and analysis of cellulosic viscose fabric incorporating *Rosa rubiginosa* extraction and pre-mordanting approaches. Mater Today Commun 38:108068. https://doi.org/10.1016/j.mtcomm.2024.108068
97. Adeel S, Younis M, Mia R, Özomay M, Mirnezhad S, Imran M (2024) Environmentally friendly greener approach toward sustainable application of amaltas brown pod-based natural colorant for wool yarn dyeing. J Eng Fibers Fabr 19:15589250241235652. https://doi.org/10.1177/15589250241235653
98. Kamboj A, Jose S, Singh A (2022) Antimicrobial activity of natural dyes—a comprehensive review. J Nat Fibers 19:5380–5394. https://doi.org/10.1080/15440478.2021.1875378
99. Xia W, Li Z, Tang Y, Li Q (2023) Sustainable recycling of café waste as natural bio resource and its value adding applications in green and effective dyeing/bio finishing of textile. Sep Purif Technol 309:123091. https://doi.org/10.1016/j.seppur.2022.123091
100. Dulo B, De Somer T, Phan K, Roosen M, Githaiga J, Raes K, De Meester S (2022) Evaluating the potential of natural dyes from nutshell wastes: sustainable colouration and functional finishing of wool fabric. Sustain Mater Technol 34:e00518. https://doi.org/10.1016/j.susmat.2022.e00518
101. Zannat A, Uddin MN, Mahmud ST, Mia R, Ahmed T (2024) Natural dyes and pigments in functional finishing. In: Islam S (ed) Renewable dyes and pigments. Elsevier, pp 271–287
102. Bhandari NL, Pokhrel B, Bhandari U, Bhattarai S, Devkota A, Bhandari G (2020) An overview of research on plant based natural dyes in Nepal: scope and challenges. J Agric Nat Res 3:45–66. https://doi.org/10.3126/janr.v3i2.32328
103. Fonseca FD, Symochko L, Pinheiro MNC (2024) Grape pomace (*Vitis vinifera* L.) waste valorization: assessing its potential as a sustainable natural dye for textiles applications. Sustainability 16:3167. https://doi.org/10.3390/su16083167
104. Provin AP, Dutra ARDA, De Sousa E, Silva Gouveia ICA, Cubas EALV (2021) Circular economy for fashion industry: use of waste from the food industry for the production of biotextiles. Technol Forecast Soc Chang 169:120858. https://doi.org/10.1016/j.techfore.2021.120858
105. Leal Filho W, Dinis MAP, Liakh O, Paço A, Dennis K, Shollo F, Sidsaph H (2024) Reducing the carbon footprint of the textile sector: an overview of impacts and solutions. Text Res J. https://doi.org/10.1177/00405175241236971

106. To MH, Uisan K, Ok YS, Pleissner D, Lin CSK (2019) Recent trends in green and sustainable chemistry: rethinking textile waste in a circular economy. Curr Opinion Green Sustain Chem 20:1–10. https://doi.org/10.1016/j.cogsc.2019.06.002
107. Volf I, Popa VI (2018) Integrated processing of biomass resources for fine chemical obtaining. In: Biomass as renewable raw material to obtain bioproducts of high-tech value. Elsevier, pp 113–160
108. Archroma (2024). https://www.archroma.com/
109. Ama Herbal (2024). https://amaherbal.com/
110. PILI (2024). https://www.pili.bio/. Accessed 10 May 2024
111. Woolmark (2024). https://www.woolmark.com/
112. IndiDye (2024). https://www.indidye.com/
113. DyeRecycle (2024). https://www.dyerecycle.com/
114. Sodhani Biotech (2024). https://naturaldyes.io/
115. SAGES (2024). https://www.sageslondon.com
116. Colorifix (2024). https://colorifix.com/
117. Rossi T, Silva PMS, De Moura LF, Araújo MC, Brito JO, Freeman HS (2017) Waste from eucalyptus wood steaming as a natural dye source for textile fibers. J Clean Prod 143:303–310. https://doi.org/10.1016/j.jclepro.2016.12.109
118. Periyasamy AP, Militky J (2020). Sustainability in textile dyeing: recent developments. In: Muthu S, Gardetti M (eds) Sustainability in the textile and apparel industries. sustainable textiles: production, processing, manufacturing & chemistry. Springer, Cham, pp 37–79

Chapter 7
Chromatography of Nature: Bio-Mordants and the Evolution of Sustainable Textile Coloration

Rezaul Karim Khan Alave, Syeda Fariha Jannat Duti, K M Noman Al Hasan, Rony Mia, Gazi Farhan Ishraque Toki, and Taosif Ahmed

Abstract This chapter focuses on the importance of bio-mordants, exploring their significance in reshaping the landscape of sustainable textile coloration. Bio-mordants, nature's alternative to traditional chemical mordants, are introduced and defined, unraveling their unique role as catalysts in fabric dyeing processes. The chapter highlights various sources of bio-mordants, elucidating their diverse properties and introducing them into the intricate chemical structures that distinguish them. A detailed exploration of the extraction processes employed to harness these bio-mordants from natural sources adds depth to the understanding of their sustainable origins. The application of bio-mordants in textile coloration takes center stage, as the chapter navigates through their use on different fabrics, encompassing both natural and synthetic fibers. The performance of bio-mordants in the context of natural dyeing is thoroughly investigated, showcasing their efficacy in achieving vibrant and eco-friendly colors. A crucial aspect explored in this chapter is the underlying mechanism behind the interaction between bio-mordants and fabric coloration. This intricate relation between natural catalysts and textile fibers is unraveled, shedding

R. K. K. Alave
Shanghai Frontier Science Research Center for Modern Textiles, College of Textiles, Donghua University, Shanghai, People's Republic of China

S. F. J. Duti
National Institute of Textile Engineering and Research, University of Dhaka, Dhaka, Bangladesh

K. M. Noman Al Hasan
School of Textile Science and Engineering, Zhejiang Sci-Tech University, Hangzhou, People's Republic of China

R. Mia (✉)
Department of Textile Engineering, Chemistry and Science, Wilson College of Textiles, North Carolina State University, Raleigh, NC, USA
e-mail: mroni_mia@yahoo.com

G. F. I. Toki · T. Ahmed
Nanotechnology Center, School of Fashion and Textiles, The Hong Kong Polytechnic University, Kowloon, Hong Kong

M. Shahid et al. (eds.), *Advancements in Textile Coloration*,
https://doi.org/10.1007/978-981-96-5091-0_7

light on the dyeing processes that define the success of bio-mordants in achieving sustainable and aesthetically pleasing outcomes. Therefore, this chapter not only defines and explores the properties of bio-mordants but also provides a comprehensive understanding of their extraction methods, application across diverse fabrics, and the intricate mechanisms that underpin their role in the evolution of sustainable textile coloration.

Keywords Bio-mordant · Natural dyes · Natural mordant · Green production · Sustainability

7.1 Introduction

The dyeing of textiles has seen various advances, particularly in the last several years. The utilization of natural dyestuff derived from plants is the first step in these developments [1]. Dyeing procedures involving natural dyestuffs sometimes involve the use of metal salts, also known as mordants [2, 3]. During the dyeing process, a mordant attaches dyestuff to fabrics or materials. It creates a complex of coordination with the colorant, which the dyestuff binds. The mordant function in adhering the dye to the substance is reflected in the term's origin from the Latin term "mordere," meaning "to bite" [4]. To stick to fibers, mordants first create a coordination combination with dyes. By strengthening the dye's binding to the fabric, this method prolongs the dye's strength and makes it more colorfast [5, 6]. Before synthetic dyes were invented, bio-mordants were commonly used to color nature-originated fibers, including linen, wool, silk, and cotton. However, as synthetic dyes were invented, their use have reduced. In recent years, interest in bio-mordants has been reignited as a result of increasing demands on manufacturers to offer more eco-friendly alternatives to petrochemical dyes. The use of heavyweight metal salts, for example, mordants, to impart hue in addition to ensuring color fastness, is necessary because many bio-mordants are not very suitable for use in textiles. When dyeing with natural materials, there are three options: simultaneous/meta-, post, and pre-mordanting [7]. Natural dyes can be classified as either substantive or non-substantive, depending on their intended application. Substantive colors, such as indigo, orchil, and turmeric, do not require any sort of fabric pretreatment. On the other hand, non-substantive dyestuffs like fustic, cochineal, madder/alizarin, or logwood color materials that have already been mordanted, or add color to a dyebath that already contains a mordant. There are three distinct types of pretreatments: direct (e.g., turmeric, safflower), acid (e.g., saffron, lac), and basic (e.g., berberine) [8]. Some natural mordant colorants, for instance, alizarin, logwood, fustic, and cochineal, are monogenetic, meaning they give the same color no matter the mordant. Conversly, some polygenetic mordants exhibits the capability to produce a variety of colors [8]. The chemical structures of natural colors allow us to classify them into several broad categories. These include: saffron, annatto, barberry, henna, sandal, alroot, favonoid, benzoquinone, henna, French marigold, and oyam [9, 10].

Similar to metallic mordants, bio-mordant compounds are utilized for the same purpose. So far, various bio-mordant varieties have been utilized, including oak galls, timber from oak, thuja, pomegranate rinds, sumach, guava, gallnut, myrobalan, rosemary, tannin, chlorophyll, banana leaves, tannic acid, valex, amla, milk, urine, natural lodhra, roots of *Rumex hymenosepalus*, green tea, chitosan, lemon juice, black tea, eggshell, tartaric acid, etc. [11]. When bio-based mordants interact with textiles like wool, silk, and cotton, strong biomolecules of bio-mordants enhance these textiles functional qualities [12]. Because of the additional hydrogen bond, these molecules give the material firm and stable tones when added to the material before or after dyeing. Furthermore, ion transfer within the system is responsible for the formation of a unique binding between the functional sites of the substance (−OH = cellulose; −NHCO = wool or silk) besides the binding site in the molecule, which in turn contributes to strong, non-fading bonds [13, 14]. Bio-mordants are safer for the environment than synthetic ones because they originated naturally, are non-toxic, and sustainable. In addition, for the most part, bio-mordants aren't harmful to people or the environment, comparatively, heavy metal mordants are dangerous [15].

For many years, researchers studying natural dyes have relied on the Acacia shrub by means of a natural mordant. Mansour et al. [16] used Acacia extraction to serve as a natural mordant as well as *Sticta coronata* for a bio-based dyestuff to color silk garments. According to the findings, silk fabric exhibits a beautiful shade of violet when dyed with *Sticta coronate* [16]. Turmeric, gallnut, and pomegranate peel tannin extracts were used as natural mordants to dye wool in an alternative investigation. Bio-mordants can substitute metal salts in the coloring of wool, according to recent findings, and their color strength and fastness characteristics are identical to those of metallic mordants. Natural-derived mordants similarly yield a different color range than projected, which is an interesting development [17]. According to a study by Kushwaha et al. [18], the use of babul bark (*Acacia nilotica)* to act by means of natural mordant increased the color yield (K/S) while dyeing pineapple fabric using coral jasmine natural dye. Both its rubbing and wash fastness were excellent. The material specimen that included the maximum amount of dye and mordant-5% of babul shell and 20% owf coral jasmine-exhibited exceptional UV protection, the antioxidant activity of up to 99.37%, and antibacterial activity of up to 98.23%. The bio-mordant, dye supply, and fiber are all part of a complicated framework. Some research has employed the bio-mordant properties of citric acid, which is present in lemon juice or peel. Another study examined the silk material dyeing capabilities of Parijat (*Nyctanthes arbor-tristis*) flowers using metal mordants in addition to natural mordants such as pomegranate, tannin, and lemon juice [19]. Many studies have investigated the possibility of using onion peel as a natural color. Still, some people have had success using onion peel as an organic mordant. Amutha et al. [20] found that some plants, including the leaves of banana, can be utilized as organic dyes to color cotton. While Onions in addition to pomegranate rinds, can be utilized as bio-mordants. Reportedly, appropriate color fastness has been attained in dyeing textiles, bio-mordants are vital because they offer an eco-friendly and non-toxic alternative. To form a strong binding, bio-mordants engage with the dye molecules as well as the fibers of the cloth [20]. This connection makes sure the dye sticks to the fabric better, which makes the color last longer and is less likely to fade. Biological materials, including plants and minerals, constitute the basis of bio-mordants. Because of this, the dyeing process is less harmful to the environment than when synthetic chemicals are used.

The main focus of this chapter is to explore the concept of using bio-mordants in sustainable textile coloration, under the theme of "Chromatography of Nature". The chapter aims to provide an initial understanding of this topic to present a comprehensive summary of the many types of mordants, together with their applications and limitations of chemical mordants, within this context. The following section briefly introduces bio-mordants, including their categorization and source. Moreover, it focuses specifically on bio-mordants and their role in advancing sustainable textiles. Finally, examine many advantageous examples of bio-mordants that possess chemical makeup.

7.2 Mordant

Dye itself is less important than a mordant. The ideal bulk mordant would also have a reasonable price point, generate a respectable color output under realistic dyeing conditions, and not negatively impact the fiber's physical characteristics or the dyes' fastness [21]. In addition, neither the processing nor the use of the colored textile material should be associated with any toxic or carcinogenic effects. Accordingly, a mordant is a material strongly adsorbed to the fiber and dye. Adjective hues dye the mordant, not the fiber, according to purists; however, no one ever says this because the difference is so subtle and can only be seen under a microscope [22, 23]. A further chemical link can be formed between the natural colorants and a mordant, which can be bonded to the fiber. By aiding in the absorption and fixation of natural dyes, it enhances the fastness characteristics of dyed materials by preventing bleeding along with the fading of colors. A variety of methods can lead to the formation of this complex, including pre-mordanting, post-mordanting, in addition with meta mordanting [24]. In the former two cases, the mordant is applied before and after the dyeing process, whereas in the latter one, the dye and mordant are applied simultaneously [25].

7.2.1 *Types of Mordants*

The choice of mordant has an impact on both the resulting shade of the dye and the dye's ability to resist fading. Generally, there are two types of mordant, they are bio-mordant and chemical-mordant.

7.2.1.1 Bio-Mordants

Plants that contain metals, tannin, tartaric acid, and tannic acid are bio-mordants. According to reports, they offer adequate dyeing and fastness qualities while being bio- and environmentally-friendly alternatives to metal mordants [26]. Tannin, metals, tannic acid, tartaric acid, that contain plants, biowastes, and by-products are all examples of bio-mordants in this setting. A vast number of plants contain tannin, an astringent vegetal component. Some of the many components of plants are their bark, wood, fruits, pods, leaves, roots, and galls. Tannin is a phenolic hydroxyl group-containing polyphenolic substance with a high molecular weight (500–3000) that occurs naturally and can crosslink proteins along with other macro-molecules when exposed to water. At 220 °C, tannin melts, and its chemical formula is $C_{76}H_{52}O_{46}$. Its molecular weight is 1701.22 [27]. Natural dyeing using biowastes, biological materials, and leftovers from many industries (e.g., food, beverage, lumber, and agriculture) has recently been the subject of extensive investigation. There are a lot of studies being conducted to find ways to make these wastes useful and environmentally safe [28]. According to reports, they offer adequate dyeing and fastness qualities while being bio- and environmentally-friendly alternatives to metal mordants. This view highlights the fact that natural colors are necessarily more safer or better for the environment than their synthetic counterparts [29].

7.2.1.2 Chemical Mordants

In common parlance, "natural" implies totally risk-free and environmentally benign. This view ignores the fact that natural colors aren't necessarily safer or better for the environment than their synthetic counterparts. Before making a definitive statement about the eco-friendliness of a product or process, it is required to do a lifetime assessment, a method that studies the environmental characteristics and the potential effects throughout the product's life [30, 31]. Metallic mordants are frequently considered to be environmentally harmful. On the other hand, the mordant's concentration is key. To meet stringent environmental standards, it is necessary to calculate how much mordant is deposited onto the fabrics and how much ends up in the wastewater [29]. Alternatives to traditional metallic mordants have been the subject of several recent investigations. The following metallic mordants have recently been found: manganese sulfate ($MnSO_4{\cdot}5H_2O$), magnesium sulfate ($MgSO_4{\cdot}7H_2O$), stannic chloride ($SnCl_4{\cdot}5H_2O$), aluminum sulfate ($Al_2(SO_4)_3$, nickel sulfate ($NiSO_4{\cdot}6H_2O$), cobalt sulfate ($CoSO_4{\cdot}6H_2O$), stannous sulfate ($SnSO_4$), ferric chloride ($FeCl_3$), ferrous chloride ($FeCl_2$), zinc sulfate ($ZnSO_4{\cdot}7H_2O$), and stannous sulfate ($SnSO_4$) are newly found metallic mordants [32, 33]. Metal mordants are further classified into two categories: dulling and brightening. Copper sulfate and ferrous sulfates are dulling mordants, while alum, potassium dichromate, and tin (stannous chloride) are brightening mordants [34].

7.2.2 Types of Mordanting

Mordants can be applied in one of three ways, depending on when they are to be used. Figure 7.1 shows the schematic diagram of the different mordanting processes.

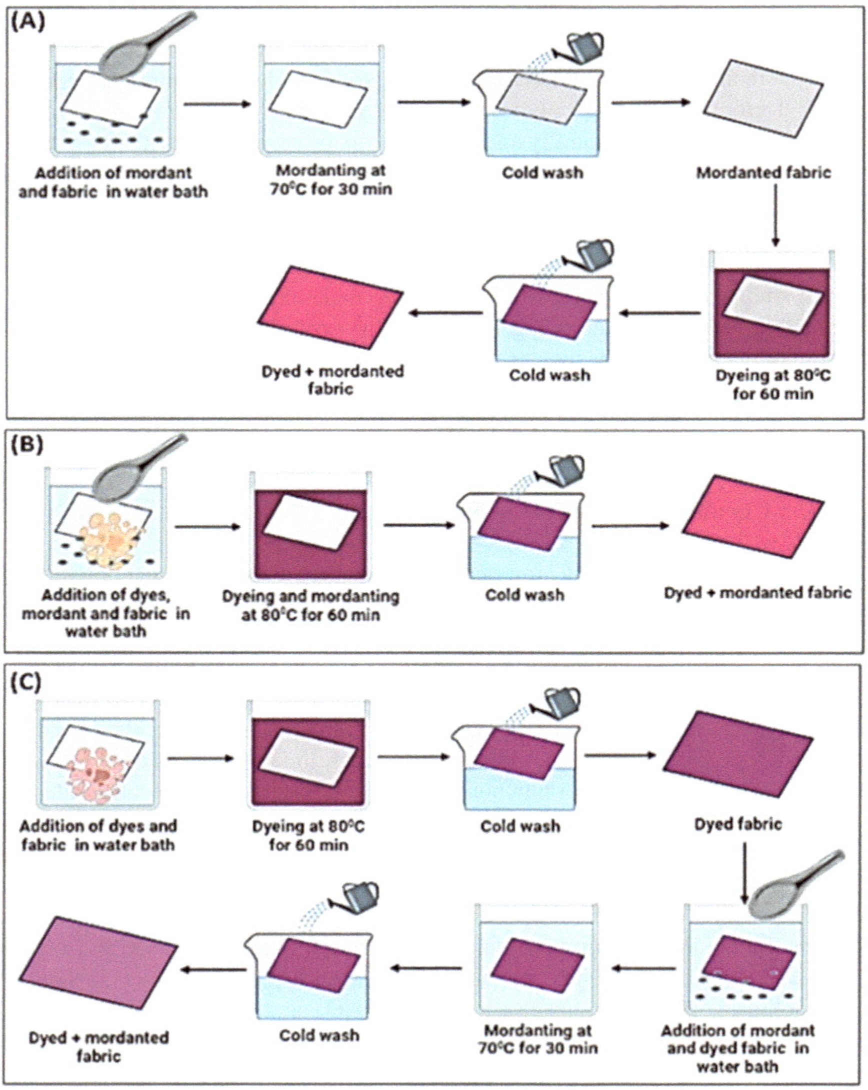

Fig. 7.1 Dyeing process flowchart showing **a** pre-mordanting, **b** meta-mordanting, and **c** post-mordanting steps. Republished with permission from Springer [21]. Copyright, Springer 2024

- Premordanting
- Metamordanting or simultaneous mordanting
- Postmordanting

7.2.2.1 Premordanting

The term "pre-mordanting" describes exactly what it is: applying mordants to the fabric before dyeing it. This is a regular occurrence since cotton and cellulosic fibers do not react well to many natural dyes when they are un-mordanted. This sort of mordanting is necessary for producing beautiful hues even when using natural dyes on animal fibers, such as cochineal. 'Kalamkari' from Andhra Pradesh and 'Sanganeri' from Rajasthan are two well-known traditional Indian printing techniques that use this mordanting procedure and natural dyes [23]. The pre-mordanting processes are schematically shown in Fig. 7.1a.

7.2.2.2 Metamordanting or Simultaneous Mordanting

The meta-mordanting technique, also known as simultaneous mordanting, involves dyeing and mordanting in the same bath. Adding mordant to the dye solution at the beginning of the dyeing process allows for the simultaneous dyeing and mordanting of cotton and cellulosic [35]. Wool and other animal fibers may require additional mordant added to the dye solution near the conclusion of the dyeing process after the color has been fully absorbed into the fabric. This method shortens the dyeing time by reducing the number of processes. For some dyes, this procedure results in deeper hues, but for others, the color yield might be lower owing to dye and mordant loss during the creation of dye-mordant complexes in the dye bath, which could lead to uneven dyeing. This method is best suited for smaller batches because the mordant cannot be recycled [23]. The meta-mordanting processes are schematically shown in Fig. 7.1b.

7.2.2.3 Postmordanting

A distinct bath is used to treat the fabric with a mordant in the post-mordanting technique following the dyeing process. In the last stage, the color is refined to perfection. This is a common method for applying iron salts to create the colors grey and black. Immersing the colored material in a mordant solution is a common step in the post-mordanting process. The dyed fabric's final color and characteristics are affected by the soak time and mordant solution concentration. The last step is to create a shade of color, different fabrics, dyes, and intended color results can affect how well post-mordanting works [36, 37] The post-mordanting processes are schematically shown in Fig. 7.1c.

7.2.3 Advantages of Bio-Mordant

The use of bio-mordants is better for the environment than that of synthetic ones. The creation of natural colorants does not require the use of harmful chemicals because they exist naturally. It is possible to employ some natural dyes as both colors and antimicrobial coatings in textiles because of their antibacterial activity [38–42]. Additionally, when applied to fabrics, some natural colorants absorb ultraviolet (UV) light, which means they shield the skin from harmful rays. Natural colorants have been the subject of extensive research due to their potential as antibacterial and UV protective coatings. Cotton knits treated with natural colors to prevent sun damage and eliminate microorganisms were detailed in a study by Ibrahim et al. [43]. Even after multiple washes in the washing machine, the results showed that the UV defense and antimicrobial effects remained. Rheum emodi, Gardenia yellow, and curcumin dyes were the subject of a comparable investigation by Zhou et al. [44] on silk fabric. All three of the natural dyes tested protected the silk cloth from ultraviolet light and killed microbes, according to the results. When consumed, certain bio-mordants, including carmine, which is present in lipsticks, do not pose any health risks [45]. When compared to synthetic mordants, bio-mordants are superior due to their many useful properties. Their environmental friendliness, low risk, affordability, compatibility with natural dyes, and backing of traditional methods make them ideal for eco-friendly textile coloring [21, 46]

7.2.4 Disadvantages of Chemical Mordant

There are still certain issues with chemical mordants, even though they are more technologically advanced and cost-effective than natural mordants. Chemical mordants are made using resources that are harmful to the environment and cannot be replenished [47]. Until non-renewable hydrocarbon resources become available, synthetic colorants will not have a future. It is believed that azo dyes and other synthetic colorants contain metabolites that cause cancer. In addition, a lot of energy is needed during the creation of synthetic colorants [48]. Toxic chemical mordants include tin, copper, and chromium. Serious health complications, including skin irritation, respiratory problems, cancer, and organ damage, can result from prolonged exposure to these compounds. Serious water contamination can result from the discharge of waste materials that include chemical contaminants into bodies of water. Contaminating drinking water supplies and harming marine life are two potential outcomes of these chemicals. Pollution from chemical mordants persists for a long time since they are not biodegradable. Ecosystems might be permanently damaged by their tenacity [49, 50].

7.3 Sources of Bio-Mordant

Bio-mordant agents can be derived from a wide variety of herbal sources. Colors are fixed onto fabrics, color fastness is improved, and hue is occasionally enhanced by using bio-mordants, which are natural chemicals produced from animals, plants, and microorganisms [51]. If it's concerned about the impact on human health and the environment caused by synthetic mordants, it may want to look into bio-mordants as an alternative [52]. The biological sources of these natural mordants vary, yet they all contribute something special to the dyeing process. Sources of bio-mordant are animals, plants, and microalgae [53].

7.3.1 Plant Source

Some of the most commonly employed natural mordant ingredients in the natural coloring processes from 1993 to 2023, particularly in the last 5 years, were Acacia, lemon, pomegranate peel, myrobalan, tannic acid, Aloe vera, oak, turmeric, henna, and tamarind [54]. Also, scientists have found over 80 different plant sources that can be employed as bio-mordants in natural dye systems [26]. Figure 7.2 shows below mentioned bio-mordants from different plant sources.

7.3.1.1 Tannin

A variety of naturally producing polyphenols goes by the term "tannin," which comes from the French term "tanin" meaning "tanning substance" [55]. When describing the method of turning animal skins into leather using extracts from various plant species, Seguin initially introduced the name "tannin" in 1796. A vast number of plants contain tannin, an astringent vegetal component. Some of the many components of plants are their bark, wood, fruits, pods, leaves, roots, and galls. Tannin is a phenolic hydroxyl group-containing polyphenolic substance with a high molecular weight (500–3000) that occurs naturally and can crosslink proteins along with other macromolecules when exposed to water. Tannin has a melting point of 220 °C, a molecular weight of 1701.22, and the chemical formula $C_{76}H_{52}O_{46}$ [27]. The term tannin does not refer to a specific chemical. They encompass a vast category of organic compounds that exhibit a wide range of variations in chemical make-up and interactions. Tannins serve a protective purpose in barks and heartwoods of plants, as a result of their protein precipitation nature, they inhibit several enzymes. Tannins can find in many different plant species; some examples include, treripod (65%), chestnut (30%), myrobalan (30–35%), tara (43–51%), and gall nuts (50–70%) [56, 57].

Fig. 7.2 Sources of plant-based bio-mordant, **a** leaves of acacia (Adapted from: Wikimedia Commons, https://commons.wikimedia.org/wiki/File:Acacia_-_leaf_detail_%286652577893%29.jpg), **b** acorns (Adapted from: Wikimedia Commons, https://commons.wikimedia.org/wiki/File:Quercus_robur_acorns_in_Tuntorp_1.jpg), **c** *Eurya acuminata* (Adapted from: Wikimedia Commons, https://commons.wikimedia.org/wiki/File:Eurya_japonica_1004.JPG), **d** *Entada spiralis* (Adapted from: Flickr, https://www.flickr.com/photos/reulim/49915367416), **e** flower of *Peganum harmala* L. (Adapted from: Wikimedia Commons, https://commons.wikimedia.org/wiki/File:Peganum_harmala_20140604.jpg), **f** flower of *Justicia carnea* L. (Adapted from: Wikimedia Commons, https://commons.wikimedia.org/wiki/File:Justicia_carnea_Jakobinia_r%C3%B3%C5%BCowawa_2023-10-31_02.jpg), **g** Manjakani (*Quercus infectoria*) (Adapted from: Wikimedia Commons, https://commons.wikimedia.org/wiki/File:Quercus_infectoria_ssp_veneris_kz2.jpg), and **h** leaves of *Musa basjoo* (Adapted from: Wikimedia Commons, https://commons.wikimedia.org/wiki/File:Musa_basjoo_13zz.jpg)

7.3.1.2 Acacia

More than 1,350 species of *Acacia* are known to exist in the world today. New research on the *Acacia* genus suggests that the species may be roughly classified into five distinct groups. It is not easy to classify and recognize *acacia* species. Hydrolyzable tannins, flavonoids, nonprotein amino acid terpenes (particularly diterpenes, essential oils, triterpene genins, phytosterol, in addition to saponins), gums, alkaloids, and amines have all been found in several *Acacia* species [58]. For many years, researchers studying natural dyes have relied on the *Acacia* shrub as a bio-mordant (Fig. 7.2a). Mansour et al. [16] used *Sticta coronata*, an organic source-derived dyestuff, and *acacia* extract, a natural mordant, to dye silk garments. The results show that the *Sticta coronata* color turned silk into a beautiful shade of violet. Wool was dyed using tannin extracts from gallnuts, pomegranate rind, and *Acacia arabica* bark in an alternative investigation. Rather et al. [17] reported that bio-mordants can replace metal salts in wool dyeing because they produce color gamuts that are quite different from what is expected of a mordant. Additionally, they are just as effective as conventional metal mordants in terms of fastness properties and color strength.

7.3.1.3 Acorns

The powdered acorns can be purchased at grocery stores or health food stores, or they can be found under oak trees during the autumn (Fig. 7.2b). To acquire the full-color intensity, make a powder from the ground, shelled acorns, and soak it in water for a few days. Colours ranging from pale beiges to greys to teal blues are created by acorns [23, 59].

7.3.1.4 Eurya Acuminata

Attenuate, serrulate, narrowly oblongelliptic leaves that measure 5–7.6 cm in length characterize the tall evergreen shrub *Eurya acuminata* (Fig. 7.2c). Its range includes the mountainous rain forests of Bangladesh in addition to Vietnam, Taiwan, China, Japan, Sri Lanka, Thailand, and India. Many compounds, such as ellagic acid, caffeine, anthrocyanins, favone, favonols, euryanoside, chrysoeriol, and β-D-glucopyranoside, were discovered in the chemical examination of many species of *Eurya* [60]. According to Vankar et al. [61], a bio-mordant and natural dye made of *Eurya acuminata* and *Rubia cordifolia* can be used to color cotton fabric in an eco-friendly way. Even the fastness characteristics yield beneficial results in this case, and the claim goes on to say that utilizing a bio-mordant increases the colorimetric data regarding dyeing and also improves the fastness qualities.

7.3.1.5 *Entada Spiralis* Ridl

A woody climber native to Malaysia, *Entada spiralis* Ridl (Leguminoceae) has a long history of medicinal and cosmetic uses, including the treatment of syphilis and insect bites, as well as for making shampoo and soap (Fig. 7.2d). According to Aboaba et al. [62], numerous studies have demonstrated that Entada sp. possesses antibacterial and antifungal properties for a variety of fungal infections. One research by Chairat et al. [63] looked at how an organic mordant solution made through the roots of *Entada spiralis* Ridl and an extract from dried-out mangosteen fruit peels (*Garcinia mangostana L.*) affected the coloring consequence of silk and cotton yarn. There is evidence that pre-mordanting colored cotton using the organic *Entada spiralis* mordant solution, as opposed to standard mordants, raised the K/S value.

7.3.1.6 Harmal (*Peganum Harmala* L.)

Plants of the glabrous, perennial Harmal (*Peganum harmala* L., group Zygophyllaceae) thrive on sandy soils, semi-arid climates, and steppes in the eastern Mediterranean (Fig. 7.2e). Based on analyses of the extracts' molecular structures, quinazoline alkaloids, and beta-carboline are the plant's most important components [64]. In a study conducted by Azeem et al. [65], natural dyes from logwood, zeera, harmal, and turmeric were employed as natural mordants for coloring woolen fabrics using ultrasonic energy. The claim is that fabrics colored with different natural mordants have significantly reddish, brighter, and yellowish colors.

7.3.1.7 *Justicia Carnea* L

The flowering plant *Justicia carnea* L. is native to several regions of Africa (Fig. 7.2f). According to Onyeabo et al. [66], the plant comprises chemical compounds, for instance, phenols, terpenoids, and flavonoids. Jabar et al. [13] studied the use of natural mordant substances to color cotton fabric using organic-derived dyestuff. The agents included *Vernonia amygdalina* leaves, *Moringa oleifera* L. bark, *Justicia carnea* L. leaves, and *Azadirachta indica* L. bark. According to reports, depending on the mordant used, various colors can be achieved, ranging from chocolate to yellow.

7.3.1.8 Manjakani (*Quercus Infectoria*)

According to Ab Rahman et al. [67], *Quercus infectoria* (Fig. 7.2g) has a long history of medicinal usage due to the belief that it contains significant levels of bioactive substances for example gallic acid, tannin, syringic acidic substances, the acid ellagic, P-sitosterol, amentofavone, as well as hexamethyl ether. In an additional investigation, the wool and silk textiles were dyed using myrrh (*Commiphora molmol*) extract as a natural dyestuff and manjakani (*Quercus infectoria*) and sumac (*Rhus coriaria*) as natural mordants. The results show that samples dyed with sumac are more vibrantly colored than those dyed using manjakani [68].

7.3.1.9 *Musa Basjoo* Leaf

The family Musaceae includes the *Musa basjoo* plant (Fig. 7.2h). *Musa basjoo* is known to contain several compounds that can interact with specific fabric macromolecules and dyestuff at the same time, including anthraquinone, cyanide, favonoids, phenols, amine, coumarins, and coumarin [69]. Traditional dyeing techniques have long made use of *Musa basjoo's* leaves, barking, and fruit rinds to color wool and silk. By the tannin found in *Musa basjoo* plants as a bio-mordant allows for the functional dyeing of wool and silk materials while also attaining ecological coloring [70].

7.3.1.10 Other Bark Sources

Some birches and willows, as well as the bark of some trees, including the European larch, Norway spruce, Picea abies, and Larix decidua, are occasionally tanned. Russian leather has a distinct aroma that comes from an essential oil found in birch bark, which is why the bark is so beloved there. Willow bark is good for tanning gloves because it makes light, supple, and malleable leather. The significance of certain tropical barks is low. The bark of the Western Australian eucalyptus tree, known as a mallet, has a tannin concentration ranging from 35 to 50% [71]. Crucial in India, the bark of the *Cassia auriculata* tree is known as Avaram. An orange-yellow dye found in the tanekaha bark of the New Zealand tree *Phyllocladus trichomanoides* makes it ideal for use as gloves [72].

7.3.2 Animal Source

Animals play a special function as natural colorants; mordants derived from animals are occasionally chosen over synthetic ones due to their lower toxicity and less environmental impact [73]. Figure 7.3 shows the bio-mordants from animal sources.

7.3.2.1 Cochineal

One of the oldest dyes used by the Aztecs and later by the Spaniards, cochineal was a replacement for Kermes dye. The Spanish called it *Grana cochininilla* because the two colors are structurally identical but the carminic acid content in cochineal is twelve times higher [74]. A bug called *Dactylopiuscoccus*, which feeds on prickly *coccus cacti* (*Nopaleacoccinellifera*), is responsible for producing this dye (Fig. 7.3a). The insects' dried carcasses are then used to make the dye. Central and South American countries where it is most common include Mexico and Brazil [75]. Other names for it include Indian fig, opuntia, and nopal. At present, though, Peru and the Canary Islands are the primary places to get this dye. The female insects that inhabit cacti feed on sap and deposit hundreds of eggs, each of which contains color in the form of a white fluid. Their skin is brownish-red in color. The structure provided

Fig. 7.3 Source of animal-based bio-mordant, **a** cochineal insect (Adapted from: Flickr, https://www.flickr.com/photos/vahemart/29330777632), **b** kermes insect (Adapted from: Wikimedia Commons, https://commons.wikimedia.org/wiki/File:Kermes_quercus_20160712_002.jpg), and **c** lac insect (Adapted from: Wikimedia Commons, https://commons.wikimedia.org/wiki/File:Lac_Insects_on_a_Babybonnet_Branch_-_Flickr_-_treegrow_(2).jpg)

shows that the fluid comprises glycerylmyristae (a fat), coccerin (cochineal wax), and 9–10% carminic acid as a colorant. Carminic acid's excellent fastness properties make it useful in foodstuffs, pharmaceuticals, and the coloring of natural fibers [76]. Due to its increased stability in the presence of light and oxidants, carminic acid does not degrade as easily over time. Colors such as pink, crimson, scarlet, and blue can be achieved by using different mordants on fabric.

7.3.2.2 Kermes

Ankle worms, or "kermes grains" as they are more commonly known, are tiny worms that resemble seeds from vegetables (Fig. 7.3b). Ingredients: Kermes bug powder and Kermes color. Various ancient civilizations in Europe, Asia, and Africa have relied on it for dyeing textiles [77]. Dry bodies of female scaled insects from the genus Kermes contain the red color. Both *Kermes ilicis* and *vermilio* are species of oak native to the Mediterranean and Asia; the former thrives on *Quercus scoccifera* and the latter on Durmast oak (*Quercus ilex*). In addition to feeding on oak sap, this insect also lives on oak leaves and stems. Once the mother has laid eggs, the eggs can be killed with vinegar fumes, and the color can be extracted by boiling the eggs in water. The colorant's yellowish-red hue is caused by the presence of kermesic acid, an aglycon of carminic acid. It dissolves in hot water. Still, it has flavokermesic acid, albeit in negligible amounts [78, 79]. The use of tin produces a scarlet red, while alum produces a scarlet with a bluer cast, and so on. Many mordants give it its many colors. A more affordable alternative to Tyrian purple, it is thought of as a dye.

7.3.2.3 Lac Insect

Another red dye, lac dye, has been utilized since prehistoric times and is one of the oldest insect dyes [80]. The Indian term "Lakh" is the origin of the English word "lac," and it depicts how hundreds of thousands of insects are used to make a tiny amount of dye. This dye can be found on the decaying branches and twigs of rain trees like *Samaneasaman* and *Pithecolobiumsaman* in Cambodia, India, Thailand, the Moluccas, and Sumarta. It is made from the resinous secretion of *Coccuslaccae*, which includes *Lacciferlacca* and *kerrialacca* (Fig. 7.3c). These trees are members of the *Tachardiidae* family. The discharge contains a sticky substance that, when extracted with water, forms a lac color. On mordanting with tin, it also provides a fiery crimson hue. Laccaic acid D is the only structure that has only one carboxyl component; all the others contain di-carboxylic acid. Wool and silk are dyed with this carboxyl compound because of its acidic character. Several fatty acids, erythrolaccin, terpenic acid, aleuritic acid, and other insoluble but minorly significant components are also present [81].

7.3.3 *Microalgae*

Microalgae are easily identifiable by the pigments they create through light harvesting, which gives them their distinctive color. As a component of their photosynthetic system, microalgae create pigments, which are secondary metabolites that add color and help with photosynthesis and photoprotection. There are three main categories into which pigments fall: phytoproteins, carotenoids, and chlorophylls [82–84]. Figure 7.4 shows the bio-mordants from microalgae sources.

7.3.3.1 Chlorophylls

As the primary green pigment involved in photosynthesis, chlorophyll (Chl) is abundant in both marine and terrestrial environments (Fig. 7.4a). Around 1.2 billion tonnes of chlorophyll are produced each year by plants, cyanobacteria, and algae [82, 84–86]. It is from their absorption spectra that it can identify the five primary classes of chl pigments: a, b, c, d, and f. Chlorophyll pigments often show their peak absorption in the 450–475 nm and 630–675 nm wavelength bands, respectively [83]. Color hues

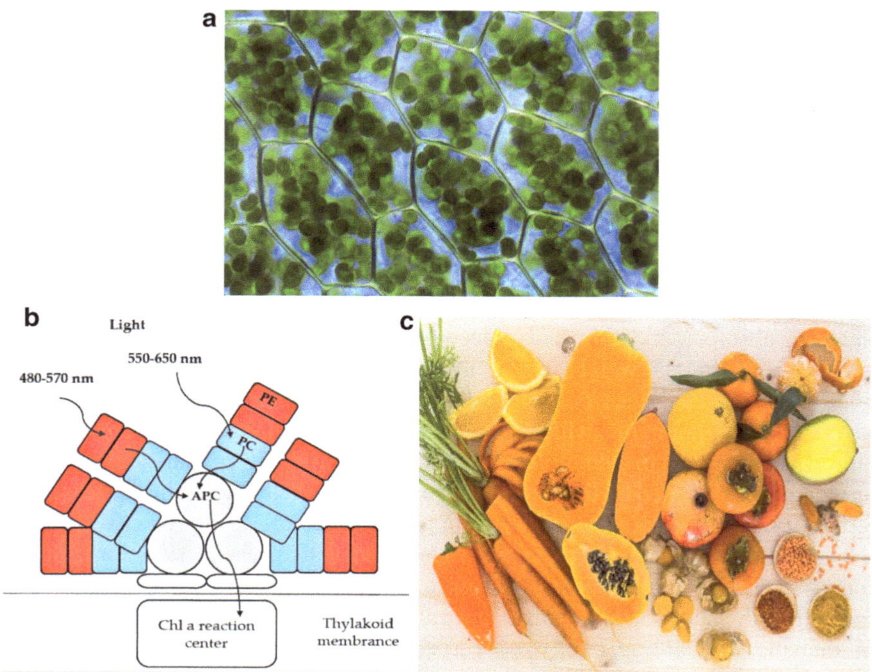

Fig. 7.4 Source of microalgae bio-mordant, **a** chlorophylls, **b** The role of phycobiliproteins in light harvesting as a schematic structure, Republished with permission from Frontiers in Pharmacology [89]. Copyright, CC-BY terms, 2021, and **c** sources of carotenoids

vary among chl pigments because of their unique absorption spectra. The pigments known as Chl a, b, c, d, and f vary in hue from blue green to brilliant green to yellow-green to forest green to emerald green. Various species of microalgae have various intracellular pigment compositions; however, chlorophyll-a pigment is present in the central reaction center of all microalgae groups.

7.3.3.2 Phycobiliproteins

Cyanobacteria and Rhodophyta create phycobiliproteins, which are protein-based auxiliary photosynthetic pigments that are hydrophilic, extremely luminous, and have vibrant colors [84, 87]. Within the cell, these pigments have assembled into phycobilisome complexes that link the chloroplast's thylakoid membranes [88]. Phycobiliprotein pigments work in tandem with chl-a pigment to improve light harvesting; their absorption spectra fall within the visible light spectrum, between 500 and 650 nm. Because of their non-toxicity and exceptional fluorescence capabilities, phytobiliproteins find employment in a wide range of industrial processes. These biliproteins serve as both a biological colorant and a chelating agent due to their molecular makeup. Their ability to neutralize reactive oxygen species allows them to lessen oxidative stress (Fig. 7.4b). Phycobiliproteins are therefore believed to have antioxidant characteristics.

7.3.3.3 Carotenoids

The yellow, orange, and red hues of carotenoids are a result of their status as lipophilic biogenic pigments. Even non-photosynthetic fungus and bacteria produce pigments belonging to this category [90, 91]. Photosynthetic organisms include plants and algae. The sources of carotenoids is shown in Fig. 7.4c. Carotenoids are essential for photosynthesis because they act as photoprotectors, allowing plants to absorb more light and reducing heat buildup [92]. Up to 8–14% of a microalgae biomass's weight can include carotenoids [93]. Carotenoids are classified as either carotenes or xanthophylls based on their chemical structure. In contrast to xanthophylls, which also include hydrogen, carotenes only contain carbon and hydrogen in their molecular structure. Conversely, primary and secondary carotenoids are defined by the functional qualities of carotenoids. The processes that occur during photosynthesis include primary carotenoids. As a protective strategy, cells produce secondary carotenoids when faced with stressful situations. Subsequently, secondary carotenoids possess a multitude of bioactivities that benefit human health, including antioxidant, anticancer, neuroprotective, anti-inflammatory, antidiabetic, and cardioprotective properties [93].

7.4 Examples of Bio-Mordants

Bio-mordants show that the textile sector can be a leader in sustainable and environmentally friendly operations by lowering the environmental footprint of dyeing without sacrificing color quality or durability [94].

*7.4.1 Mango (***Mangifera Indica** *L.) Tree Bark*

The mango tree, or *Mangifera indica* as it is officially named is shown in Fig. 7.5a, has an extensive past of remedial usage in both conventional and alternative medical systems. Around the globe, you may find hundreds of distinct mango varieties. The shape, size, flavor, flesh color, and skin tone, of mango fruits can differ from light yellow to gold to green to orange, all dependent on the cultivar [95].

7.4.1.1 Chemical Composition of *Mangifera Indica* L.

The plant's many biochemical constituents, together with polyphenols, triterpenoids, and flavonoids. The main bioactive components of mangiferin, a xanthone glycoside, are isomangiferin, tannins, and gallic acid derivatives shown in Fig. 7.5b [96].

7.4.1.2 Advantages of *Mangifera Indica* L.

Jayalakshmi et al. [53] employed ratanjot and manjistha as natural dyestuff to colorant woolen yarn, while mango peel and pomegranate were utilized as natural mordants. In another study, Wangatia et al. [97] investigated using mango tree bark as mordant

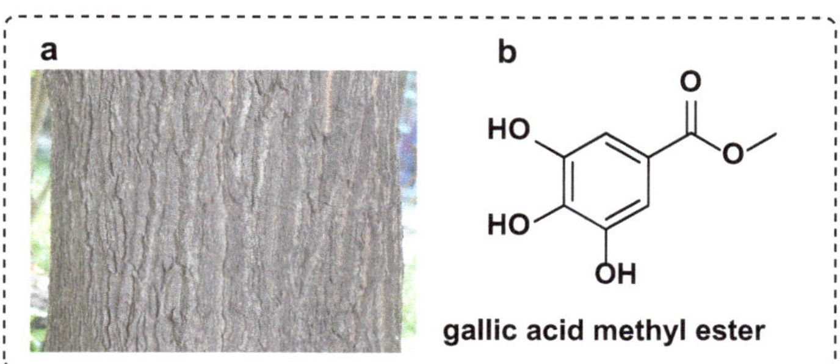

Fig. 7.5 *Indica Mangifera* L., **a** bark of the plant (Adapted from: Wikimedia Commons, https://commons.wikimedia.org/wiki/File:Mangifera_indica_0004.jpg) and **b** chemical composition as a biomordant

substances on cotton, both before and after the mordanting process, to produce environmentally friendly dyes. Using a natural dyestuff produced from bitter plants before mordanting produces weaker colors and less colorfastness than post-mordanting with mango. Mango peel is being used as an organic mordant by several researchers, including Bhandari et al. [98]. As natural mordants, they used aloe vera with mango bark extract. To color the cloth, they used the bark of *Castanopsis indica* as a dye. It has been shown that the colors that end up becoming brown can be affected by the sort of mordant that is employed.

*7.4.2 Papaya (*Corica Papaya *L.)*

Carica papaya is a member of the *Caricaceae* family and one of 21 species in the genus *Carica*; its fruit is known as a papaya, as shown in Fig. 7.6a. The initial region of domestication was Mesoamerica, which encompasses the present-day southern Mexican and Central American regions. Tropical regions of several nations are ideal for its cultivation. In 2022, Indian papayas accounted for 38% of the global total [99, 100].

7.4.2.1 Chemical Composition of *Corica Papaya* L.

Papain along with chymo-papain are two major active components of papaya, according to Silva et al. [101]. Along with phenolic compounds like rutin, erulic acid, as well as caffeic acid, papaya fruit contains carotenoids including lycopene, β-cryptoxanthin, as well as β-carotene [102] qre shown in Fig. 7.6b.

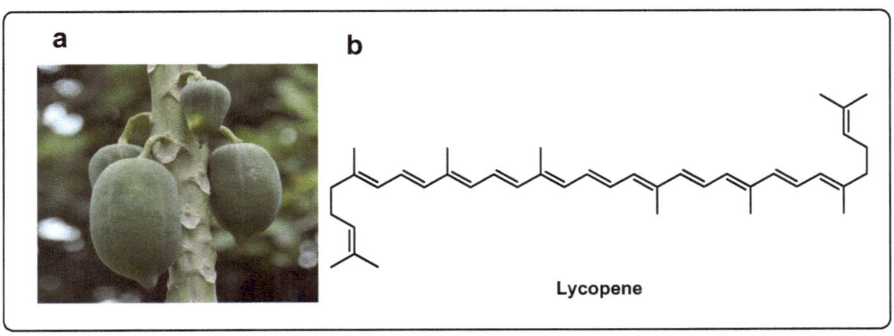

Fig. 7.6 Papaya (*Corica papaya* L.) **a** fruits (Adapted from: Freerange, https://freerangestock.com/photos/171070/papayas-growing-on-tree-in-nature.html), and **b** chemical composition

7.4.2.2 Advantages of *Corica Papaya* L.

Papaya is a safer alternative to synthetic mordants since it breaks down naturally and does not contain any hazardous or heavy metals. A study conducted by Rosyida et al. [103], discovered that metallic mordants, papaya fruit extract (a bio-mordant agent), and jackfruit wood extract may be used to color cotton garments. Green, purple, and grey color tones with sufficient fastness attributes may supposedly be created with the right mordant chemicals [104].

7.4.3 *Orange (*Citrus Sinensis *L.) Peel*

The original habitat of the orange tree is in the regions of southern China, northeast India, and Myanmar. It wasn't until 314 BC that the sweet orange made its literary debut in China [105]. Planting orange trees is widespread because of the excellent fruit they provide. Raw or cooked, the orange tree fruit is delicious and aromatic whether eaten in its natural state or with the juice or peel. Global orange output in 2022 was 76 million tonnes, with 22% coming from Brazil, 22% from China, and 12% from India [106]. Figure 7.7a shows a picture of an orange peel.

7.4.3.1 Orange (*Citrus Sinensis* L.) Peel Chemical Composition

Citrus sinensis's efficiency as a bio-mordant is due in part to its mineral content, essential oils, flavonoids, and organic acids. *Citrus* fruits include three types of flavonoids: flavones, and flavonols, flavanones. *Citrus* species mostly include hesperidine, naringin, narirutin, and eriocitrin as flavonoids [107]. To enhance the dye molecules' binding to fabric fibers, acids that are organic, flavonoids, as well as minerals can create complexes with them. As shown in Figure 7.7b, a combination

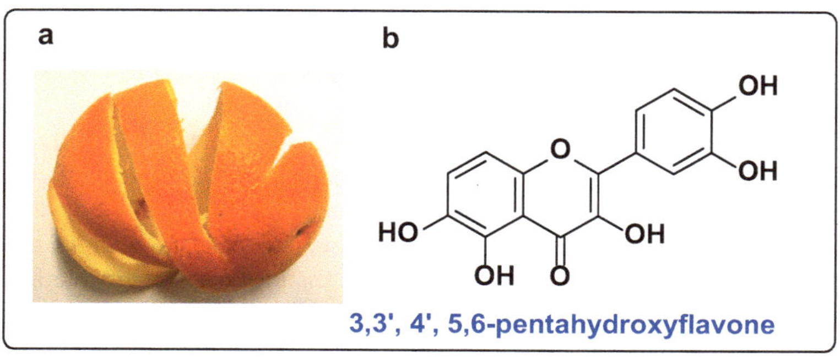

Fig. 7.7 Orange (*Citrus sinensis* L.) peel, **a** physical appearance (Adapted from: Flickr, http://flickr.com/photos/fdecomite/6298546165) and **b** flavone-based chemical composition

of hesperosiodes, naringoside, and ecryodietyoside make up the citrin, a class of citroflavonoids found in 7.6% of the peel. Citrus peel is a known source of limonoid glycosides [108].

7.4.3.2 Advantages of Orange (*Citrus Sinensis* L.) Peel

Several benefits are associated with natural dyeing when orange (*Citrus sinensis* L.) peel is used as a bio-mordant. The compound includes pectin, several organic acids (including citric and ascorbic acid), flavonoids (including hesperidin and naringin), essential oils (including limonene and linalool), and more. Through their interactions with fibers and dyes, these chemicals help make mordanting work. In a work, Adeel et al. [52] used natural dyes for wool dyeing, including tannic acid, acacia bark, coral jasmine (*Nyctanthes arbortristis*), orange and pomegranate peels, and acacia. According to them, depending on the mordant ingredient, it is possible to obtain green and grey color tones with fastness values that are high enough [109].

*7.4.4 Arjun Bark (*Terminalia Arjuna*)*

An enormous extensive tree with a typically buttressed trunk, *Terminalia arjuna* (family Combetaceae) is native to South Asia and is used for a variety of medicinal purposes. Commonly known as the "Arjun tree," the Arjuna is an extensively planted tree in India [110]. Antioxidant, anti-inflammatory, and antibacterial effects are only a few of its numerous medicinal uses. By strengthening and conditioning the cardiac muscles, arjuna not only reduces the likelihood of heart disease but also helps the heart function normally. In addition to reducing hypertension, the arjuna tree has strong antihypertensive characteristics [111]. For the greatest advantages in situations of cardiac difficulties, it is recommended to drink cooked arjuna bark in milk one or two times daily (Fig. 7.8a).

7.4.4.1 Arjun Bark (*Terminalia Arjuna*) Chemical Composition

Arjun bark's primary potential coloring components, baicalein, and ellagic acid, impart a yellowish to brownish-red hue to cotton fabric [112]. The bark of *Terminalia arjuna* contains glycosides, triterpenoids, and a large variety of polyphenolic flavonoids, such as gallic acid, arjunolone, bicalien, flavones, and many more [113]. The ability of Arjun dye, like other natural colorants, to form acceptable hydrogen bonds with appropriate functional groups that are active of the fiber substrate or adequate coordinating structures with metal ions (mordants) is seen in Fig. 7.8b [114].

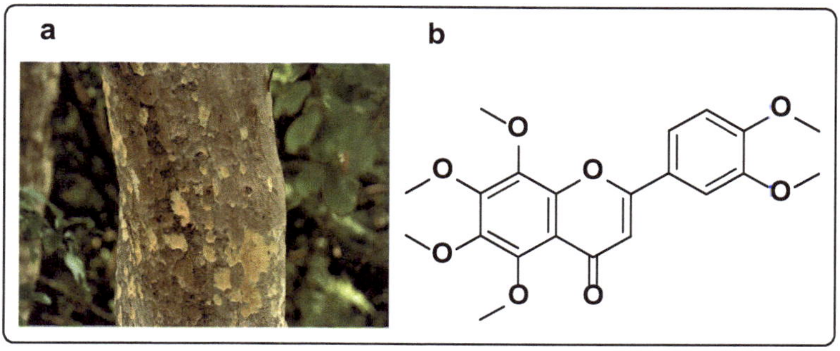

Fig. 7.8 Arjun bark, **a** physical appearance (Adapted from: Wikimedia Commons, https://commons.wikimedia.org/wiki/File:Terminalia_arjuna_W_IMG_2895.jpg) and **b** flavonoid-based chemical composition

*7.4.5 Chestnuts (*Castanea Sativa *Mill.)*

The chestnut tree, shown in Fig. 7.9a, scientifically known as *Castanea sativa* Mill., is an environmentally beneficial plant that serves several purposes, including timber, nuts, and tannin. The distribution is mostly in the Northern Hemisphere, including Southern Europe (from Turkey to the Atlantic Islands), Asia (primarily in China, Korea, and Japan), and the United States of America. A valuable resource for producing nations, *Castanea sativa* Mill. is a member of the Fagaceae family and is abundant throughout southern Europe and Asia [115]. A great deal of bioactive material-rich byproducts, including chestnut leaves, blossoms, shells, and burs, are produced during the processing of chestnuts. Its high time to identify these wastes smart usability to reduce the waste from the enviroment [116].

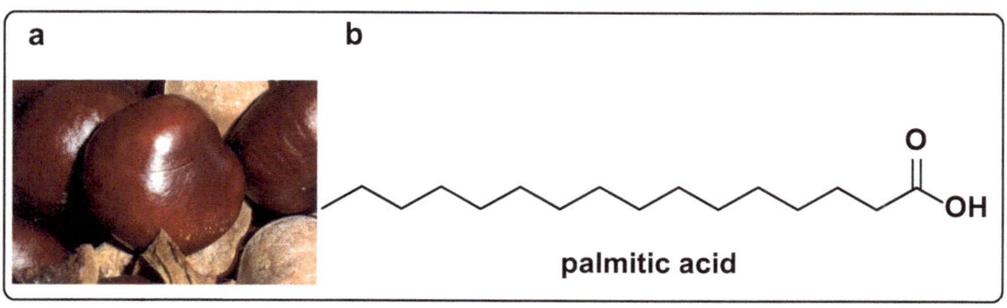

Fig. 7.9 Chestnuts (*Castanea sativa* Mill.) fruit, **a** physical appearance (Adapted from: Picryl, https://picryl.com/media/chestnut-autumn-fruits-189c61), **b** palmitic acid as a chemical composition

7.4.5.1 Chestnuts (*Castanea Sativa* Mill.) Chemical Composition

Despite claims to the contrary, a fatty acid analysis of chestnuts shows that they are rather high in unsaturated fats. The composition shown in Fig. 7.9b of the oil is as follows: 30–40% monounsaturated (cis-oleic acid, C18:1), 50–70% polyunsaturated, as well as 10–20% saturated (palmitic acid, C16:0). Extracts of tannins from chestnuts-wax, bark, and flesh-, belong to the family of hydrolyzable tannins [117].

7.4.5.2 Advantages of Chestnuts (*Castanea Sativa* Mill.)

The natural dyestuff sources that might be utilized to color vegetable fibers (cotton and jute) include *Helichrysum italicum Roth* (Roth), *Rubia peregrina* L. (wild madder), *Daphne gnidium* L. (daphne), *Lavandula stoechas* L. (wild lavender), and *Cynara scolymus* L. (artichoke). While chestnut tannins did help *Lavandula* and *Helichrysum* plant dyes adhere better to potassium alum, they did little to enhance the fabric's UV protection. Instead of tannins, potassium alum was shown to increase the UV protection capabilities of flax samples colored with Helichrysum extract [118]. The dye can be protected from oxidative breakdown by the antioxidants in chestnuts, making the color last longer. Dye absorption and color fixing are both aided by the acidic environment that tannins naturally provide [119].

*7.4.6 Turmeric (*Curcuma Longa *L.)*

Turmeric is one plant shown in Fig. 7.10a, that serves dual purposes as a bio-mordant and a natural color. Its native Asia, especially China, and India, are major producers [120]. Turmeric is an evergreen herb that belongs to the Zingiberaceae family, which is often known as the ginger family. There are a lot of metabolites in it, such as curcuminoid, oil, flavonoids, phenolics, protein, and a lot of alkaloids. It also contains a few important amino acids [121].

7.4.6.1 Turmeric (*Curcuma Longa* L.) Chemical Composition

Turmeric, which is rich in curcuminoids, a class of phenolic compounds, is the most popular natural color. Figure 7.10b shows the structure of curcumin. The aryl-C7-aryl structure is shared by a class of compounds called diarylheptanoids (sometimes called diphenylheptanoids), to which curcuminoids belong [122]. The primary active components of turmeric are these yellow pigments, which are commonly utilized as culinary colorants. Turmeric rhizomes typically contain 3–15% of these polyphenols, with curcumin being the primary component. Curcumin, whose molecular formula is $C_{21}H_{20}O_5$, was first isolated in 1815. Other names for it include diferuloyl methane and 1,6-heptadiene-3,5-dione-1,7-bis(4-hydroxy-3-methoxyphenyl)-(1E,6E) [123].

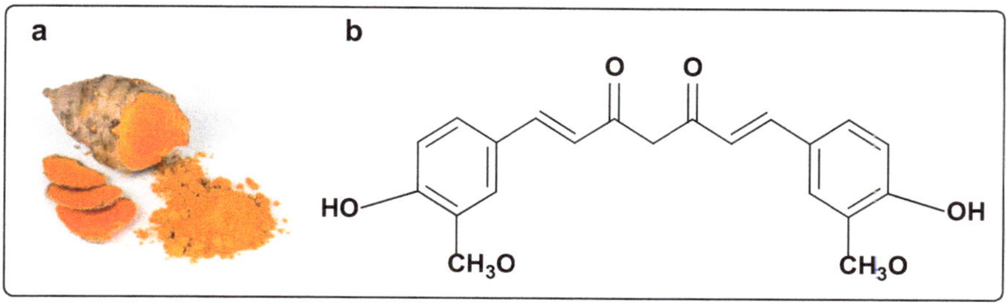

Fig. 7.10 Turmeric (*Curcuma longa* L.), **a** physical appearance (Adapted from: Freerange, https://freerangestock.com/photos/122627/turmeric-root-and-powder-displayed-together.html), and **b** curcumin as a chemical composition

7.4.6.2 Advantages of Turmeric (*Curcuma Longa* L.)

When used as a mordant to cotton, silk, polyamide, or acrylic fibers, turmeric extract might affect the fabric's dyeability. Ferrous sulfate is a common mordant, and the colors it creates are often redder and brighter than those of other mordants [124]. To get a yellow hue in cotton, it can be mordanted with tannin, zinc chloride, or aluminum sulfate, or left unmordanted. When cotton samples were mordanted with tannin, aluminum sulfate, and ferrous sulfate, the colorfastness was found to be increased, both in terms of washability and hot pressing resistance [125]. Sunscreen products made from it are well known for their anti-inflammatory, anti-cancer, anti-inflammatory, antibacterial, antiparasitic, antimutagenic, along with anti-inceptive qualities [94].

7.5 Conclusion

The adoption of sustainable practices in the textile dyeing industry is essential for maintaining ecological balance and improving societal well-being. The resurgence of natural dyes, coupled with the use of bio-mordants, represents a significant step toward eco-friendly and sustainable dyeing methods. Bio-mordants, derived from natural materials, provide a safer and more environmentally responsible alternative to synthetic mordants, reducing pollution and conserving natural resources. By integrating natural solutions with advanced techniques the industry can achieve more efficient and vibrant dyeing processes while minimizing its ecological footprint. This blend of nature and technology not only promotes sustainability but also offers innovative possibilities for the future of textile coloration. The shift toward sustainable dyeing practices is a promising direction for the preservation of natural assets, ensuring that the textile industry contributes positively to the environment and supports a healthier planet for generations to come.

Declaration The authors declare that they have no known competing financial interests or personal relationships that could have appeared to influence the work reported in this chapter. Generative AI and AI-assisted technologies have been used in the writing process to improve the readability and language of the work only. After using this tool, the authors reviewed and edited the content as needed and take full responsibility for the content of the publication.

References

1. Mia R, Mofasser A, Bhat MA, Howlader MI, Sayed NI, Bakar MA (2024) Utilizing neem wood waste as a natural dye in the development of bio-based functionalized colored jute fabric. Sustainable Chemistry and Pharmacy 38:101489. https://doi.org/10.1016/j.scp.2024.101489
2. Cunningham AB, Maduarta IM, Howe J, Ingram W, Jansen S (2011) Hanging by a thread: natural metallic mordant processes in traditional Indonesian textiles. Econ Bot 65(3):241–259. [Online]. http://www.jstor.org/stable/41408235
3. Ahmed T et al (2024) Chapter 8-new trends in printing applications of natural dyes and pigments. In: Ul Islam S (ed) Renewable dyes and pigments. Elsevier, pp 139–163
4. Chakraborty J (2015) Fundamentals and practices in colouration of textiles. CRC Press
5. Thavapalan S (2019) The meaning of color in ancient Mesopotamia. Brill
6. Bakar MA et al (2023) Exploring the potential of mahogany extract as a natural dye for the coloration of jute fabric. Heliyon 9(9). https://doi.org/10.1016/j.heliyon.2023.e19464
7. Pisitsak P, Hutakamol J, Thongcharoen R, Phokaew P, Kanjanawan K, Saksaeng N (2016) Improving the dyeability of cotton with tannin-rich natural dye through pretreatment with whey protein isolate. Ind Crop Prod 79:47–56. https://doi.org/10.1016/j.indcrop.2015.10.043
8. Yameen M et al (2024) Sustainable appraisal of lipstick tree seeds (Bixa orellana)-based bixin natural orange colorant for green mordanted silk fabrics and wool yarns. Sci Prog 107(2):00368504241242282. https://doi.org/10.1177/00368504241242282
9. Roy Choudhury AK (2013) Green chemistry and the textile industry. Textile Progress 45(1):3–143. https://doi.org/10.1080/00405167.2013.807601
10. Habib N et al (2024) Green extraction and application of yellow natural curcumin colorant from Curcuma aromatica rhizomes for silk dyeing. Sci Rep 14(1):13032. https://doi.org/10.1038/s41598-024-63927-7
11. Batool F, Iqbal N, Adeel S, Azeem M, Hussaan M, Mia R (2024) Sugar beet (Beta vulgaris L.) leaves as natural colorant for cotton dyeing using an ecofriendly approach toward industrial progress. Sci Prog 107(3):00368504241271737. https://doi.org/10.1177/00368504241271737
12. Hosseinnezhad M, Gharanjig K, Imani H, Razani N (2022) Green dyeing of wool yarns with yellow and black myrobalan extract as bio-mordant with natural dyes. J Nat Fibers 19(10):3893–3915. https://doi.org/10.1080/15440478.2020.1848735
13. Jabar JM, Owokotomo IA, Ogunsade AF (2022) Sustainable dyeing of cotton fabric with mangiferin: roles of microwave-rays and bio-mordants on fabric colorimetric and fastness properties. Sustain Chem Pharm 29:100822. https://doi.org/10.1016/j.scp.2022.100822
14. Botteri L, Miljković A, Glogar MI Influence of cotton pre-treatment on dyeing with onion and pomegranate peel extracts. Molecules 27(14):4547. https://doi.org/10.3390/molecules27144547
15. Mia R, Habib S, Assiri MA, Liu H, Mahmud S (2023) Coloration and antibacterial treatment of aramid fabric: a comprehensive review. Surf Innov 12(3–4):130–146. https://doi.org/10.1680/jsuin.23.00065
16. Mansour HF, Heffernan S (2011) Environmental aspects on dyeing silk fabric with sticta coronata lichen using ultrasonic energy and mild mordants. Clean Technol Environ Policy 13(1):207–213. https://doi.org/10.1007/s10098-010-0296-2

17. Rather LJ et al (2016) Ecological dyeing of Woolen yarn with Adhatoda vasica natural dye in the presence of biomordants as an alternative copartner to metal mordants. J Environ Chem Eng 4(3):3041–3049. https://doi.org/10.1016/j.jece.2016.06.019
18. Kushwaha A, Singh S, Chaudhary K (2023) Eco-friendly multifunctional dyeing of pineapple using Nyctanthes Arbortristis dye and Acacia Nilotica bio-mordant. Sustain Chem Pharm 34:101146. https://doi.org/10.1016/j.scp.2023.101146
19. Sundrarajan M, Gandhi RR, Suresh J, Gowri S, Selvam S (2012) Natural dyeing of silk fabric using eco-friendly mordants. Asian J Chem 24(7):3109–3112
20. Amutha K, Annapoorani SG (2019) Dyeing of banana/cotton union fabric with natural dye and mordants. Int J Appl Res 5(8):24–26
21. Repon MR et al (2024) Textile dyeing using natural mordants and dyes: a review. Environ Chem Lett 22(3):1473–1520. https://doi.org/10.1007/s10311-024-01716-4
22. Bancroft WD, Mordants I (1992) J Phys Chem 26(5):447−470. https://doi.org/10.1021/j150221a003
23. Singh S, Singh DR (2018) Application of natural mordants on textile. Int J Appl Home Sci 5(1):252−260
24. Mia R et al (2022) Natural dye extracted from Triadica sebifera in aqueous medium for sustainable dyeing and functionalizing of viscose fabric. Clean Eng Technol 8:100471. https://doi.org/10.1016/j.clet.2022.100471
25. Vankar PS (2000) Chemistry of natural dyes. Resonance 5(10):73–80
26. Benli H (2024) Bio-mordants: a review. Environ Sci Pollut Res 31(14):20714–20771. https://doi.org/10.1007/s11356-024-32174-8
27. Ramakrishnan K, Selvi S, Shubha R (2006) Tannin and its analytical techniques. Indian Chem Eng 48(2):88
28. Ismal Ö (2016) Patterns from nature: contact printing. J TEXTILE Assoc 77(2):81–91
29. İşmal ÖE, Yıldırım L (2019) 3-metal mordants and biomordants. In: Shahid ul I, Butola BS (eds) The impact and prospects of green chemistry for textile technolog. Woodhead Publishing, pp 57–82
30. Mia R, Das S, Banna BU, Ahmed T, Bakar MA (2023) Enhancing antibacterial properties of organic cotton fabric using mahogany wood waste. Case Stud Chem Environ Eng 100387
31. Mia R, Bakar MA, Islam MR, Ahmed T (2023) Eco-friendly coloration from mahogany wood waste for sustainable dyeing of organic nonwoven cotton fabric. Results Eng 17:101032. https://doi.org/10.1016/j.rineng.2023.101032
32. Shahid M, Shahid ul I, Mohammad F (2013) Recent advancements in natural dye applications: a review. J Clean Prod 53:310–331. https://doi.org/10.1016/j.jclepro.2013.03.031
33. Yusuf M, Khan SA, Shabbir M, Mohammad F (2017) Developing a shade range on wool by madder (Rubia cordifolia) root extract with gallnut (Quercus infectoria) as biomordant. J Nat Fibers 14(4):597–607. https://doi.org/10.1080/15440478.2016.1240641
34. Prabhu K, Bhute AS (2012) Plant based natural dyes and mordants: a review. J Nat Prod Plant Resour 2(6):649–664
35. Mia R, Sk MS, Oli ZBS, Ahmed T, Kabir S, Waqar MA (2021) Functionalizing cotton fabrics through herbally synthesized nanosilver. Clean Eng Technol 4:100227. https://doi.org/10.1016/j.clet.2021.100227
36. Yusuf M, Shabbir M, Mohammad F (2017) Natural colorants: historical, processing and sustainable prospects. Nat Prod Bioprospecting 7(1):123–145. https://doi.org/10.1007/s13659-017-0119-9
37. Sheikh J, Jagtap PS, Teli MD (2016) Ultrasound assisted extraction of natural dyes and natural mordants vis a vis dyeing. Fibers Polym 17(5):738–743. https://doi.org/10.1007/s12221-016-5031-0
38. Han S, Yang Y (2005) Antimicrobial activity of wool fabric treated with curcumin. Dye Pigment 64(2):157–161. https://doi.org/10.1016/j.dyepig.2004.05.008
39. Singh R, Jain A, Panwar S, Gupta D, Khare SK (2005) Antimicrobial activity of some natural dyes. Dye Pigment 66(2):99–102. https://doi.org/10.1016/j.dyepig.2004.09.005

40. Khan MI et al (2011) Assessment of antimicrobial activity of Catechu and its dyed substrate. J Clean Prod 19(12):1385–1394. https://doi.org/10.1016/j.jclepro.2011.03.013
41. Khan SA et al (2012) Antimicrobial activity of wool yarn dyed with Rheum emodi L. (Indian Rhubarb). Dye Pigment 95(2):206–214. https://doi.org/10.1016/j.dyepig.2012.04.010
42. Toki GFI et al (2024) Sustainable coloration and analysis of cellulosic viscose fabric incorporating Rosa rubiginosa extraction and pre-mordanting approaches. Mater Today Commun 38:108068. https://doi.org/10.1016/j.mtcomm.2024.108068
43. Ibrahim NA, El-Gamal AR, Gouda M, Mahrous F (2010) A new approach for natural dyeing and functional finishing of cotton cellulose. Carbohydr Polym 82(4):1205–1211. https://doi.org/10.1016/j.carbpol.2010.06.054
44. Zhou Y, Zhang J, Tang RC, Zhang J (2015) Simultaneous dyeing and functionalization of silk with three natural yellow dyes. Ind Crop Prod 64:224–232. https://doi.org/10.1016/j.indcrop.2014.09.041
45. Choudhury AKR (2018) Eco-friendly dyes and dyeing. Adv Mater Technol Environ 2:145–176
46. Periyasamy AP (2022) Natural dyeing of cellulose fibers using syzygium cumini fruit extracts and a bio-mordant: a step toward sustainable dyeing. Sustain Mater Technol 33:e00472. https://doi.org/10.1016/j.susmat.2022.e00472
47. İşmal ÖE, Yıldırım L (2019) Metal mordants and biomordants. In: Islam S, Butola BS (eds) The impact and prospects of green chemistry for textile technology. Elsevier. pp 57–82. https://doi.org/10.1016/B978-0-08-102491-1.00003-4
48. Saxena S, Raja ASM, Arputharaj A (2017) Challenges in sustainable wet processing of textiles. In: Muthu SS (ed) Textiles and clothing sustainability: sustainable textile chemical processes. Springer Singapore, Singapore, pp 43–79
49. Hassan RM, Ibrahim N, Hassan NA, Idris N, Nor NM (2023) Effects of metal ion mordant on different fabrics dyed with coconut husk dye. AIP Conf Proc 2625(1):020008. https://doi.org/10.1063/5.0129648
50. Chakraborty S, Chakraborty L (2022) Aspects of mordant dyeing of textile fibers In: Yusuf M, Shahdi M (eds) Emerging technologies for textile coloration. CRC Press, pp 37–56
51. Vajpeyee P, Singh S, Maity S, Pandit P, Singha K (2022) Aspects of mordants and metal complex dyes. In: Textile dyes and pigments: a green chemistry approach. pp 393–409. https://doi.org/10.1002/9781119905332.ch20
52. Adeel S, Ahmad S, Habib N, Fazal ur R, Mia R, Ahmed B (2022) Coloring efficacy of Nyctanthes Arbortristis based yellow natural dye for surface-modified wool. Ind Crop Prod 188:115571. https://doi.org/10.1016/j.indcrop.2022.115571
53. Jayalakshmi I, Amsamani S (2007) Bio-mordant for wool. Man-Made Text India 50(7):267–270
54. Naqvi SAR, Adeel S, Mia R, Hosseinnezhad M, Rather LJ, Imran M (2024) Modern ecofriendly approach for extraction of luteolin natural dye from weld for silk fabric and wool yarn dyeing. Sustain Chem Pharm 39:101554. https://doi.org/10.1016/j.scp.2024.101554
55. Khanbabaee K, van Ree T (2001) Tannins: classification and definition. Nat Prod Rep 18(6):641–649. https://doi.org/10.1039/B101061L
56. Mueller-Harvey I (2001) Analysis of hydrolysable tannins. Anim Feed Sci Technol 91(1):3–20. https://doi.org/10.1016/S0377-8401(01)00227-9
57. Chavan RB, Nalankilli G (1991) Concurrent sublimation transfer printing and finishing of silk. Text Res J 61(8):461–466. https://doi.org/10.1177/004051759106100805
58. Seigler DS (2003) Phytochemistry of Acacia—sensu lato. Biochem Syst Ecol 31(8):845–873. https://doi.org/10.1016/S0305-1978(03)00082-6
59. Ozdemir MB, Karadag R (2023) Anatolian acorn oak's economic potential in the application to the textile and leather industries. Text & Leather Rev 6:320–332
60. Faisal T, Ahsan M, Choudhury JA, Azam A (2017) Phytochemical and biological investigations of Eurya acuminata (Theaceae). Dhaka Univ J Pharm Sci 15(2):151–154
61. Vankar PS, Shanker R, Mahanta D, Tiwari SC (2008) Ecofriendly sonicator dyeing of cotton with Rubia cordifolia Linn. using biomordant. Dye Pigment 76(1):207–212. https://doi.org/10.1016/j.dyepig.2006.08.023

62. Aboaba O, Smith S, Olude F (2006) Antibacterial effect of edible plant extract on Escherichia coli 0157: H7. Pak J Nutr 5(4):325–327
63. Chairat M, Bremner JB, Chantrapromma K (2007) Dyeing of cotton and silk yarn with the extracted dye from the fruit hulls of mangosteen,Garcinia mangostana linn. Fibers Polym 8(6):613–619. https://doi.org/10.1007/BF02875998
64. Moloudizargari M, Mikaili P, Aghajanshakeri S, Asghari MH, Shayegh J (2013) Pharmacological and therapeutic effects of Peganum harmala and its main alkaloids (in eng). Pharmacogn Rev 7(14):199−212. https://doi.org/10.4103/0973-7847.120524
65. Adeel S et al (2019) Microwave-assisted sustainable dyeing of wool fabric using cochineal-based carminic acid as natural colorant. J Nat Fibers 16(7):1026–1034. https://doi.org/10.1080/15440478.2018.1448317
66. Onyeabo C, Achi NK, Ekeleme-Egedigwe CA, Ebere CU, Okoro CK (2017) Haematological and biochemical studies on Justicia carnea leaves extract in phenylhydrazine induced-anemia in albino rats. Acta Sci Pol Technol Aliment 16(2):217–230
67. Ab Rahman N, Harisun Y (2015) Quantification of gallic acid and tannic acid from Quercus infectoria (Manjakani) and their effects on antioxidant and antibacterial activities. Pertanika J Sci & Technol 23(2)
68. Al Alamoudi HA, Salem AA (2019) Application of myrrh extract as an eco-friendly dye and antimicrobial agent on wool and silk fabrics part 1: dyeing with myrrh extract. J Nat Fibers 16(7):1006–1014. https://doi.org/10.1080/15440478.2018.1447417
69. Liu F et al (2021) The complete chloroplast genome and characteristics analysis of Musa basjoo Siebold. Mol Biol Rep 48:7113–7125
70. Tai Z, Chen A, Qin B, Cai L, Xu Y (2014) Chemical constituents and antioxidant activity of the Musa basjoo flower. Eur Food Res Technol 239:501–508. https://doi.org/10.1007/s00217-014-2244-6
71. Miranda I, Lima L, Quilhó T, Knapic S, Pereira H (2016) The bark of Eucalyptus sideroxylon as a source of phenolic extracts with anti-oxidant properties. Ind Crops Prod 82:81–87. https://doi.org/10.1016/j.indcrop.2015.12.003
72. Hayes RA, Piggott AM, Smith TE, Nahrung HF (2014) Corymbia phloem phenolics, tannins and terpenoids: interactions with a cerambycid borer. Chemoecol 24:95–103. https://doi.org/10.1007/s00049-014-0153-6
73. Pranta AD, Rahaman MT (2024) Extraction of eco-friendly natural dyes and biomordants for textile coloration: a critical review. Nano-Struct & Nano-Objects 39:101243. https://doi.org/10.1016/j.nanoso.2024.101243
74. Khan M, Al-Mamun M, Al-Amin M, Moniruzzaman M, Hasan M (2015) Dye-sensitized solar cell using used semiconductor glass and natural dye: towards alternative energy challenge. Int J Renew Energy Resour 5(2):38–44
75. Serrano A, Sousa MM, Hallett J, Lopes JA, Oliveira MC (2011) Analysis of natural red dyes (cochineal) in textiles of historical importance using HPLC and multivariate data analysis. Anal Bioanal Chem 401(2):735–743. https://doi.org/10.1007/s00216-011-5094-0
76. Serrano A, Sousa M, Hallett J, Simmonds MSJ, Nesbitt M, Lopes JA (2013) Identification of Dactylopius cochineal species with high-performance liquid chromatography and multivariate data analysis. Analyst 138(20):6081–6090. https://doi.org/10.1039/C3AN00052D
77. Al-Ghadeer HA (2010) Acute ocular complications from self-administered topical kermes. Middle East Afr J Ophthalmol 17(4). [Online]. https://journals.lww.com/mejo/fulltext/2010/17040/acute_ocular_complications_from_self_administered.21.aspx
78. Doménech-Carbó A, Doménech-Carbó MT, Calisti M, Maiolo V (2010) Identification of naphthoquinonic and anthraquinonic dyes via sequential potential steps applied to the voltammetry of microparticles methodology. J Solid State Electrochem 14(3):465–477. https://doi.org/10.1007/s10008-009-0899-x
79. Karadag R, Yurdun T, Dolen E (2010) Identification of natural red dyes in 15–17th centuries ottoman silk textiles (kaftans, brocades, velvets and skullcaps) by HPLC with diode array detection. Asian J Chem 22(9):7043

80. Lluveras-Tenorio A et al (2017) Spectroscopic and mass spectrometric approach to define the Cyprus orthodox icon tradition-the first known occurrence of Indian lac in Greece/Europe. Microchem J 131:112–119. https://doi.org/10.1016/j.microc.2016.12.005
81. Ramani R, Pal G (2016) Lac marketing and crop economics. In: Sharma KK, Monobrullah M, Mohanasundaram A, Ramani R (eds) Beneficial insect farming. ICAR Indian Institute of Natural Resins and Gums, pp 133–152
82. Srivastava A, Kalwani M, Chakdar H, Pabbi S, Shukla P (2022) Biosynthesis and biotechnological interventions for commercial production of microalgal pigments: a review. Bioresour Technol 352:127071. https://doi.org/10.1016/j.biortech.2022.127071
83. Pagels F, Salvaterra D, Amaro HM, Guedes AC (2020) Chapter 18-pigments from microalgae. In: Jacob-Lopes E, Maroneze MM, Queiroz MI, Zepka LQ (eds) Handbook of microalgae-based processes and products. Academic Press, pp 465–492
84. Christaki E, Bonos E, Florou-Paneri P (2015) Chapter 14-innovative microalgae pigments as functional ingredients in nutrition. In: Kim S-K (ed) Handbook of marine microalgae. Academic Press, Boston, pp 233–243
85. Khanra S, Mondal M, Halder G, Tiwari ON, Gayen K, Bhowmick TK (2018) Downstream processing of microalgae for pigments, protein and carbohydrate in industrial application: a review. Food Bioprod Process 110:60–84. https://doi.org/10.1016/j.fbp.2018.02.002
86. García-Vaquero M, Brunton N, Lafarga T (2021) Chapter 7-microalgae as a source of pigments for food applications. In: Lafarga T, Acién G (eds) Cultured microalgae for the food industry. Academic Press, pp 177–198
87. Lauceri R, Chini Zittelli G, Torzillo G (2019) A simple method for rapid purification of phycobiliproteins from Arthrospira platensis and Porphyridium cruentum biomass. Algal Res 44:101685. https://doi.org/10.1016/j.algal.2019.101685
88. İlter I, Akyıl S, Demirel Z, Koç M, Conk-Dalay M, Kaymak-Ertekin F (2018) Optimization of phycocyanin extraction from Spirulina platensis using different techniques. J Food Compos Anal 70:78–88. https://doi.org/10.1016/j.jfca.2018.04.007
89. Abidizadegan M, Peltomaa E, Blomster J (2021) The potential of cryptophyte algae in biomedical and pharmaceutical applications (in English). Front Pharmacol, Rev 11. https://doi.org/10.3389/fphar.2020.618836
90. Fernandes AS et al (2021) Bioaccessibility of microalgae-based carotenoids and their association with the lipid matrix. Food Res Int 148:110596. https://doi.org/10.1016/j.foodres.2021.110596
91. Liu C, Hu B, Cheng Y, Guo Y, Yao W, Qian H (2021) Carotenoids from fungi and microalgae: a review on their recent production, extraction, and developments. Bioresour Technol 337:125398. https://doi.org/10.1016/j.biortech.2021.125398
92. Gui L, Xu L, Liu ZY, Zhou ZG, Sun Z (2022) Carotenoid-rich microalgae promote growth and health conditions of Artemia nauplii. Aquaculture 546:737289. https://doi.org/10.1016/j.aquaculture.2021.737289
93. Gallego R et al (2022) Anti-inflammatory and neuroprotective evaluation of diverse microalgae extracts enriched in carotenoids. Algal Res 67:102830. https://doi.org/10.1016/j.algal.2022.102830
94. Rahman MM, Kim M, Youm K, Kumar S, Koh J, Hong KH (2023) Sustainable one-bath natural dyeing of cotton fabric using turmeric root extract and chitosan biomordant. J Clean Prod 382:135303. https://doi.org/10.1016/j.jclepro.2022.135303
95. Jabar JM, Ogunsade AF, Odusote YA, Yılmaz M (2023) Utilization of Nigerian mango (Mangifera indica L) leaves dye extract for silk fabric coloration: influence of extraction technique, mordant and mordanting type on the fabric color attributes. Ind Crops Prod 193:116235. https://doi.org/10.1016/j.indcrop.2022.116235
96. Shah KA, Patel MB, Patel RJ, Parmar PK (2010) Mangifera indica (mango) (in eng). Pharmacogn Rev 4(7):42–8. https://doi.org/10.4103/0973-7847.65325
97. Wangatia LM, Tadesse K, Moyo S, Wangatia LM, Tadesse K, Moyo S (2015) Mango bark mordant for dyeing cotton with natural dye: fully eco-friendly natural dyeing. Int J Text Sci 4(2):36–41

98. Bhandari NL, Shrestha S, Bhandari G, Parajuli N, Silwal SB, Adhikari R (2021) Extraction of dye from Castanopsis indica for its use in textile dyeing and medicinal purpose with natural mordant. Egypt J Chem 64(11):6681–6690. https://doi.org/10.21608/ejchem.2021.48519.2991
99. Giordani R, Moulin A, Verger R (19991) Tributyroylglycerol hydrolase activity in Carica papaya and other lattices. Phytochem 30(4):1069–1072. https://doi.org/10.1016/S0031-9422(00)95174-4
100. McKee RA, Smith H (1986) Purification of proteinases from Carica papaya. Phytochem 25(10):2283–2287. https://doi.org/10.1016/S0031-9422(00)81680-5
101. Silva JD et al (2007) Papaya (Carica papaya L.) biology and biotechnology. Tree For Sci Biotechnol 1(1):47–73
102. Rivera-Pastrana DM, Yahia EM, González-Aguilar GA (2010) Phenolic and carotenoid profiles of papaya fruit (Carica papaya L.) and their contents under low temperature storage. J Sci Food Agric 90(14):2358–2365. https://doi.org/10.1002/jsfa.4092
103. Rosyida A, Suranto S, Masykuri M, Margono M (2022) Optimisation of the process of dyeing cotton fabric with natural dyes from jackfruit wood by immersion without heating. Pigm Resin Technol 51(6):663–673. https://doi.org/10.1108/PRT-03-2021-0034
104. Trimboli P, Mouly S (2022) Pharmacokinetics and clinical implications of two non-tablet oral formulations of L-thyroxine in patients with hypothyroidism. J Clin Med 11(12):3479. https://doi.org/10.3390/jcm11123479
105. Chen Z-T, Chu H-L, Chyau C-C, Chu C-C, Duh P-D (2012) Protective effects of sweet orange (Citrus sinensis) peel and their bioactive compounds on oxidative stress. Food Chem 135(4):2119–2127. https://doi.org/10.1016/j.foodchem.2012.07.041
106. Toki GFI et al (2024) Sustainable coloration and analysis of cellulosic viscose fabric incorporating Rosa rubiginosa extraction and pre-mordanting approaches. Mater Today Commun 108068. https://doi.org/10.1016/j.mtcomm.2024.108068
107. Hegazy A, Ibrahium M (2012) Antioxidant activities of orange peel extracts. World Appl Sci J 18(5):684–688
108. Toki GFI et al (2024) Eco-friendly sustainable dyeing of lycra viscose fabric with allium sativum using natural and metal mordants. Fibers Polym 25:3887–3900. https://doi.org/10.1007/s12221-024-00710-5
109. Montero-Calderon A, Cortes C, Zulueta A, Frigola A, Esteve MJ (2019) Green solvents and ultrasound-assisted extraction of bioactive orange (Citrus sinensis) peel compounds. Sci Rep 9(1):16120. https://doi.org/10.1038/s41598-019-52717-1
110. Pandey AK, Mandal AK (2012) Sustainable harvesting of terminalia arjuna (roxb.) wight & arnot (Arjuna) and Litsea glutinosa (Lour.) Robinson (Maida) bark in Central India. J Sustain For 31(3):294–309. https://doi.org/10.1080/10549811.2011.583865
111. Gangadevi V, Muthumary J (2009) Taxol production by Pestalotiopsis terminaliae, an endophytic fungus of Terminalia arjuna (arjun tree). Biotechnol Appl Biochem 52(1):9–15. https://doi.org/10.1042/BA20070243
112. Adeel S et al (2019) Utrasonic assisted sustainable dyeing of mordanted silk fabric using arjun (Terminalia arjuna) bark extracts. Environ Prog & Sustain Energy 38(s1):S331–S339. https://doi.org/10.1002/ep.13048
113. Mia R, Toki GFI, Kamal SAB (2024) Nanoscale coating for flavonoid-based natural colorants. In: Ahmed S, Adeel S (eds) Nanoscale textile coatings for enhanced performance. Springer, pp 171–207
114. Benhelal E et al (2019) ACEME: synthesis and characterization of reactive silica residues from two stage mineral carbonation process. Environ Prog & Sustain Energy 38(3):e13066. https://doi.org/10.1002/ep.13066
115. De Vasconcelos MC, Bennett RN, Rosa EA, Ferreira-Cardoso JV (2010) Composition of European chestnut (Castanea sativa Mill.) and association with health effects: fresh and processed products. J Sci Food Agric 90(10):1578–1589. https://doi.org/10.1002/jsfa.4016
116. Pinto D, Braga N, Silva AM, Costa P, Delerue-Matos C, Rodrigues F (2020) Chapter 6-chestnut. In: Galanakis CM (ed) Valorization of fruit processing by-products. Academic Press, pp 127–144

117. Vázquez G, Fontenla E, Santos J, Freire MS, González-Álvarez J, Antorrena G (2008) Antioxidant activity and phenolic content of chestnut (Castanea sativa) shell and eucalyptus (Eucalyptus globulus) bark extracts. Ind Crop Prod 28(3):279–285. https://doi.org/10.1016/j.indcrop.2008.03.003
118. Grifoni D et al (2014) UV protective properties of cotton and flax fabrics dyed with multifunctional plant extracts. Dye Pigment 105:89–96. https://doi.org/10.1016/j.dyepig.2014.01.027
119. Conedera M et al (2021) How future-proof is sweet chestnut (Castanea sativa) in a global change context? For Ecol Manage 494:119320. https://doi.org/10.1016/j.foreco.2021.119320
120. Naveed R et al (2022) Microwave assisted extraction and dyeing of cotton fabric with mixed natural dye from pomegranate rind (Punica granatum L.) and turmeric rhizome (Curcuma longa L.). J Nat Fibers 19(1):248–255. https://doi.org/10.1080/15440478.2020.1738309
121. Verma RK, Kumari P, Maurya RK, Kumar V, Verma R, Singh RK (2018) Medicinal properties of turmeric (Curcuma longa L.): a review. Int J Chem Stud 6(4):1354–1357
122. Chatzinasiou L, Booker A, MacLennan E, Mackonochie M, Heinrich M (2019) Turmeric (Curcuma longa L.) products: what quality differences exist? J Herb Med 17:100281. https://doi.org/10.1016/j.hermed.2019.100281
123. Chempakam B, Parthasarathy VA (2008) Turmeric. In: Parthasarathy VA, Chempakam B, Zachariah TJ (eds) Chemistry of spices. Cambridge, CABI, pp 97–123
124. Hosen MD, Rabbi MF, Raihan MA, Al Mamun MA (2021) Effect of turmeric dye and biomordants on knitted cotton fabric coloration: a promising alternative to metallic mordanting. Clean Eng Technol 3:100124. https://doi.org/10.1016/j.clet.2021.100124
125. Toprak-Cavdur T, Uysal S, Gisbert-Paya J (2022) Dyeing recycled cotton fibers using Curcuma longa and Pterocarpus santalinus natural dyes and bio-mordant chitosan. J Nat Fibers 19(16):2105469. https://doi.org/10.1080/15440478.2022.2105469

Chapter 8
Sustainable Fashion Through Naturally Tie-Dyed Textiles

Md Mahbubul Alam, Md. Shah Nawaz Mazumder, A Z M Mofasser, Nadim Ibn Sayed, and Rony Mia

Abstract When the context is sustainable fashion, ensuring the environment's safety and human health is considered to hold the utmost significance. Tie-dyeing is one of the most recent and popular resist-dyeing techniques corroborating sustainable fashion. This chapter offers an overview of sustainable fashions with tie-dyeing techniques such as spiral tie-dyeing, binding and knotting, folding and clamping, crumple tie-dyeing, ombre tie-dyeing, knotting and tying, stitching and binding, and striped tie-dyeing, where natural dyes can be used. Here, comparative analyses between conventional and recently emerging techniques have been presented. The chapter provides information on the sustainable alternatives to the tie-dyeing processes that consume excessive amounts of water. Besides, a comprehensive description has been provided of the process of redesigning second-hand clothes by tie-dye.

Keywords Sustainable fashion · Tie-dyeing · Redesign · Natural dyes

8.1 Introduction

The concept of sustainable fashion comprises perspectives like social and economic, including the prime concerns regarding its adverse environmental impacts. The approach to sustainable fashion encourages both manufacturers and consumers to create an ecosystem of fashion that will make the planet safer for human beings [1, 2]. To meet the demand of consumers, clothing manufacturers tend to pursue low production costs while ignoring environmental regulatory systems, especially in developing

M. M. Alam · Md. S. N. Mazumder · A. Z. M. Mofasser · N. Ibn Sayed
Department of Textile Engineering, European University of Bangladesh, Mirpur, Dhaka, Bangladesh
e-mail: nadimibnsayed@yahoo.com

R. Mia (✉)
Department of Textile Engineering, Chemistry and Science, Wilson College of Textiles, North Carolina State University, Raleigh, NC, USA
e-mail: mroni_mia@yahoo.com

M. Shahid et al. (eds.), *Advancements in Textile Coloration*,
https://doi.org/10.1007/978-981-96-5091-0_8

textile-producing countries [1, 3]. Consumers are also concerned about environmental sustainability; thus, demand for eco-friendly textile products is increasing [4, 5].

Tie-dyeing is a technique of dyeing fabrics or clothing that is done by binding or tying fabrics in order to produce eye-catching and creative patterns that result in colorful, unique designs [6]. Tie-dyeing offers sustainable fashion by using less water in the whole process, giving a new life to old clothes, and reducing waste through recycling. More than a fashion trend, tie-dyeing embodies a sustainable ethos, allowing individuals to align their aesthetic choices and customization with environmental responsibility and conservation efforts [7]. Furthermore, using sustainable materials such as natural dyes, chemicals, tying materials, and dyeing techniques makes tie-dyeing more sustainable in fashion [8].

Tie-dyeing is dyed with basic techniques like knotting, folding, binding, clapping, and stitching, where desired patterns are produced with different methods. This basic process involves garments or fabric being rinsed and washed after dyeing. Spiral, crumple, stripped are some prominent techniques that have been derived from this method. For the sustainability aspect, there are continuous improvements in the technique and materials of tie-dyeing and also in the finishing process [9].

Elimination of the use of petroleum-based synthetic dyes in tie-dyeing processes has nowadays become one of the most important requirements to ensure ecological balance and health safety [10, 11]. Natural dyes are sustainable alternatives for toxic and carcinogenic synthetic dyes. To ensure environmental and health safety, not only the colorants but also the chemical auxiliaries need to be harmless. In the conventional processes of producing naturally tie-dyed apparels, metal-containing mordants are typically used which have a negative impact to some extent on human health and aquatic life [12]. Therefore, investigations are needed to innovate new tie-dyeing methods where metal-free mordants can be implemented to fix the natural dyes on the surface of textile substrates.

This chapter offers a bird's eye view of sustainable fashions through naturally tie-dyed textiles. Based on history, the materials and techniques of tie-dyeing have been gathered here with conceptual schematic presentations. An exclusive review of the natural dyes and mordants incorporated in tie-dyeing methods has been provided and discussions have been made regarding the future scopes for developing novel sustainable methods to manufacture fashionable tie-dyed apparel. A comprehension of the impacts of tie-dyeing techniques on the aesthetics of textiles has been given. In the final part of this chapter, discussions on the market value of tie-dyed apparels are made to emphasize the demand for sustainable tie-dyed textiles.

8.2 History of Tie-Dyeing

Tie-dyeing was introduced in the Sui Dynasty in China in the fifth century AD where the designs included brighter colors with small circles and lines. Asians introduced a resist technique of tie-dyeing named *shibori* in Japan in the eighth century AD.

This resist technique was accomplished by two main steps to prevent dye from penetrating certain areas: first, elaborate patterns were stitched and gathered tightly; second, the fabric was wrapped around a core of rope or wood and bound with string or thread [7]. *Bandhani* is another type of tie-dyeing where the desired design is created by plucking the cloth with fingernails to form bindings and resist dye in knot patterns. This was one of the oldest methods of tie-dyeing and was initially discovered at 4000 B.C. back in western India and still widely being practiced [13]. For centuries, the Hausa communities of West Africa have preserved the cultural practice of tie-dyeing, utilizing indigo dyes in pit-dyeing processes, especially in the Kano region of Nigeria. In Yoruba culture, tie-dyeing is called *adire*, a clothing made using resist-dyeing techniques, mainly with indigo dye, and worn by the Yoruba people of southwestern Nigeria. The practice began in the early 20th century and was mostly done by women. Tie-dyeing was introduced to the U.S. in 1909 by Professor Charles P. Pellow from Columbia University, who delivered a lecture and live demo of his technique using tie-dyed muslin fabric. Before the 1960s, techniques like *Shibori* and *batik* were used occasionally. Tie-dyeing remained trendy till the late 1960s, inspired by rockstars like Janis Joplin and John Sebastian, who dyed their own clothes. It gained huge popularity in the U.S., especially with the arrival of affordable commercial dyes, which made it easy for people to turn clothes into colorful designs [14].

8.3 Natural Dyes Used in Tie-Dyeing

Natural colorants can be collected from vegetable, animal, mineral, and microbial sources. Nevertheless, the use of natural dyes of vegetable and animal origin has been reported in numerous literature [15, 16] for tie-dyeing of textiles. In Indonesia, natural colorants extracted from various parts of plants are used in tie-dye. As cherry leaves contain two major color-bearing components, flavonoid, and tannin, the dyes extracted from cherry leaves are preferred for tie-dyeing [17]. Moreover, *cherry leaf extracts* exhibit antioxidant and antimicrobial properties due to polyphenols, saponins, and triterpenoids apart from flavonoids and tannins. Mangiferin, a natural compound containing carbonyl and hydroxyl groups as chromophore and auxochrome, can be collected from *mango leaves*. Mango peels may also be a source of natural color which can provide brown color [18–20]. Curcuminoid groups (curcumin, bis-demetoxy curcumin, and demetoxy curcumin) present in *turmeric* are able to impart yellow-orange color to a textile substrate. Brazilin, found from *barks of brazilwood,* is a potential natural colorant that can be used for tie-dyeing as it can impart a red color [21]. In Indonesia and Malaysia, tannins (brown) and anthocyanins (red) are reported to be extracted from the peels of mangosteen fruit [22]. Desta et al. [23] reported that cotton fabric panels can be treated in tie-dyeing technique using the natural dyes extracted from the *petals of marigold flowers* and Kenyan *muuku bark*. Arpana et al. [24] tie-dyed silk kurtis using extracts from *kilmora*, *brahmi*, *arjun*, *madder* and *golden drop*. *Sappanwood barks*, *purple cabbage* and *basil leaves* can impart orange, purple and light green colors respectively [25]. According to the investigation of Gausa et al. [26], *logwood*, *sunflower*, *henna*, *woad* and *saffron* are vegetable sources of natural colorants as well. Machado et al. [27] showed a way to tie-dye cotton fabrics with the natural color extracted from acorns of red, white and pin oak trees using metal rusts as mordant. Tie-dyeing of fabrics with the colorants obtained from animal sources like *cochineal*, *tekhelet* (snail), and *tyrain purple* is familiar in West Africa [26].

8.3.1 Advantages of Natural Dyes

Inspired by the worldwide consciousness regarding the environment and human health, many countries have imposed statutory regulations on the usage of carcinogenic and toxic dyes derived from petroleum-based resources. This is the fact that is grabbing the attention of modern researchers to explore environment-friendly substitutions of using synthetic dyes [10]. Researchers are spotlighting the natural sources dyes extraction and utilizing them as an option that is better than harmful synthetic dyes from the perspective of health and the environment [28].

8.3.1.1 Environmental Benefits of Using Natural Dyes in Tie-Dyeing

Notwithstanding that synthetic dyes have superior brightness, fastness properties, usability and shade-reproducibility compared to natural counterparts, they carry insoluble contaminants that are eventually released into the environment after the dyeing process [29]. The developing countries, to be specific, do not have well-developed effluent treatment systems and they discharge the harmful dyeing effluents with insufficient treatment [30]. According to the information provided by Hassaan et al. [31], 80,000 to 120,000 tons of synthetic dyes are annually discharged into the ecosystem all over the world. Such uncontrolled disposal of high amounts of synthetic dyes has severe consequences on aquatic life. Typically, around 300 mg/L dye residues are present in the effluent, whereas, only 1 mg/L or even lesser concentration of dyes can spoil the transparency of water. In opaque water, light cannot pass through and hence the photosynthesis processes of aquatic plants are hindered. Unused synthetic dyes are responsible for the degradation of the quality of dissolved oxygen and alteration of the level of pH [30]. The contemporary tie-dyeing processes allow the use of substantial amounts of synthetic dyes that are non-renewable petrochemical products. Extreme issues related to the environment can occur owing to the discharge of amine-containing wastewater after the application of synthetic dyes in wet processes like tie-dyeing [10, 32]. Unlike synthetic dyes, natural dyes are both biodegradable and collected from renewable sources making them sustainable [33].

8.3.1.2 Health Benefits of Using Natural Dyes in Tie-Dyeing

Azo-containing synthetic dyes can not only create cancer, and allergy in the skin, but also they can hamper the human reproductive system. Consequently, stringent regulation has been dictated by the European Union against the use of synthetic dyes having high azo content. Disperse dyes which are used to dye hydrophobic fibers (like polyester), are capable of creating allergy as well [11]. East African people practice tie-dyeing in an uncontrolled way in their backyards posing a detrimental impact on the state of their health. Inappropriate disposal of synthetic dyes and their essential auxiliaries can cause skin discomfort, breathing difficulties, and allergic responses [34]. Some natural dyes not only impart color to a substrate but also offer some functional attributes, such as antioxidant properties [35], antimicrobial properties [36], and ultraviolet resistance [37], aromatic properties and insect repellency [38].

8.3.2 The Science Behind Tie-Dyeing with Natural Dyes

As natural dyes have insufficient substantivity to textile fibers, the dyeing process typically involves treatment with suitable mordanting agents like ferrous sulphate [39], copper sulphate, aluminum sulphate [40], potash alum [41], zinc sulphate, potassium dichromate, stannic chloride, stannous chloride, tannic acid, tannin [42] and so on. The objective of mordanting is to form a dye-fiber complex by establishing connections between the functional groups of dyes and the fibers and hence improving color fastness [43, 44]. Metal mordants have an impact on the hues of the dyed materials.

Mordants can be applied before, during or after the dyeing process. According to Vankar et al. [45], three types of compounds can be used as mordants such as metallic salts, tannins and oils. Based on the ability to brighten the shade of the dyed fabric, mordants are of two types. $K_2Cr_2O_7$, alum, and $SnCl_2$ are said to be brightening mordants for their ability to produce bright shade. Ferrous sulphate and copper sulphate fall into the category of dulling mordants as their use offers dullness in the shade of naturally dyed fabrics.

Since the mordants are normally metal salts, sometimes they can pose harmful impacts on the environment [12]. The application of metallic salts in the tie-dyeing process has been reported in many research [23, 44, 46, 47]. Oils derived from natural sources (stearic acid, palmitic acid, oleic acid, etc.) can act as mordants if they are treated with concentrated sulphuric acid and metals. Concentrated sulphuric acid imparts sulphonic acid groups to the oils which have superior metal binding capacity. The incorporation of metals converts the sulphonic acid groups into metallic sulphonate groups in which the bound metal is capable of creating metal complexes with the functional groups of natural dyes. The use of tannins can solve this issue as they are natural products found in plenty of vegetable sources like bark, fruits, fruit pods, galls, leaves and roots of plants and they do not contain metals in their chemical structure. Tannins refer to a vast class of high molecular weight organic compounds bearing hydroxyl groups capable of creating cross-linkages between fiber-forming polymers [42]. Machado et al. [27] used tannin-iron complexes to endow a variety of colors to the tie-dyed fabrics. In Nigeria, sustainable tie-dyeing is carried out using mordants made from wood ash or a blend of bulrush millet and wood [26].

Bio-mordanting has not gained popularity in the tie-dyeing processes yet, however, apart from reducing the utilization of synthetic dyes, the use of bio-mordants is of significance in resolving environmental and health issues.

8.4 Types of Tie-Dyeing Techniques

Tie-dyeing is a manual method where desired patterns are produced on textiles by tying required parts of the fabric or yarn so that they will not soak dye and the united portion will absorb the dye. Tie dyeing is usually done by direct dyeing where the fabric is plunged right into the dyeing bath for coloring [48]. During dyeing, different shapes, designs, and structures are obtained by different tying processes. Splashing dyeing, pulling dyeing, and hanging dyeing, known as direct dyeing, are also applied in tie-dyeing. Tie dyeing is also done by ice dyeing, reverse dyeing, immersion, and low-water immersion [49]. Desired patterns are produced by spiral, crumple, ombre, stripes, folding and clamping, knotting and tying, knotting and binding, stitching, and binding.

8.4.1 Tie-Dyeing Materials

In tie-dyeing, several tools and materials are used, including fabrics or garments, along with fiber-reactive dyes. Assorted fiber-reactive dyes, fabric items, soda ash, laundry detergent, rubber bands, squeeze bottles, wire racks, gloves, and dust masks are needed for the tie-dyeing process which needs to fulfill some criteria as per dyeing requirements.

- **Fiber reactive dye**: Assorted colored fiber reactive dyes are usually selected for tie-dyeing, which have the optimal properties to react with the fiber easily and show a good color fastness. Most garments are dyed with man-made dye but the use of natural dyes is increasing day by day because of the sustainability aspect.
- **Fabric item**: Mostly, finished garments like shirts, T-shirts, socks, or sweatshirts, usually made from cotton and have clean surfaces to produce the desired design are tie-dyed.
- **Soda ash**: Before dyeing garments, they are soaked in soda ash solution for 15 min during the preparatory process to prepare them for dyeing.
- **Laundry detergent**: Synthrapol or mild laundry detergent is used for washing the garment to clean the impurities at the preparatory process and excess dyes at the washing.

- **Tying materials**: Rubber, polyethylene, or cotton threads are usually used because these materials have the proper strength to overcome the repeated process of dyeing chemicals and temperature and are flexible for easy wrapping.
- **Squeeze bottles**: Bottles with squeezes are needed to be applied assorted colors on the garment.
- **Wire racks**: Wire racks are needed to dry the washed garments.
- **Gloves**: To protect hands, gloves are used which keep hands safe from chemicals.
- **Dust Masks**: During the dyeing process dust masks are used to protect from dust.

Besides, these different materials are used as per the requirements of the dyeing process [50].

8.4.2 *Spiral Tie-Dyeing*

Spiral tie-dyed apparels consist of a complex classical design. After dyeing, the garments consist of a spiral design. Reactive dyes are more suitable for cotton fabrics for tie-dyeing. Usually, brighter and more durable dyes are used in this technique. Fabrics are washed with detergent before dyeing and also soaked for 15 min in a soda ash solution. To produce the spiral design, the fabrics need to be rolled up by twisting clockwise or anti-clockwise while pinching a small desired section or center of the fabrics lying on a flat surface, as illustrated in Fig. 8.1a and b. After rolling, the fabrics need to be secured by rubber bands, passing rubber over the center of the spiral and dye as desired, as shown in Fig. 8.1c and d. Spiral design tie-dyed fabric is obtained after rinsing, washing, and drying as presented in Fig. 8.1e [51].

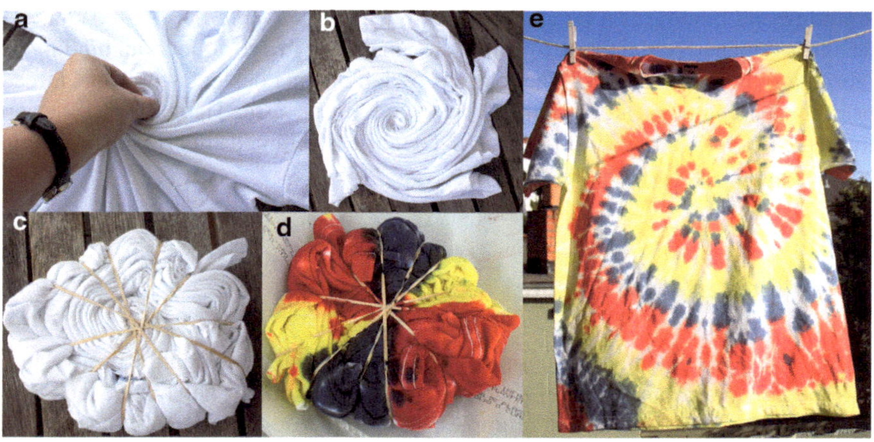

Fig. 8.1 Spiral tie-dyeing. **a** Pinching at the fabric. **b** Rolling fabric. **c** Binding with rubber. **d** Dye applying and **e** Finished tie-dyed apparel. Adapted from Flickr [https://www.flickr.com/photos/wendycopley/albums/72157621965922012/]

Fig. 8.2 Knotting and binding technique. **a** Knotting and binding fabric and **b** Finished tie-dyed fabric. Adapted from Flickr [https://www.flickr.com/photos/wendycopley/albums/72157621965922012/]

8.4.3 Tie-Dyeing by Knotting and Binding

Knotting and binding is a famous tie-dyeing technique where the desired shape or design is obtained on the fabric by immersing the fabric in dye liquor after binding the fabric into the desired shape. The fabric is tied in distinct designs with rope, thread, or rubber in knotting and binding, as presented in Fig. 8.2a. Thereafter, fabrics are immersed in dye liquor. The desired tie-dyed fabric is obtained after rinsing, washing, and drying, as shown in Fig. 8.2b [9].

8.4.4 Crumple Tie-Dyeing

In crumple tie-dyeing the fabric is also prepared as spiral tie-dyeing by pre-washing and soaking in soda ash solution. Crumple tie dyeing is an easy and unique technique where the fabric needs random crumples. In crumple tie-dyeing fabric needs to be crumpled randomly and tied with rubber springs, as shown in Fig. 8.3a and b. Random design tie-dyed fabrics are obtained after applying dye and finishing as presented in Fig. 8.3c and d [52].

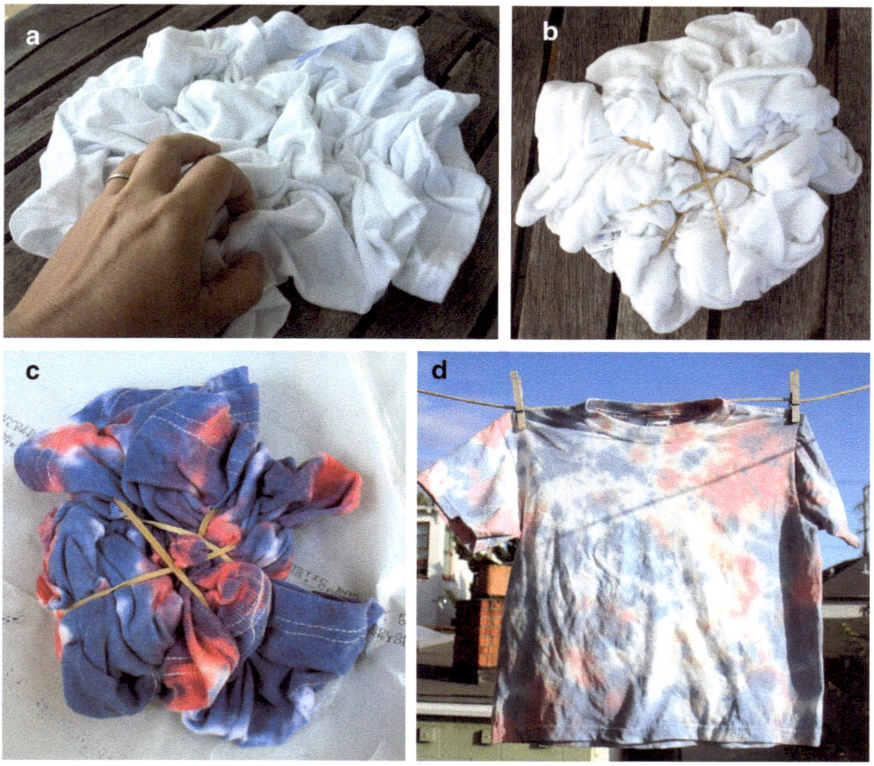

Fig. 8.3 Crumple tie-dyeing. **a** Crumpling fabric, **b** Tying fabric, **c** Applying dye, and **d** Finished crumpled tie-dyed apparel. Adapted from Flickr [https://www.flickr.com/photos/wendycopley/albums/72157621965922012/]

8.4.5 *Folding and Clamping Technique*

Folding and clamping technique provides layers of color in fabric with 3D design. In this technique, fabrics are folded in different designs using clips, wooden sticks, and blocks, as presented in Fig. 8.4a. Sharp designs are gained after immersing in the dye bath and also after rinsing, washing, and drying (Fig. 8.4b) [9].

8.4.6 *Ombre Tie-Dyeing*

In ombre tie-dyeing process, the fabric is prepared in the same way as spiral tie-dyeing. In the beginning, dyes are squeezed at the bottom or top of the fabric, and then it is brushed with a wet brush upward or downward the dye as shown in Fig. 8.5a,

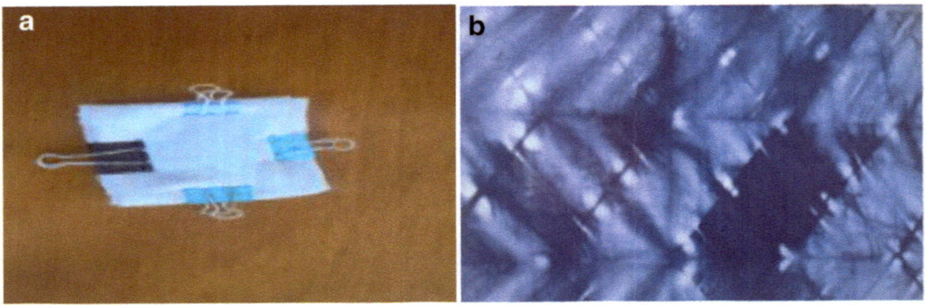

Fig. 8.4 Folding and clamping. **a** Folding and clamping fabric and **b** Finished tie-dyed fabric. Republished with permission from fibres & textiles in Eastern Europe [9], Copyright, CC-BY terms, 2023

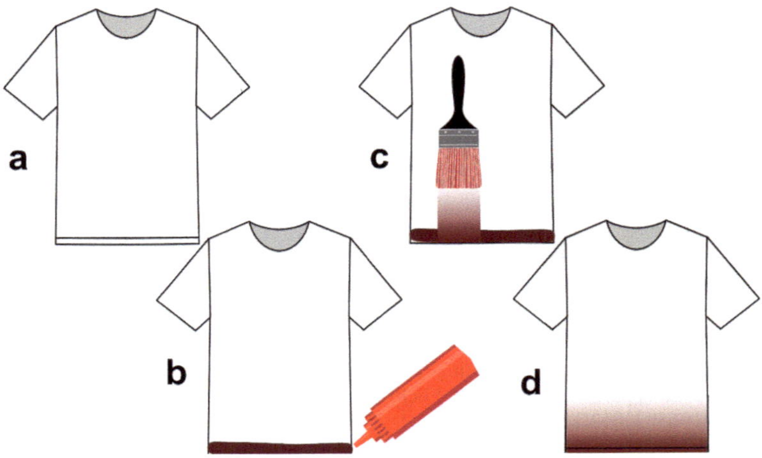

Fig. 8.5 Ombre dyeing, **a** Ready to dye plain garment, **b** Dye applying, **c** Dye movement, and **d** Finished apparel

b, and c. Finished apparel is obtained after washing and drying, as presented in Fig. 8.5d [9].

8.4.7 Knotting and Tying

The knotting and tying coloration technique produces smooth patterns with vibrant colors. The fabric is folded and knotted in various desired parts to produce the selected pattern by dyeing as illustrated in Fig. 8.6a. Finished products are produced with the desired design by completing all finishing processes as presented in Fig. 8.6b [9].

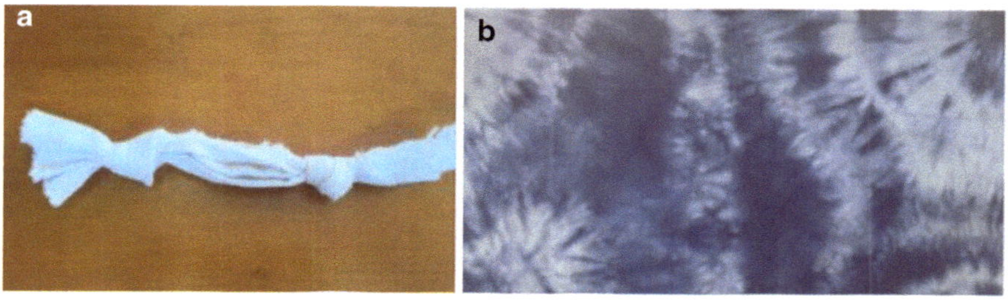

Fig. 8.6 Knotting and tying technique. **a** Applying knot and tying at the desired pattern and **b** Finished tie-dyed apparel. Republished with permission from fibres & textiles in Eastern Europe [9], Copyright, CC-BY terms, 2023

8.4.8 Striped Tie-Dyeing

Striped designs are formed in fabric by accordion folding and applying alternative colors. The striped design could be vertical, horizontal, or diagonal which is produced by a series of 1-inch accordion-foldings. Folded fabrics are secured by applying rubber bands in every 1 to 2 inches and also different colored dyes are applied to create a stripe design as presented in Fig. 8.7a, b and c. Striped-designed apparel is produced by finishing washing and drying as illustrated in Fig. 8.7d [9].

8.4.9 Stitching and Binding

Stitching and binding are done by folding and boundling by stitches which produces diverse and vivid patterns. In this technique, apparels are folded and bounded by stitching according to the desired design after which the bounded apparel is immersed in the dye bath to produce the selected design as presented in Fig. 8.8a. Finished apparel is obtained by finishing washing and drying with the selected design (Fig. 8.8b) [9].

8.5 Setting and Finishing Naturally Tie-Dyed Textiles

The tie-dyed fabric needs to go through some finishing process to prevent bleeding or fading of the colors. Rinsing, washing, drying, ironing, and setting processes are the finishing processes to set the color of tie-dyed fabric which will prevent color bleeding and fading [53, 54].

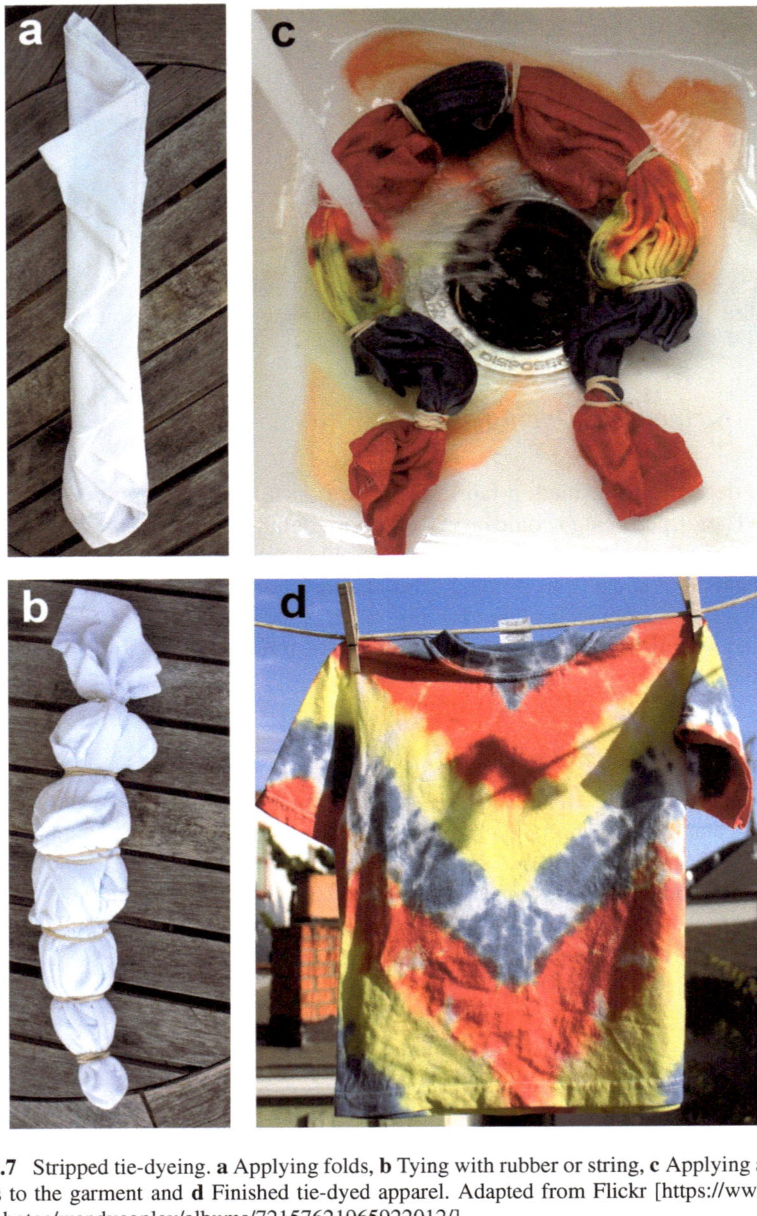

Fig. 8.7 Stripped tie-dyeing. **a** Applying folds, **b** Tying with rubber or string, **c** Applying assorted colors to the garment and **d** Finished tie-dyed apparel. Adapted from Flickr [https://www.flickr.com/photos/wendycopley/albums/72157621965922012/]

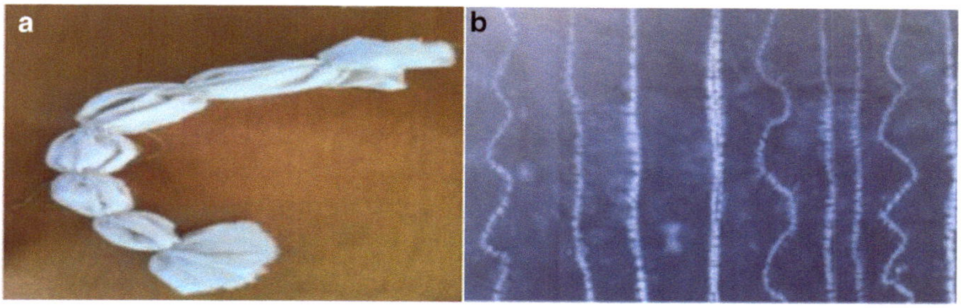

Fig. 8.8 Stitching and binding technique. **a** Stitching the desired design, and **b** Finished tie-dyed apparel. Republished with permission from fibres & textiles in Eastern Europe [9] , Copyright, CC-BY terms, 2023

8.5.1 Rinsing and Washing

During the rinsing process, the rubber or ties are placed under cold running water, after that the ties and rubber need to be removed, and continue rinsing under cool water or warm water till the water gets clear.

Completing the rinsing process, the dyed fabric needs to go through a washing process where tie-dyed fabric is washed with synthrapol detergent in hot water which removes unfixed dyes. To prevent muddy-look in tie-dyed fabric, assorted color-dyed multiple garments should not be washed together. However, multiple garments could be washed together until they are dyed the same colors [55].

8.5.2 Drying and Ironing

Tie-dyed garments need to go through the drying process after washing and rinsing to maintain the colors and design. Tie-dyed garments are dried using the air drying method, which is more suitable for maintaining pattern and color. The garment could be dry by hanging on the rack or lying the garment on a towel. To prevent distortion of the pattern and shrinkage, a heat dryer should be used, and also avoid drying in the sunlight to prevent fading.

To provide a smooth and wrinkle-free appearance, tie-dyed fabric is calendered after drying. Tie-dyed garments must be dyed inside out at a proper temperature with gentle press and smooth moving motions. Tie-dyed garments must be stored in a cool, dry place while using acid-free tissue paper between accurate folds to maintain the pattern [54].

8.5.3 *Care and Maintenance*

To preserve the colors and lifespan of tie-dyed garments, the application of harsh chemicals or bleach, rubbing or scrubbing should be avoided during washing. Hand washing is preferable over machine washing while using a gentle detergent to avoid twisting, which needs to be hung by the hem, or using plastic hangers with padded inserts. The stain needs to be treated immediately by stain remover or mild detergent and water [52].

8.6 Sustainable Fashion by Tie-Dye

In recent years, there has been a growing concern for environmental consciousness, prompting industries like fashion to adopt more sustainable approaches. This evolution has reignited interest in tie-dye, now with a stronger emphasis on environmental responsibility [7]. Incorporating eco-friendly principles in tie-dyeing processes holds the potential to exert a substantial ecological influence. Traditional tie-dyeing practices commonly utilize synthetic dyes and significant quantities of water, leading to environmental pollution and excessive water usage. Adopting eco-friendly practices enables crafting stunning tie-dyed patterns while ensuring a sustainable future for future generations. Furthermore, incorporating tie-dyed merchandise into inventory management strategies aids waste reduction and sustainability promotion, facilitating the efficient distribution of vibrant, eco-conscious clothing [8].

8.6.1 *Water Management by Tie-Dye*

Efficient water management is essential for environmentally sustainable tie-dyeing processes. Instead of employing water-intensive rinsing methods, alternative techniques are being explored such as ice dyeing or steam fixation to minimize water usage during the dyeing process. Employing such techniques conserves water resources without compromising on creating vivid and unique designs [8].

8.6.2 *Redesigning Second-Hand Cloth by Tie-Dye*

Young et al. [56] developed a collection of ready-to-wear garments, distinct from conventional mass-produced apparel, utilizing post-consumer men's suits and sweaters, thus defining it as post-consumer recycled clothing. Redesigning garments involves a detailed process exceeding the basic repairs or modifications by tailors to reduce functional issues or enhance the garment's fit [57]. The integration of tie-dyeing into the sustainable fashion narrative is largely attributed to its role in repurposing used garments. As opposed to conventional consumption patterns, individuals are now embracing methods to refurbish and reimagine their existing clothing collection, thus fostering a more sustainable approach to fashion [7]. Redesigning cloth used by tie-dyeing explores the option of upcycling existing garments or textiles rather than acquiring new materials for tie-dyeing projects. Utilizing upcycling techniques offers a practical solution to decreasing textile waste and reducing the ecological footprint associated with fabric production [8].

8.7 Effect of Tie-Dyeing on Textile Properties

Durability is a crucial aspect of quality and longevity of textile products. Tie-dyeing processes, being integral to textile production, significantly impact durability. Investigations have been employed to asses the impacts of tie-dyeing techniques on various characteristics of textiles, including wear resistance, tensile strength, and colorfastness [9].

8.7.1 Effect of Tie-Dyeing Techniques on the Water Resistance and Stability of Textiles

The term stability in textiles denotes the capability to endure wear and tear without deformation, whereas water resistance characterizes their ability to withstand washing cycles without compromising their original qualities. Tie-dyeing techniques express a substantial effect on enhancing the water-resistance properties and long-term stability of textile substrates. Meng et al. [9] demonstrated that using distinct tie-dyeing methods yields diverse impacts on durability and water resistance. Textile water resistance and stability are substantially influenced by tie-dyeing techniques, primarily attributed to textile structure and fiber orientation alterations. Through in case of the folding and clamping method, textile water resistance and stability are enhanced in contrast to that of the stitch and bind dyeing technique that compromises these attributes; however, the direct dyeing method exerts minimal influence on the water repellency and durability of fabrics [9].

8.7.2 Effect of Tie-Dyeing Techniques on the Abrasion Resistance and Tensile Strength of Textiles

Textile durability is commonly evaluated based on two important factors, such as tensile strength and resistance to abrasion. Different tie-dyeing techniques exert considerable effects on abrasion and tensile strength performance. Wear resistance and tensile strength are primarily influenced by the structural transformations induced by tie-dyeing techniques. Enhanced fiber alignment achieved through the folding and clamping dyeing technique minimizes distortion and breakage in textiles by promoting a more parallel orientation of fibers. Utilization of the knotting and binding dyeing method induces fiber tilting, augmenting surface area and inter-fiber friction, thereby enhancing textile abrasion resistance. Contrariwise, the stitching and binding technique induces entanglement and fiber distortion, causing heightened friction and reduced abrasion resistance and tensile strength in textiles [9].

8.7.3 Mechanisms of Tie-Dyeing Techniques in Improving Textile Durability

Changes in textile structure, largely influenced by tie-dyeing techniques, are identified as key mechanisms in enhancing textile durability. The folding and clamping dyeing method promotes fiber alignment, minimizing distortion and breakage, consequently improving textile strength, resistance to abrasion, and overall durability. Furthermore, it contributes to textile stability, maintaining their original shape and size integrity. On the other hand, the stitching and binding method induces fiber intertwining and distortion, exacerbating the friction between fibers. This phenomenon leads to a reduction in the abrasion resistance, tensile strength, and overall durability of textiles. Moreover, this technique compromises the stability of textiles, resulting in a loss of shape and size stability. The knotting and binding dyeing method induces fiber reorientation, leading to an augmented fiber surface area that fortifies textile durability against abrasion. Nevertheless, this technique does not yield significant improvements in textile tensile strength or stability [9].

8.8 Impacts of Tie-Dyeing Techniques on the Aesthetics of Textiles

The aesthetics of textiles is considered as visual properties of the textile products, including visual appearance, surface texture, tactile experience, and coloration and tie-dyeing techniques have immense influence on the aesthetics of the textile products.

8.8.1 Appearance and Hand-Feel of Textiles

Applying different techniques of tie-dyeing, different hand-feel, and appearance are achieved in textiles. Meng et al. [9] studied the impacts of different tie-dyeing methodologies on the hand-feel and appearance of textiles. They found that among the popular techniques of tie-dyeing, the direct dyeing technique results in textiles with superior hand-feel and even appearance. On the contrary, if the textiles are dyed with stitching and binding techniques, the products show a relatively rough appearance with an inferior hand feel. Alterations strongly influence the aesthetic and tactile features of tie-dyed textiles in fiber structure and alignment. In particular, the direct dyeing method induces a more parallel fiber arrangement, resulting in a smoother and softer hand feel with a uniform surface texture. The stitching and binding method creates entanglement and fiber distortion that leads to an uneven surface texture with a conceded subsided visual appearance and hand feels [9].

8.8.2 Texture and Color of Textiles

Tie-dyeing techniques offer unique approaches to design textiles that enable the creation of innovative textured patterns that overcome limitations associated with flat design patterns and unvarying color applications. Beyond color variations, tie-dyeing techniques demonstrably affect the texture of textiles. Research has shown that distinct tie-dyeing methods produce unique textural outcomes. For example, the direct dyeing technique imparts well-defined parallel textures onto the fabrics. On the other hand, the stitching and binding method produces a crisscrossed textural effect. In contrast, the folding and clamping technique exhibits minimal influence on textile texture [9, 58].

8.9 Commercialization of Tie-Dyed Fashion

Tie-dyeing techniques gained popularity as a fashion trend in the 1960s and 70s. Although the trend faded, its resurgence in 2019 was stimulated by celebrities like Justin Bieber and Beyoncé embracing and displaying the style [59]. At the same time, major fashion brands like Stella McCartney, Proenza Schouler, and Prada promoted tie-dyed fashions in their inventory. The trend is now being adopted by niche brands too [60].

8.10 Conclusion

Clothings are traditionally tie-dyed with synthetic dyes, which can be replaced with natural dyes where metal salts are used as mordants. The usage of natural dyes and bio-mordants is increasing in terms of sustainability. Also, different dyeing methods are being invented to make it more sustainable, which are derived from basic techniques such as knotting, tying, clamping, and stitching, where the rinsing and washing processes are improving to make it sustainable. This chapter concluded the tie-dyeing of textiles naturally and sustainably through different types of tie-dyeing methods. The different textile properties and the aesthetics of textiles change over the variations in the tie-dyeing techniques. Tie-dyed textiles were popular in some particular regions of the world in the past, now, their popularity is spreading rapidly among consumers with the incorporation of sustainability.

Declaration The authors declare that they have no known competing financial interests or personal relationships that could have appeared to influence the work reported in this chapter. Generative AI and AI-assisted technologies have been used in the writing process to improve the readability and language of the work only. After using this tool, the author reviewed and edited the content as needed and take full responsibility for the content of the publication.

References

1. Devaki E, Suganya M, Sreelakshmi S (2018) Sustainable fashion–a review. In: Parthiban M, Srikrishnan MR, Kandhavadivu P, Sustainability in Fashion and Apparels. WPI Publishing, pp 84–93
2. Chowdhury M et al (2022) A feasibility study to analyze the behavior of heat settings on the cleaner production of knitted fabrics. Clean Eng Technol 7:100429
3. Nagurney A, Yu M (2012) Sustainable fashion supply chain management under oligopolistic competition and brand differentiation. Int J Prod Econ 135(2):532–540
4. Chan TY, Wong CW (2012) The consumption side of sustainable fashion supply chain: Understanding fashion consumer eco-fashion consumption decision. J Fash Mark Manag: Int J 16(2):193–215. https://doi.org/10.1108/13612021211222824
5. Rahman MM et al (2023) CAD technology aided design and evaluation of performance parameters of warp knitted seamless garment. Int J Interact Des Manuf 17(2):787–799
6. Barasa N, Olal FO (2020) Creative designing of fabric patterns by tie-dyeing with synthetic dyes. Int J Innov Res Dev 9(2):186–191
7. The renaissance of tie-dye: a sustainable twist. https://endless-summer-nz.com/blogs/endless-magazin/the-renaissance-of-tie-dye-a-sustainable-twist
8. Going green with tie dye: earth-friendly ways to create stunning designs. https://www.factorytwofour.com/going-green-with-tie-dye-earth-friendly-ways-to-create-stunning-designs/
9. Meng X, Yang C, Zhu C (2023) Impact of tie-dye techniques on the durability and aesthetic appeal of textiles. Fibres & Text East Eur 31(5):43–55. https://doi.org/10.2478/ftee-2023-0044
10. Naveed R et al (2022) Microwave assisted extraction and dyeing of cotton fabric with mixed natural dye from pomegranate rind (Punica granatum L.) and turmeric rhizome (Curcuma longa L.). J Nat Fibers 19(1):248–255
11. Ostad-Ali-Askari K (2022) Management of risks substances and sustainable development. Appl Water Sci 12(65). https://doi.org/10.1007/s13201-021-01562-7

12. Sk S, Mia R, Haque MA, Shamim AM (2021) Review on extraction and application of natural dyes. TextE & Leather Rev 4(4):218–233. https://doi.org/10.31881/TLR.2021.09
13. Belfer N (1992) Batik and tie dye techniques. Courier Corporation
14. Okoli-Akirika N, Chukueggu C (2023) Redesigning the surface of tie–dyed fabrics for a new fashion trend in ABA, ABIA state. Awka J Fine Appl Arts 9(2):68–84
15. Banna BU, Mia R, Hasan MM, Ahmed B, Shibly MAH (2023) Ultrasonic-assisted sustainable extraction and dyeing of organic cotton fabric using natural dyes from Dillenia Indica Leaf. Heliyon 9(8):e18702. https://doi.org/10.1016/j.heliyon.2023.e18702
16. Mia R et al (2022) Natural dye extracted from Triadica sebifera in aqueous medium for sustainable dyeing and functionalizing of viscose fabric. Clean Eng Technol 8:100471
17. Ahmed T, et al (2024) New trends in printing applications of natural dyes and pigments. In: Islam S (ed) Renewable Dyes Pigments. Elsevier, pp 139–163
18. Zannat A, Uddin MN, Mahmud ST, Mia R, Ahmed T (2024) Natural dyes and pigments in functional finishing. In: Islam S (ed) Renewable dyes and pigments. Elsevier, pp 271–287
19. Islam SR et al (2024) Natural dyes and pigments as a source of medicine. In: Islam S (ed) Renewable dyes and pigments. Elsevier, pp 177–232
20. Al Fahad MA et al (2024) Pharmaceutical applications of natural dyes and pigments. In: Islam S (ed) Renewable dyes and pigments. Elsevier, pp 165–175
21. Zia KM, Adeel S, Aslam H, Khosa MK, Zuber M (2019) Influence of ultrasonic radiation on extraction and green dyeing of mordanted cotton using neem bark extract. J Ind Eng Chem 77:317–322
22. Kusumawati N, Santoso AB, Sianita MM, Muslim S (2017) Extraction, characterization and application of natural dyes from the fresh Mangosteen (Garcinia mangostana L.) peel. Int J Adv Sci, Eng Inf Technol 7(3):878–884
23. Karen Desta A, John K, Josphat Igadwa M (2019) Wash fastness properties of pre-mordanted tie-dyed cotton fabrics dyed with muuku bark and marogold flowers natural dyes. Ann Univ Oradea Fascicle Texts, Leatherwork 20(1)
24. Arpana K, Surabhi M (2018) Development of silk Kurtis with tie and dye technique using natural dyes. Asian J Home Sci 13(1):175–179
25. Mayusoh C (2015) The art of designing, fabric pattern by tie-dyeing with natural dyes. Procedia Soc Behav Sci 197:1472–1480
26. Gausa S, Abubakar E (2015) Tie-dye (Adire) among the Jukun people. Mgbakoigba: J Afr Stud 4:1–13
27. Machado C, Oliynyk AO, Silverman JR (2022) Tie-dyeing with foraged acorns and rust: a workshop connecting green chemistry and environmental science. J Chem Educ 99(6):2431–2437
28. Samanta AK, Agarwal P (2009) Application of natural dyes on textiles
29. Bide M (2014) Sustainable dyeing with synthetic dyes. In: Roadmap to sustainable textiles and clothing: eco-friendly raw Materials, technologies, and processing methods. pp 81–107
30. Slama HB et al (2021) Diversity of synthetic dyes from textile industries, discharge impacts and treatment methods. Appl Sci 11(14):6255
31. Hassaan MA, El Nemr A, Hassaan A (2017) Health and environmental impacts of dyes: mini review. Am J Environ Sci Eng 1(3):64–67
32. Uddin F (2021) Environmental hazard in textile dyeing wastewater from local textile industry. Cellulose 28(17):10715–10739
33. Saxena S, Raja A (2014) Natural dyes: sources, chemistry, application and sustainability issues. In: Roadmap to sustainable textiles and clothing: eco-friendly raw materials, technologies, and processing methods. Springer, pp 37–80
34. Jnr EB-A, Acquaye R (2022) Mitigating risks in inappropriate handling of chemicals by small-scale textile producers in Sekondi-Takoradi, Ghana. J Text Sci Technol 8(1):1–14
35. Baseri S (2020) Eco-friendly production of anti-UV and antibacterial cotton fabrics via waste products. Cellulose 27:10407–10423
36. Mia R, Mofasser A, Bhat MA, Howlader MI, Sayed NI, Bakar MA (2024) Utilizing neem wood waste as a natural dye in the development of bio-based functionalized colored jute fabric. Sustain Chem Pharm 38:101489. https://doi.org/10.1016/j.scp.2024.101489

37. Grifoni D, Bacci L, Zipoli G, Albanese L, Sabatini F (2011) The role of natural dyes in the UV protection of fabrics made of vegetable fibres. Dyes Pigm 91(3):279–285
38. Pizzicato B, Pacifico S, Cayuela D, Mijas G, Riba-Moliner M (2023) Advancements in sustainable natural dyes for textile applications: a review. Molecules 28(16):5954
39. Haji A (2019) Dyeing of cotton fabric with natural dyes improved by mordants and plasma treatment. Prog Color, Color Coat 12(3):191–201
40. Chhipa MK, Srivastav S, Mehta N (2017) Study of dyeing of cotton fabric using peanut pod natural dyes using Al2SO4, CuSO4 and FeSO4 mordanting agent. Int J Environ Agri Res 3(2):36–44
41. Haar S, Schrader E, Gatewood BM (2013) Comparison of aluminum mordants on the colorfastness of natural dyes on cotton. Cloth Text Res J 31(2):97–108
42. Prabhu K, Bhute AS (2012) Plant based natural dyes and mordants: a review. J Nat Prod Plant Resour 2(6):649–664
43. Ma X, Wei Y, Wang S, Zuo X, Shen B (2020) Sustainable ultrasound-assisted ultralow liquor ratio dyeing of cotton fabric with natural turmeric dye. Text Res J 90(5–6):685–694
44. Bigambo P, Wellah H, Ngaga H, Kimbokota S, Mrango M (2023) Turmeric dyeing of cotton fabrics using tie-dyeing techniques. Tanzan J Sci 49(5):1064–1078
45. Vankar PS (2000) Chemistry of natural dyes. Resonance 5(10):73–80
46. Jung JS (2017) Making of natural dyeing scarves by tie-dyeing technique. In: MATEC web of conferences, vol 108. EDP Sciences, p 03006
47. Samik S, Santoso AB, Kusumawati N (2019) Diversification of "NAJWA" Hijab staining using tie-dye method based on natural dyes. In: National seminar on chemistry 2019 (SNK-19). Atlantis Press, pp 17–21
48. Aspland J (1991) Direct dyes and their application. Text Chem Color 23(11):41–45
49. 20+ Tie-dye techniques for all skill levels. https://www.chandye.com/folding-and-dyeing-techniques/
50. Stearns S 17+tie dye patterns and folding techniques. https://sarahmaker.com/tie-dye-patterns/
51. How to tie dye 4 ways. https://chalkfulloflove.com/blogs/news/how-to-tie-dye-4-ways
52. 25 free tie dye patterns and folding techniques. https://www.craftpassion.com/tie-dye-patterns/
53. Tie-dye technique. https://anuprerna.com/story-details/tie-dye-technique/17835
54. What do you do after tie dying a shirt? https://alanbalen.com/blogs/tops/what-do-you-do-after-tie-dying-a-shirt
55. How to Tie Dye–Easy Techniques for Beginners. https://sarahmaker.com/how-to-tie-dye/
56. Young C, Jirousek C, Ashdown S (2004) Undesigned: a study in sustainable design of apparel using post-consumer recycled clothing. Cloth Text Res J 22(1–2):61–68
57. Janigo KA, Wu J (2015) Collaborative redesign of used clothes as a sustainable fashion solution and potential business opportunity. Fash Pract 7(1):75–97
58. Juniati N (2018) Indonesian tie-dye exploration with fabric manipulation. https://repository.ubaya.ac.id/34784/
59. Tie-dye is back for 2019 and this is why. https://www.harpersbazaar.com/uk/fashion/shows-trends/a25945717/tie-dye-is-back-for-2019-and-this-is-why/
60. Trend alert: tie-dye. https://members.asicentral.com/news/web-exclusive/april-2019/trend-alert-tie-dye/

Chapter 9
Microbial Colorants: Past, Present and Future

Saptarshi Maiti and Ravindra V. Adivarekar

Abstract Currently, as far as coloration is concerned, the textile industry is dominated by synthetic colors. Synthetic dyes emerged into this world due to an accidental discovery of a dye named '*Mauveine*' by 18-year-old chemist William Henry Perkin in 1856. However, these synthetic colors have adverse effects on the environment and humans. As a result, intense research in natural colors has been started, and a lot of developments are being taking place, shifting the entire focus of the textile processing sector towards *Sustainability*. Colors from natural sources can be derived from different parts of nature, like plants, animals, vegetables, fruits, and microorganisms. Microbial colors are the colored metabolites secreted by microbes under stress conditions, which can be a scintillating source of non-toxic and non-carcinogenic alternatives for human use, having a great potential to substitute synthetic colors. The present chapter discusses the evolution of such microbial colors, the latest technology involved, the production stages, the metabolic mechanism of biosynthesis, the economics involved in pigment production, its application in the textile processing sector for therapeutic and other commercial finishes, and novel strategies being undertaken by various researchers. It also touches upon the scalability of microbial colors and concludes with their future challenges and scope for further research.

Keywords Mauveine · Sustainability · Natural · Microbial · Pigment

9.1 Introduction

Color is one of the most promising characteristic features of any material. It defines their characteristics as well as their outlook to the viewers. Color is a perception, and it readily contributes to the aesthetics of any product. This is because its appearance is regarded as the primary aspect that is perceived by our senses whereas its color indicates acceptability. In today's era of fast fashion, such factors have become

S. Maiti (✉) · R. V. Adivarekar
Department of Fibres and Textile Processing Technology, Institute of Chemical Technology, Matunga (E), Mumbai, India
e-mail: maiti.sapta@gmail.com

M. Shahid et al. (eds.), *Advancements in Textile Coloration*,
https://doi.org/10.1007/978-981-96-5091-0_9

more significant, and popular sectors like textiles, food and cosmetics have enhanced the application of coloring matter for their products. Such an enhanced craze for colorants resulted in the acceptability of synthetic colorants over natural colorants for their reliability, economics and reproducibility [1]. If we look back into the history of the inception of synthetic dyes, as earlier mentioned, it started in 1856. On an accidental note, aniline purple was prepared by William Henry Perkin, and the color was named "Perkin's Mauve or Mauveine". Since then, numerous colorants have been chemically prepared, primarily utilizing petroleum resources. Natural colorants, on the other hand, have been utilized since the dawn of human civilization. The wall paintings of the cave of Altamira, Spain are one of the popular examples of it. Scientists have found those to be painted during Upper Paleolithic Age about 36,000 years ago. Colored fabrics were traced in the Indus Valley Civilization (2600–1900 B.C.), where the clothes show traces of madder dye in the remnants at Mohenjo-daro and Harrapa. The art of dyeing and the attraction of the human eye towards color also have evidence during the Bronze Age in Europe. Traces of blue pigment on very old cotton fabrics from Huaca Prieta on the north coast of Peru were positively identified as an indigoid dye (indigotin), derived from *Indigofera* spp. native to South America [2–4]. In 1884, cultivation of *Monascus* sp. was seen and utilised for manufacturing Chinese rice and rice wine both of which were red in color. The commercialisation of the first carotenoid pigment derived from *Cryptococcus* happened in 1954 followed by production of the carotenoid pigment from other species like *Rhodotorula* sp. Later, the production and development of astaxanthin from *Phaffia rhodozyma* and beta carotene from *Dunaliella salina* using natural products containing beta carotene was reported [5]. There are many other examples which can be drawn out from the history of color production from natural resources and utilisation of it has gradually grown with the civilizations. Various natural sources like plants, animals, microorganisms and minerals were used to achieve colors for various types, and they were the only source of colorants. However, with the advent of steam engines followed by industrial revolution, amount of yarn and fabric production increased multiple times. To suffice the need for colored fabric of this amount, natural sources became scarce and synthetic colors created their place in textile coloration. Their brightness, wide range of color gamut, color reproducibility, ease of application, performance properties, etc. made them the favourite of most of the dyers [6].

9.1.1 Synthetic Colorants

Synthetic colorants are being used for a long time since their inception. However, they possess certain drawbacks which are creating trouble for the environment. First of all, most of these dyes are petroleum based and therefore their resources are not abundant since they are depleting rapidly. The textile industry ranks second globally in terms of pollution and the major pollution is contributed by the dyeing process. It is predicted that about 20% of the global water pollution involves the textile dyeing processes. Such petroleum-based colorants, fixation agents (e.g. chromium and formaldehyde), finishing agents like flame retardants, etc. are the major contributors to pollution [7, 8]. The release of varieties of colorants and fixing agents post-treatment in water streams leads to primary destruction in the surrounding environment because due to the non-biodegradability and toxicity [9, 10]. Thus, there is a recent shift towards the consumption of biodegradable and eco-friendly materials and the necessity is increasing for nature-based colorants day by day. Various natural sources such as plants, insects, microorganisms and mineral ores are explored for alternatives to the synthetic colorants. The environmental demand requires the utilisation of natural dyes which have minimum risk to the ecology and colors obtained from such sources possess characteristic features like antioxidant, anti-bacterial, UV protection and anti-carcinogenic taken up from their parent sources [7, 9–12].

Textile industry predominately uses synthetic dyes because they are easily synthesised at a low cost, thermally stable, highly photostable and encompasses a wider spectrum of colors. But synthetic dyes cause a discharge of high amount of colored wastewater. Due to this not only water transparency is getting affected in the water bodies but also creating trouble for photosynthesis to take place naturally. Additionally, synthetic dyes are also harmful due to their toxicity, mutagenicity, as well as carcinogenic nature that pose threats to human health. Alongside, many irreducible synthetic colors are found to be major obstacles in the effluent treatment plants leading to the blockage of the membranes. These drawbacks of synthetic colors have diverted a huge interest towards natural colorants.

9.1.2 Natural Colorants

Natural coloring components are commonly (primarily known as pigments) recovered from animal, plant or microbial sources to impart distinct and vibrant colors. These pigments are being used as colors for ages. Synthetic dyes have the advantage of providing a full range of shades, but at the same time, they pose a threat to the environment and humans. In view of this, the sustainable textile market has once again started to divert itself toward natural colorants. Practically, so far plants are found to be the major sources of natural dyes and their extracts from roots, stems, leaves, flowers, fruits and seeds of various plant species produce colors along with therapeutic properties. The demand for natural dyes has exponentially increased to multiple folds. It is not possible to meet such demand from only plant-derived dyes due to land shrinking. Thus, there is a need to explore alternative sources of natural dyes. Insects and mineral sources are also having their own individual shortcomings based on their suitability as of now and availability in bulk is prominently missing [13]. Additionally, such sources though natural can eventually strain the eco-system due to the very high requirement of quantity in order to obtain the similar tinctorial strength produced by the synthetic colorants [14]. In most of the cases, the quality and colorfastness of such colorants are not found to be at par with the synthetic colors. Sufficient agricultural land and water are always required to grow any plant, and color extraction also requires ample volumes of water and energy. Along with it, most of the plant sources are seasonal. The major limitation of these natural dyes is their low yield after extraction.

Altogether, an alternative coloring source is of great interest to the textile and fashion sector, this has led the research arena to focus on microbial pigments to overcome such issues. Generally, EU and US consumers appear to be well informed of such problems. For example, the commercial success of underwear having natural dyes on it indicates a new positive outlook towards the usefulness of natural colorants though their synthetic counterparts are cheaper. To evaluate the techno-economic feasibility of various natural dye sources, several research projects are being undertaken.

9.1.3 Microbial Pigments

Microorganisms are considered as the potential source for bio-pigment production owing to their positive features over other natural sources in terms of availability, cost efficiency, stability, labour and easy downstream processing [15]. Varieties of bio pigments have been produced so far that are likely quinines, monascins, flavins, melanins, carotenoids, violancein, etc., from microorganisms [16]. Cultivation of microorganisms can be attained through submerged and solid-state fermentation on natural raw materials or industrial organic waste. Most of the microbial pigments not only serve as coloring compounds in Texile and Food processing, cosmetics industry but also possess antioxidant, anti-cancer, anti-inflammatory and anti-microbial activities [17].

Very specific and limited research works have been carried out considering microorganisms for color production which typically points out the scope for undergoing research in this sector. The present chapter will create a pathway for us to explore the potential of microorganisms for natural coloration.

Microorganisms are the most versatile tools in biotechnology to produce various molecules that include antibiotics, enzymes, organic acids and pigments. Recent survey has resulted in considering microorganism as a promising source of natural colors.

9.2 Pigment Production

The process of screening coloring components from microorganisms is not new. Traditionally there have been a lot of cultural processes to have utilized it, although not going into much deeper to check out the technology behind it. Two of the most common ancient and conventional microbial coloration processes are the preparation of red yeast rice in Japan and China and the preparation of black color paste popular in Kalamkari printing in India.

In ancient days, in Japan and China, some rice types were cultivated along with *Monascus purpureus* yeast because of which, the rice grains were used to screen the red purple pigment from the yeast and convert that into a colorant. The inception of this process is as old as 300 BC. Along with its use in culinary, Chinese people used to apply this in various traditional medicines.

Kalamkari printing is one of the most traditional and epic hand-printing processes carried out in our country. This method involves artistic craftsmanship using pens. The black color used in this process is prepared by the fermentation of jaggery using iron rust in a clay vessel for a period of a fortnight. The extract is then applied on naturally mordanted cellulosic fabric using wooden pens. This technique results in a bright black color and is primarily based on the utilization of microorganisms to produce colors.

Systematic extraction of pigments from microorganisms can become a major phenomenon for textile coloration in today's era. In this process, the first step is always to identify a microorganism that can produce colorant which can be performed in two ways. The culture can either be isolated from nature or any existing microorganism can be genetically modified for the same purpose. This method is also known as screening and the main objective of this step is the identification and isolation of color-generating microorganisms from non-producing ones. It involves a culture enrichment technique at an optimised incubation condition followed by the isolation of microorganisms. When a pure microorganism is obtained through screening and isolation, its gene sequencing analysis is generally performed and deposited at a registered bank [18]. The production of crude pigment can be carried out by growing microbial inoculums under the required conditions of temperature, pH and nutrients. As mentioned earlier, nourishing these microorganisms in agricultural wastes, e.g. in sugar cane bagasse due to higher nutrient content is an effective and cost-cutting step at the industrial scale of production [18–21]. However, the volume of product generated depends on the fermentation step. The required pigment is then extracted as per their solubility level and characterised for further application in various fields (Fig. 9.1).

9.2.1 Strain Collection

Microorganisms can sustain in desired environmental conditions depending on environmental stimuli like pH, temperature, salinity, humidity, their food source, aerobic nature, requirement of light, etc. In adverse conditions, they tend to produce more pigments as an action of shock and protection. Ideally, microorganisms that can produce pigment should be familiar with carbon and nitrogen sources, have pH tolerance, minerals and temperature, and produce good colors. Such microorganisms having non-pathogenic and non-hazardous nature can be easily isolated from the cell biomass and are always opted for bulk production.

Bacteria, fungi, algae and lichens are the three major sources of pigments [2, 21]. Among these sources, fungi are the well-explored source of microbial pigments.

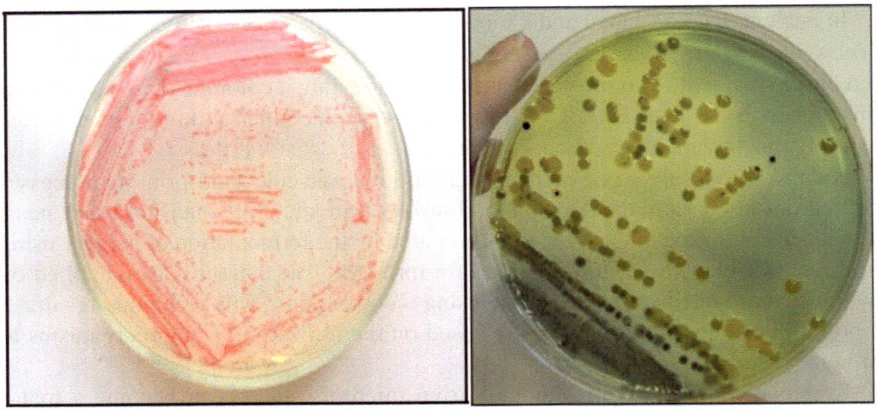

Fig. 9.1 Color-producing microorganisms

Many fungal pigments are chemically stable, non-toxic and are stable over a wide range of temperatures and pH [22–27]. However, the fermentation process requires high investment for installation. Fungi get very easily contaminated, are highly contagious and therefore its screening and growth is not easy. High production costs are also a matter of concern. Therefore, scientists are attempting to explore the possibility of industrial side streams and wastes for the fermentation processes to counterbalance the cost of production [28]. There are lots of fungi that can produce pigment from an economical substrate, mainly in the agro-residual wastes [29, 30]. Different fungi sources have been explored for the industrial production of red, yellow and orange pigments [31].

Bacteria are also one of the significant sources of microbial pigments that can be utilised in textile industries. *Actinobacteria* is the most popular pigment-producing group of bacteria. Some of the color-producing bacteria are *Flavobacterium* spp., *Bacillus* spp., *Serratia marcescens*, *Chromobacterium violaceum*, *Staphylococcus aureus*, *Agrobacterium aurantiacum*, etc. Vibrant colors can be obtained by organisms such as Pink-red, Golden Yellow, Purple, Red, Creamy and Yellow [32]. Production of these pigments has shorter life cycles as compared to fungi. Furthermore, the genetic modification of bacteria is simpler. Fungal pigments hold a higher position than the bacterial pigments in terms of commercial spectrum [33].

Algae are another type of abundantly available microorganisms engulfed with varying-colored pigments. It has been observed that macroalgae may contain phycoerythrins and carotenoids, which can become important pigments for the textile finishing processes [34, 35]. Many algal pigments are soluble in nature and therefore can be extracted without much effort. Extraction of algal pigment on large scale results in biomass disintegration, followed by organic solvent mixture treatment. The supernatant of the solvent mixture can be utilised further to extract pigment using various methods that include chromatographic techniques, phase separation, etc.

Lichens are symbiotic organisms derived from fungi and algae, known for producing fascinating metabolites with unique medicinal and functional properties. Lichens are also one of the most underexplored resources for natural colorants. They produce various depsides and depsidones. Under suitable conditions, such compounds act as precursors of various chromophores [36]. However, the area of lichen-derived natural dyes lacks systematic studies such as proper identification of coloring component(s) that are majorly responsible for shade, mechanism of dyeing and evaluation of performance properties of dyed specimen. Very limited research is being carried out on the dyeing application of lichens in textiles that has been established to date [37].

The color pigments extracted from these sources have great structural diversity. The derivatives of benzoquinone, prodiginines, pyocyanin, melanin, carotenoids, violacein, anthraquinone, azaquinones, shikonin, pyrrole dyes, phenazine dyes, etc. are among the few colorants already extracted from microbial sources and have great potential usage in textile processing sector [38–46]. A further boost to the screening of these natural pigments has been developed by their extraordinary pharmacological characteristics like anti-carcinogenic, anti-bacterial, anti-inflammatory, antioxidant, etc. which gives value addition to the aesthetics [47].

9.2.1.1 Pigments from Wild Bacteria

Marine water has proved to be an efficient source of microorganisms, and a novel rose red pigment has been extracted from *Serratia marcescens* subsp. collected from the mangrove forest soil of Mumbai, India [48]. *Pseudomonas aeruginosa* (GS-33) from marine produced Phenazine-1-carboxylic acid (PCA), which is a lemon-yellow pigment having anti-carcinogenic properties [49].

9.2.1.2 Pigments from Wild Fungi

Fungus tends to grow in varying atmospheric conditions, preferably in moist weather. In most of the cases, it has been observed that fungi grow well in wood and cause some coloration. This is also known as spalting and researchers are using it to explore pigment-producing fungi. *Scytalidium cuboideum* (red), *Scytalidium ganodermophthorum* (yellow), and *Chlorociboria aeruginosa* (green) were found to be producing pigments in different varieties of plants like pine wood, bamboo, etc. at North Pacific Region, USA [50–53].

Healthy roots of trees and plants are another good source of dark pigment-producing endophyte fungi. These endophytes can develop pigments that are bioactive secondary metabolites. *Phyllosticta capitalensis*, a melanin-producing fungi has been traced in multiple types of vegetation like both dry and wet deciduous forests, mangroves, semi-evergreen forests, etc. [54–56].

Biochar is a charcoal produced from plant matter and stored in the soil. It is generally executed by pyrolysis of biomass. Biochar and similar chemical compounds are found to be boosters for the growth of fungi, and it has been reported that aspergillin pigment (black) producing *Aspergillus niger* grow very well in these types of soils [30, 56, 57].

9.2.2 Strain Development

The technological science of engineering and improving strains of microorganisms, to improve the metabolic capacities for biotechnological applications is referred to as strain development. A recent survey says that genetically developed strains can produce pigments in a shorter span of time with several fold increase in pigment generation [58, 59]. This genetic transformation, the mutation process can be sped up using certain mutagens like Ultraviolet light, 1-methyl-3-nitro-1-nitrosoguanidine (NTG), ethyl methane sulphonate (EMS), etc. [60, 61]. In another report, microwave radiation was successfully utilised for the mutation of *Serratia marcescens* which caused considerable augmentation of production in prodigiosin pigment [62].

9.2.3 Fermentation

Fermentation is the process of chemical breakdown of substances by microorganisms, typically involving effervescence and generation of heat as a by-product. It can happen in the presence or in the absence of air. Pigments form as a secondary metabolite during fermentation. Fermentation can be categorised into two types. Solid state and Submerged fermentation (Fig. 9.2).

9.2.3.1 Submerged Fermentation

Submerged fermentation started its course in early 30s and has become very promising since then. In this process, the nutrients used for fermentation are in a liquid state that allows the microbial culture to grow throughout the solution. It can be both batch-wise as well as continuous process. This process has been utilized in many microbial color production processes (Fig. 9.3) [1, 63, 64].

9.2.3.2 Solid-State Fermentation

This process does not involve the presence of any liquid medium. Microorganisms are grown in solid media without any water, and only 15% moisture is required for their growth. These solid substrates are sparingly soluble biopolymers and are rich sources of nutrients for microorganisms [65]. It is a conventional process, and the most common reactors utilized for this process are Forced aeration, Drum, and Tower

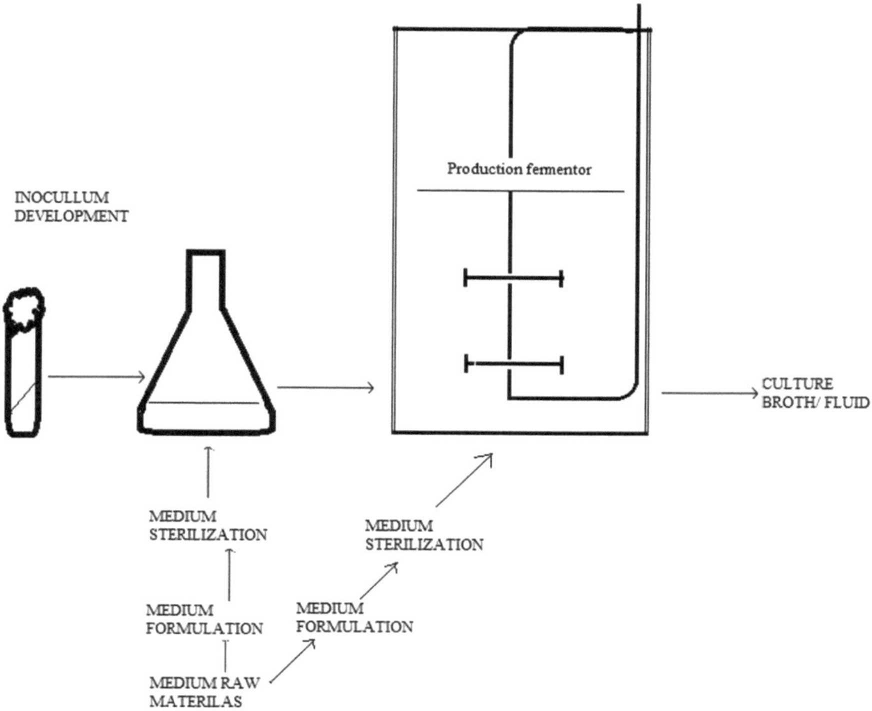

Fig. 9.2 Fermentation process of microbial extract

reactors. Agricultural materials like corn cob, jackfruit seed, etc., are reported to be used for the production of pigments from *Monuscus* spp. (Fig. 9.4) [30, 66–68].

9.3 Extraction

The fermentation process is followed by the separation of the pigments from the residue nutrients, dead mass, water, etc. This step is commonly referred to as the extraction which finally delivers the purified pigment for further processes. The extraction process of natural colorants can be widely classified into aqueous, super-critical fluid, alkali or acid assisted, microwave, ultrasonic method and solvent extraction (Fig. 9.5).

Aqueous extraction is always considered to be a sought-after process as it is an eco-friendly process, and the colored solution can directly be utilized for the subsequent dyeing process. It is to be noted that only the water-soluble colorants are suitable for such type of extraction.

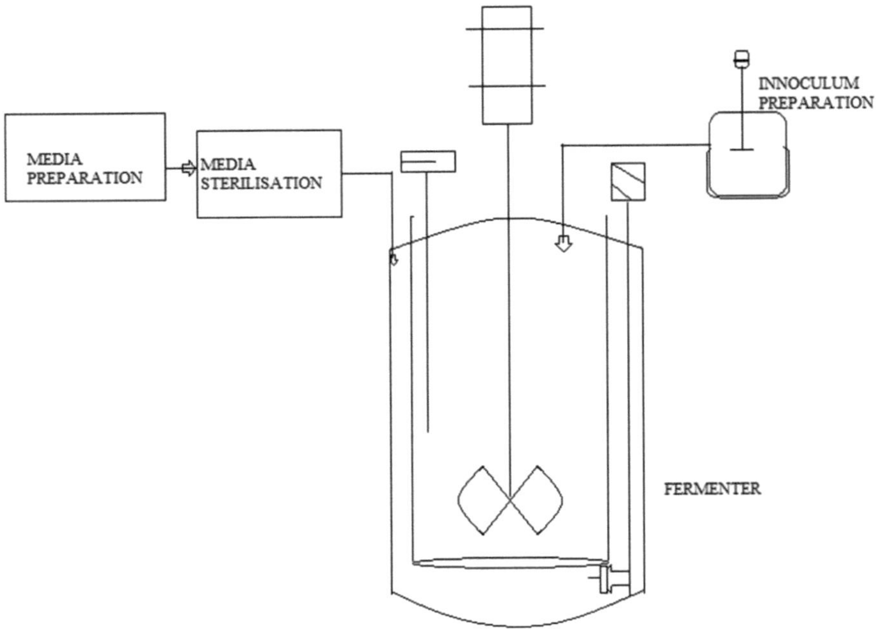

Fig. 9.3 Schematic process diagram of submerged fermentation

Enzyme extraction process uses enzymes for the separation of the dye from its origin. It usually occurs by the breakdown of the bonds between pigments and the substrate. As the enzyme is very target specific, the separation happens without any chemical degradation of the pigment and the pigment thus produced is dissolved in the extraction media.

Fermentation of roots, leaves, barks, etc., is carried out to extract the colors from the substrates. A very common example of dye extraction using the fermentation method is natural indigo. Fresh twigs and leaves are collected in a wooden vat, initially kept in warm water of around 32 °C temperature, and left for fermentation. The cellulose part of the leaves degrades, and the pigment comes out of it. The enzyme and fermentation-based processes are mainly used to extract dyes from plant sources. However, they can also be utilized to extract microbial colors from wood-based microorganisms [69].

Acid or alkali-based extraction is very much suitable for the extraction of polar color compounds from the fermented media. Alkaline extraction is especially suitable for colorants that contain polyphenolic compounds. Acidic extraction, on the other hand, is utilized for flavones-based dyes as it prevents the oxidative degradation of the molecules. This process is comparatively more promising and economical. The only limitation is the removal of alkali or acid later.

Solvent-based extraction process is very much popular mostly as it can be utilized for both the polar and non-polar colorants. It is efficient, cheap, in case of recycling

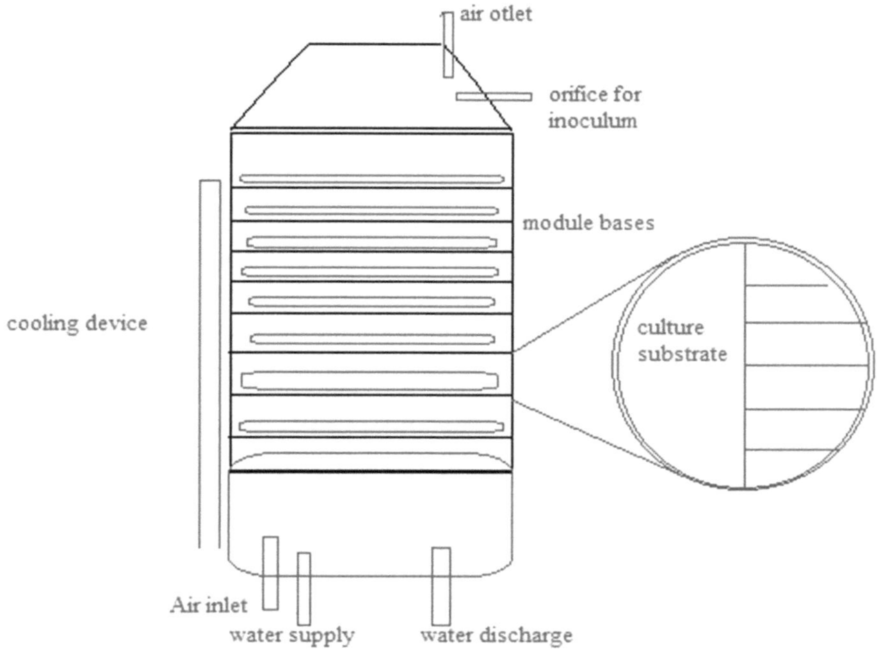

Fig. 9.4 Schematic diagram of solid-state fermentation

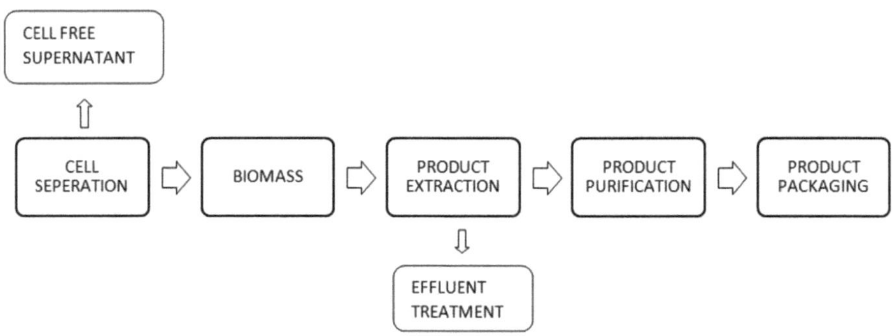

Fig. 9.5 Extraction process of the colored pigments

and reuse, it is quite eco-friendly. Some of the common organic solvents are methanol, ethanol, PET Ether, chloroform, acetone and water-alcohol mixture. This water-alcohol combination is applied to extract both water-soluble and insoluble pigments [70, 71]. Alkali or acid is also added in certain cases to hydrolyse the glycosides for a higher yield %.

Microwave and ultrasound technology can be implemented during the extraction process to shorten the time and increase productivity. These processes cause collision and agitation at the molecular level which causes pigment separation from the holding substrate much more effectively than normal mechanical agitation processes [69].

Utilization of supercritical fluids for color extraction from plant sources has also been surveyed [72]. Although this process is very efficient and eco-friendly, it requires high capital investment for installation and very limited suitability of raw materials.

9.4 Recent Trends of Microbial Pigment Production

Microorganisms in the surrounding environment are the inspiration and origin of various colors. It has been discussed throughout the chapter about those microorganisms that constitute vibrant colors. However, the important challenge is to improve the pigment production from these microorganisms so that the process becomes economically viable. One of the very clear solutions might be using a big fermenter; however, the best option is to use molecular biology to produce high-yielding microorganisms from the existing ones. The genes responsible for the biosynthesis of several pigments can be either mutated/cloned and recombinant DNA technology can be used to figure out different species with sufficient pigment yield capacity. Amgen Inc. scientists engineered a non-hazardous *E. coli* strain whose yield of producing indigo pigment (exclusively obtained from the woad plant) was very high [73, 74]. Polyketide like kalafungin is generally bright yellow in color and is a genetically modified version of *Streptomyces coelicolor*; a blue pigment actinorhodin. Alternatively, biosynthesis of actinorhodin can also produce orange or yellow–red anthraquinones by the use of genetic engineering [75].

9.5 Characterisation

The colors obtained from such microbial sources need to be characterised to know their structure, chemical nature like functional groups, chemical stability, etc. This is significant for their final application. Sometimes, a combination of various pigments is produced by the microorganisms and needs to be separately identified for a similar purpose. Above all, it is very much essential to find out whether the color has any affinity towards textiles. If the colors don't have any affinity for textiles that means they are not having any functional group or are non-polar in nature. There might be an additional step to carry out the fixation of the colorant onto textile materials physically or chemically. Therefore, it is always suggested to check their dyeability before application onto textiles. The extracted pigments can be tested under FTIR, Raman, and NMR Spectroscopy to find out the nature of the functional groups in the pigment required for coloration. Table 9.1 illustrates a list of some pigments majorly used for textiles with their chemical structures and functional groups.

Table 9.1 Reported microbial pigments with chemical structures

Sr. no.	Pigment group	Chemical structure
1	*Astaxanthin* (Purple)	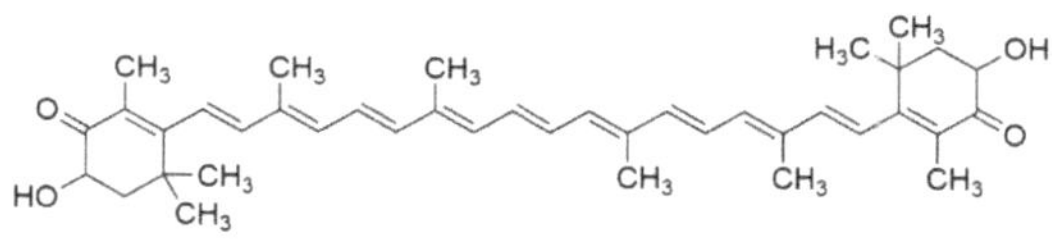
2	*Flavoglaucin* (Yellow)	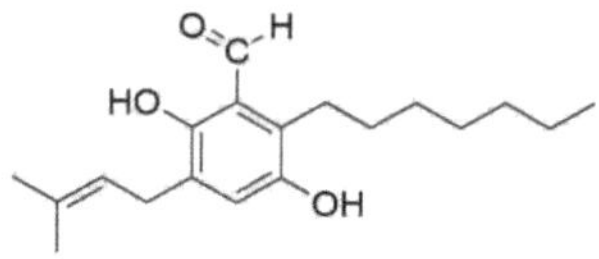
3	*Indigo* (Blue)	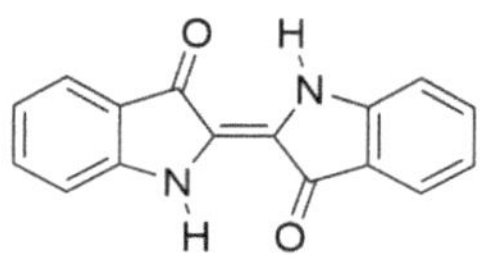
4	*Riboflavin* (Yellow orange)	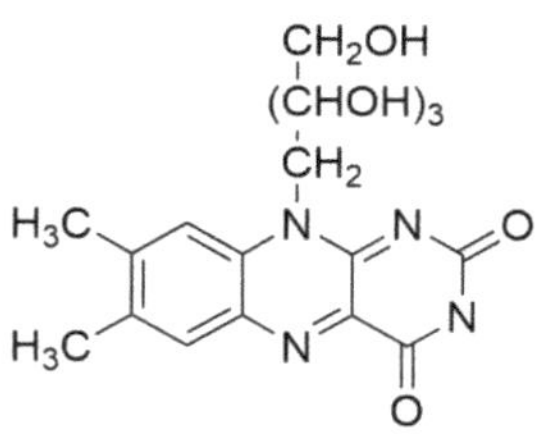
5	*Anthraquinone* (Red)	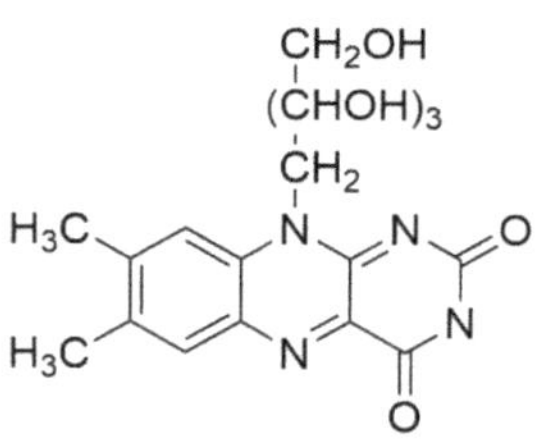
6	*Violacein* (Purple)	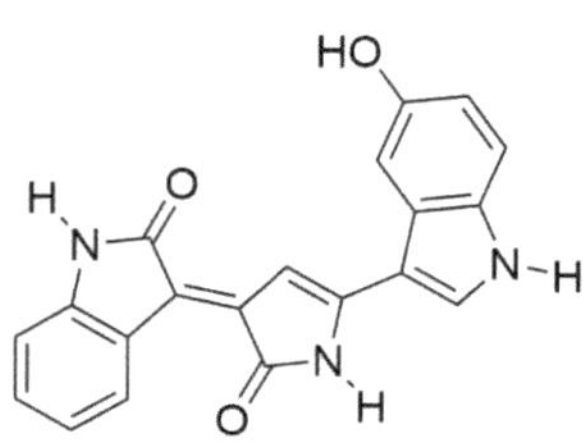

(continued)

Table 9.1 (continued)

Sr. no.	Pigment group	Chemical structure
7	*Napthaquinone* (Deep blood red)	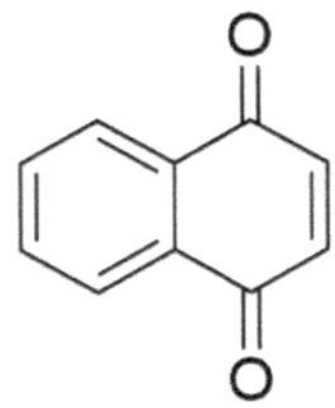
8	*Prodigiosin* (Red)	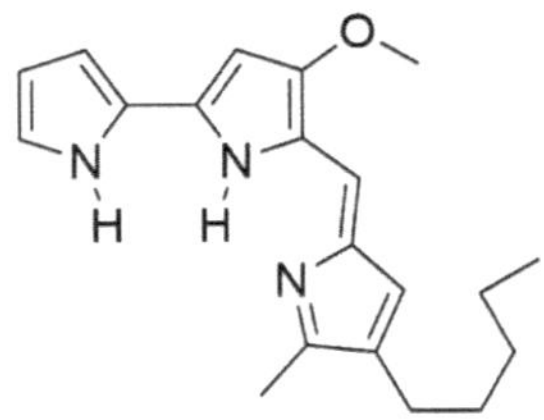
9	*Monascorubamine* (Red)	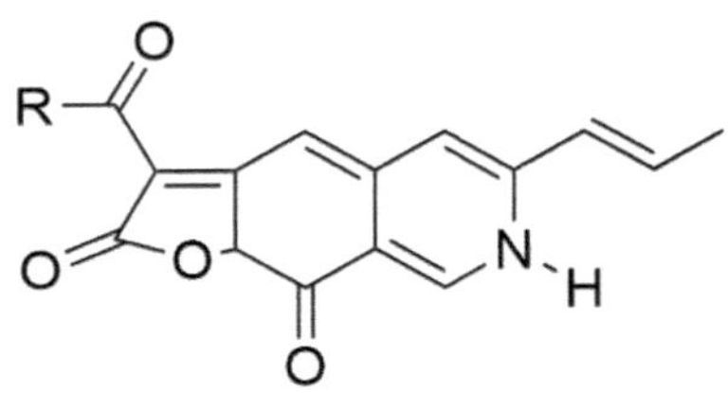
10	*Monascorubin* (Orange)	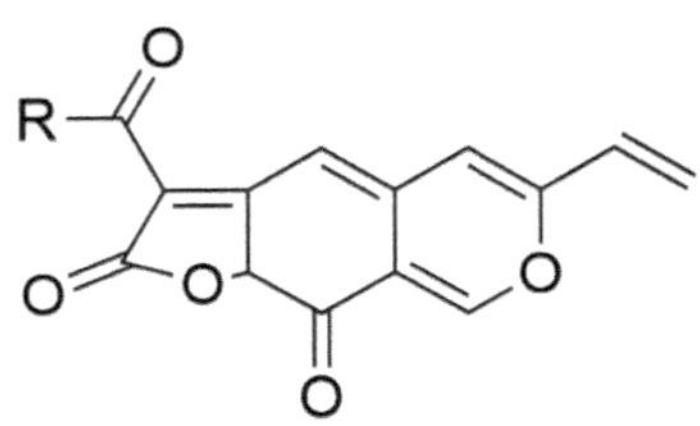
11	*Carotenoid* (Yellow)	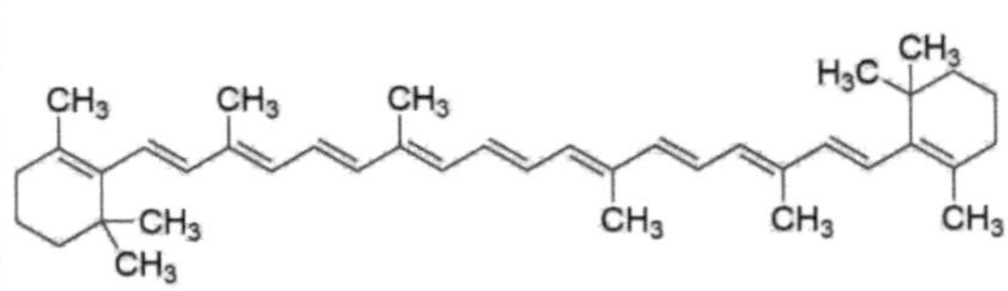

9.6 Application of Pigments

Pigments are molecules of significant importance to many industries. Their application varies from a wide spectrum of coloring agents to antioxidants and additives. These colors are mostly applied in areas of textiles, food, pharmaceuticals, cosmetics, and plastics. Pigments that are applied in textile coloration are mentioned below.

9.6.1 Textile Coloration

Colored fabrics with vibrant shades are always magnificent. It requires art, skill and knowledge of science, technology and craftsmanship for textile coloration. The choice of color selection for both a designer and customer can be sometimes challenging at times and to fulfil the demand, a dyer must have various colors that can cover the complete color gamut when used standalone or in combination. These dyes should have fair chemical and thermal stability because during dyeing, temperature and different chemicals like acid, alkali, salt and many other formulated chemicals are used. There are various classes of dyes that exist and are applied to textiles by distinctively different processes. To use the colorants in combination, it is necessary to have suitability within them and get onto fabric by a single process. All these variables make the process very complicated. This is the main challenge for coloration using natural colorants. Moreover, it is also a fact that natural colors are superior to synthetic colorants. Within natural colors, microbial dyes and pigments are slowly finding their niche and have a great potential to become commercial success in the near future. The need is to undertake intensive research with more and more discovery.

Researchers have produced violacein by the local isolation of *Chromobacterium violaceum* (Gen Bank accession no. HM132057) which is mainly found in different agricultural waste materials like sugarcane bagasse, solid pineapple waste, molasses, brown sugar as a substitute to the conventional rich medium [18]. This highly competitive, methanol and acetone-soluble pigment is stable over a wide range of pH and temperature.

Marine bacteria were bioprospected in an attempt to extract a feasible dye pigment source. *Serratia* sp. [48]. BTWJ8 was isolated which secreted prodigiosin-like pigment. This red-yellow natural pigment was evaluated for and showed stable performance activities [76]. Research has been conducted on extraction of *Vibrio* sp. strain from marine sediments that resulted in huge production of bright red pigments for the dyeing of fibres like wool, silk, wool, nylon and acrylic [44, 77]. Ahmad et al. have evaluated the impact of various pigments like prodigiosin and violacein for dyeing of silk and cotton fabrics [18, 71].

A survey was conducted for the isolation of three fungi, namely *Trichoderma virens, Alternaria alternate* and *Curvularia lunata* to isolate pigments for textile dyeing. It showed good results for dyeing silk and wool in absence of mordant.

Pigment from *Trichoderma virens* also demonstrated antifungal properties. During the molecular analysis of the pigments, it was found that they were having multiple coloring components [78]. Several *Monascus purpureus* species were cultivated on rich starch substrates and were well known to produce rubropunctamine and monascorubramine contributing to yellow and red color respectively [79].

A lot of research works are being carried out focusing on the microbial coloration of textiles some of which are illustrated below in Table 9.2.

9.7 Future Challenges

The inception of microbial colorants is very promising and eco-friendly, but there are certain hurdles that need to be overcome. The textile industry being one of the major manufacturing sectors always works under a heavy workload where there is all time battle for competition for textile production at the lowest possible time using minimum deployment of resources. This makes the coloration harder than it should be in terms of time. This may become a hindrance to using these colors. For a textile dyer, it is necessary to know the following criteria of a colorant.

- Classification of the dye,
- Substrates that can be dyed,
- A simple application procedure,
- Nature of fixation,
- Compatibility when used in combination.

Microbial colorants contain multiple functional groups and many a time, they don't possess any functional groups (pigments). Therefore, using them in a single process becomes very difficult. Along with it, there are some limitations in the production of pigments itself which are hindering the industrial scale production and coloration with microbial pigments. The primary limitations are

- Limited color gamut,
- Compatibility with one another,
- Solubility,
- Thermal and chemical stability,
- Initial high installation cost,
- Maintaining the cultural purity, etc.

9.8 Economics for Pigment Production

Public interest in microbial colorants has increased due to the awareness of health safety, and eco-friendliness, and their growth is expected to rise by 10% annually [23, 93]. Microbial source of pigment is an important source of natural colorants, but a lot of efforts are required to generate low-cost organic substrates to grow

Table 9.2 List of microorganisms for textile coloration

Sl. No.	Microorganism	Type	Pigment	Color	Application	References
1	*Chromobacterium violaceum*	Bacteria	Violacein	Purple	Pure cotton, pure silk, pure rayon, acrylic, polyester	[18]
2	*Serratia* sp.	Bacteria	Prodigiosin	Red-yellow	Wool, nylon, acrylics and silk	[48, 80]
3	*Vibrio* sp.	Bacteria	Prodiginines	Red	Wool, silk, nylon and acrylic	[44, 76]
4	*Chryseobacterium* sp.	Bacteria	Violacein	Yellow	Natural silk, linen, cotton	[18]
5	*Trichoderma virens*	Fungus	Multi-component	Yellow	Wool and silk	[81]
6	*Alternaria alternate*	Fungus	Multi-component	Brilliant olive	Wool and silk	[81]
7	*Curvularia* sp.	Fungus	Multi-component, anthraquinone	Skin melanin, Blue	Wool and silk	[81, 82]
8	*Dermocybe sanguinea*	Fungus	Anthraquinone	Yellow/Red	Wool/cellulosics	[83]
9	*Chlorociboria aeruginosa*	Fungus	Xylindein	Green	Polyester	[84–86]
10	*Scytalidium cuboideum*	Fungus	Not reported by the author	Red	Polyester	[84, 86]
11	*Scytalidium ganodermorphthorum*	Fungus	Not reported by author	Yellow	Polyester	[84, 86]
12	*Monascus purpureus*	Fungus	Monascorubramine, rubropunctamine	Orange, Red, Yellow	Wool, cotton	[87–89]
13	*P. purpurogenum*	Fungus	Not reported by the author	Yellow	Cotton	[88, 89]
14	*Thermomyces* sp.	Fungus	Not reported by the author	Yellow	Silk	[90]
15	*Phymatotrichum* sp. *(NRC 151)*	Fungus	Not reported by the author	Brown	Wool	[91]
16	*Drechslera* sp.	Fungus	Anthraquinone	Reddish	Polyamide	[82]

(continued)

Table 9.2 (continued)

Sl. No.	Microorganism	Type	Pigment	Color	Application	References
17	*Trichodenna* sp.	Fungus	Anthraquinone	Reddish	Polyamide	[82]
18	*Aspergillus* sp.	Fungus	Anthraquinone	Reddish	Polyamide	[82]
19	*Janthinobacterium lividum*	Bacteria	Violacein	Bluish-purple	Cotton, wool, silk and nylon	[92]

pigment-producing organisms. The bio-processing of microbial pigments is a tiresome job as it requires a higher level of nutrient enrichment and sterility. On the one hand, the natural/microbial colorant is gaining higher interest, but on the other hand, this may be more expensive than the synthetic colorants. It is expected in a few years, the application of microbial pigments will increase by multiple folds, and the consumers will be ready to accept it with its cost factor. Microbial pigments are comparatively costlier than synthetic colors and plant/insect-based natural colors. A standard synthetic violet (Brilliant blue FCF, E133) costs around 42–60 US $/kg while violacein costs around 5000 × 105 $/kg; Allura Red AC, E129 comes for 24–42 $/kg, annatto extract (plant-based red color) is of 80 $/kg while prodigiosin costs around 5000 × 105 $/kg. In the case of yellow color, FD&C Yellow No. 5 is comparatively costlier (500 $/kg) but still is far cheaper than carotenoid (1000 $/kg) and saffron (1400 $/kg) [66]. Such high costs include the cost of obtaining a pure nutrition source, the fermentation process, the extraction, purification, and final packaging of the product. Along with this, the low to medium yield of pigment production also adds up. However, these unjustifiably higher costs may be owing to the newer technology and high purity supply of these colors as per the requirement of other sectors. More and more research and developments will likely abate the cost to a marginal level. In addition to the present high cost, the major limitation of its use in textile coloration is that it requires an ample amount of dyes or pigments to color the substrate in comparison to its application in other industries. This brings the present market value amounting to nearly US$ 1/g, which restricts their application mainly to naturally colored garments with high-value additions. To face such challenges, there is a requirement for the development of an economical process to produce pigments that can successfully substitute the stand of synthetic dyes and pigments. Using abundantly available bio residues in solid-state fermenters might help decrease production costs to a greater extent [94].

9.9 Summary

The term 'Sustainability' is a buzzword in today's society. To attain it, the balance between society, environment, and economy is absolutely necessary. As an integral part of society and the economy, the textile industry requires more attention to follow eco-friendly processes. Finding renewable sources, utilizing such resources at optimum levels, and creating eco-friendly processes are the key drivers for sustainability. Overall, this chapter encompasses the production, characterization, and application of the pigments from microorganisms in an economical and efficient way to achieve sustainability, which has been reviewed. The significance of microbial color over other natural colorants has been discussed. The potential sources of microbial colors and their chemical structures have been summarized. Furthermore, the reported colorants used in textile materials have also been listed. The advantages, limitations, and prospects of textile coloration using microbial colorations have been lucidly explained.

References

1. Joshi VK, Attri D, Bala A, Bhushan S (2003) Microbial pigments. Ind J Biotechnol 2:362–369
2. Mallick A (2018) Interconnection between biotechnology and textile: a new horizon of sustainable technology. In: Handbook of renewable materials for coloration finishing. Wiley Online Library, pp 549–573
3. Cardon D (2007) Natural dyes: sources, tradition, technology and science. Archetype Books
4. Siva R (2007) Status of natural dyes and dye-yielding plants in India. Curr Sci 92:916–925
5. Venil KC, Zakaria ZA, Ahmad WA (2013) Bacterial pigments and their applications. Process Biochem 48:1065–1079
6. Lomax SQ, Learner T (2006) A review of the classes, structures, and methods of analysis of synthetic organic pigments. J Am Inst Conserv 45:107–125
7. Singh OP (2000) Natural dyes: the pros & cons. Indian Text J 110:42–47
8. Chattopadhyay SN, Pan NC, Roy AK, Saxena S, Khan A (2013) Development of natural dyed jute fabric with improved color yield and UV protection characteristics. J Text Inst 104:808–818
9. Jimenez-Gonzalez O, Ruiz-Espinosa H, Luna-Guevara JJ, Ochoa-Velasco CE, Luna VD, Luna-Guevara ML (2018) A potential natural coloring agent with antioxidant properties: microencapsulates of *Renealmia alpinia* (Rottb.) Maas fruit pericarp. NFS J 13:1–9
10. Gupta D, Khare SK, Laha A (2004) Antimicrobial properties of natural dyes against Gram-negative bacteria. Color Technol 120:167–171
11. Venil CK, Lakshmanaperumalasamy P (2009) An insightful overview on microbial pigment, Prodigiosin. Electronic J Biol 5:49–61
12. Tuli HS, Sandhu SS, Sharma AK, Gandhi P (2014) Anti-angiogenic activity of the extracted fermentation broth of an entomopathogenic fungus, Cordyceps militaris 3936. Int J Pharm Pharm Sci 6:581–583
13. Neves MIL, Silva EK, Meireles MAA (2019) Trends and challenges in the industrialization of natural colorants. Food Public Heal 9:33–44
14. Saxena S, Raja ASM (2014) Natural dyes: sources, chemistry, application and sustainability issues. In: Roadmap to sustainable textiles and clothing. Springer, Singapore, pp 37–80
15. Joshi V, Attri D, Bala A, Bhushan S (2003) Microbial pigments. Indian J Biotechnol 2:362–369
16. Duffose L (2006) Microbial production of food grade pigments, food grade pigments. Food Technol Biotechnol 44(3):313–321
17. Venil CK, Lakshmanaperumalsamy P (2009) An insightful overview on microbial pigment, prodigiosin. Elect J Biol 5(3):49–61
18. Ahmad WA, Ahmad WYW, Zakaria ZA, Yusof NZ (2012) Application of bacterial pigments as colorant. In: Springerbriefs in molecular science. Springer, Berlin, Heidelberg
19. Nigam JN, Kakati MC (2002) Optimization of dilution rate for the production of value added product and simultaneous reduction of organic load from pineapple cannery waste. World J Microbiol Biotechnol 18:303–308
20. Tanaka K, Hilary ZD, Ishizaki A (1999) Investigation of the utility of pineapple juice and pineapple waste material as low-cost substrate for ethanol fermentation by *Zymomonas mobilis*. J Biosci Bioeng 87:642–646
21. Creczynski-Pasa TB, Antônio RV (2004) Energetic metabolism of *Chromobacterium violaceum*. Genet Mol Res 3:162–166
22. Poorniammal R, Gunasekaran S (2015) Physical and chemical stability analysis of Thermomyces yellow pigment for food application. Int J Food Ferment Technol 5:47–52
23. Mapari SA, Nielsen KF, Larsen TO, Frisvad JC, Meyer AS, Thrane U (2005) Exploring fungal biodiversity for the production of water-soluble pigments as potential natural food colorants. Curr Opin Biotechnol 16:231–238
24. Morales-Oyervides L, Oliveira JC, Sousa-Gallagher MJ, Méndez-Zavala A, Montañez JC (2015) Effect of heat exposure on the color intensity of red pigments produced by *Penicillium purpurogenum* GH2. J Food Eng 164:21–29

25. Morales-Oyervides L, Oliveira J, Sousa-Gallagher M, Méndez-Zavala A, Montañez JC (2017) Assessment of the dyeing properties of the pigments produced by *Talaromyces* spp. J. Fungi 3:38–46
26. Hamlyn PF (1998) Fungal biotechnology. British Mycological Society Newsletter, pp 17–18
27. Gill M, Steglich W (1987) Pigments of fungi (Macromycetes). In: Fortschritte Der Chemie Org. Naturstoffe/Progress in the chemistry of organic natural products. Springer, pp 1–297
28. Panesar R, Kaur S, Panesar PS (2015) Production of microbial pigments utilizing agro-industrial waste: a review. Curr Opin Food Sci 1:70–76
29. Brar SK, Dhillon GS, Soccol CR (2014) Biotransformation of waste biomass into high value biochemicals. Springer, USA
30. Nigam PSN, Pandey A (2009) Biotechnology for agro-industrial residues utilisation: utilisation of agro residues. Springer Science & Business Media
31. Narsing Rao MP, Xiao M, Li WJ (2017) Fungal and bacterial pigments: secondary metabolites with wide applications. Front Microbiol 8:1–13
32. Usman HM, Abdulkadir N, Gani M, Maiturare HM (2017) Bacterial pigments and its significance. MOJ Bioequiv Avail 4:1–5
33. Dufossé L (2006) Microbial production of food grade pigments. Food Technol Biotechnol 44:313–323
34. Takaichi S (2011) Carotenoids in algae: distributions, biosyntheses and functions. Mar Drugs 9:1101–1118
35. Rossano R, Ungaro N, D'Ambrosio A, Liuzzi GM, Riccio P (2003) Extracting and purifying R-phycoerythrin from Mediterranean red algae Corallina elongata Ellis & Solander. J Biotechnol 101:289–293
36. Upreti DK, Tiwari P, Shukla P, Dwivedi A (2012) Natural thalli and cultured mycobiont of *Usnea ghattensis* G. Awasthi—a potential source of purple dye yielding lichen from India. Indian J Nat Prod Resour 3(4):489–492
37. Roychowdhury R, Maiti S, Adivarekar RV, Singhal RS (2024) Sustainable dyeing of silk using an acetylshikonin-based natural colourant from the lichen *Parmotrema perlatum*. Green Chem 26(2):904–917
38. Dharmaraj S, Ashokkumar B, Dhevendaran K (2009) Food-grade pigments from *Streptomyces* sp. isolated from the marine sponge *Callyspongia diffusa*. Food Res Int 42:487–492
39. Saranya R, Jayapriya J, Tamilselvi A (2012) Dyeing of silk fabric with phenazine from Pseudomonas species. Color Technol 128:440–445
40. Nagia FA, EL-Mohamedy RSR (2007) Dyeing of wool with natural anthraquinone dyes from *Fusarium oxysporum*. Dye Pigment 75:550–555
41. Durán N, Justo GZ, Ferreira CV, Melo PS, Cordi L, Martins D (2007) Violacein: properties and biological activities. Biotechnol Appl Biochem 48:127–133
42. El-Naggar NEA, El-Ewasy SM (2017) Bioproduction, characterization, anticancer and antioxidant activities of extracellular melanin pigment produced by newly isolated microbial cell factories *Streptomyces glaucescens* NEAE-H. Sci Rep 7:1–19
43. El-Fouly MZ, Sharaf AM, Shahin AAM, El-Bialy HA, Omara AMA (2015) Biosynthesis of pyocyanin pigment by *Pseudomonas aeruginosa*. J Radiat Res Appl Sci 8:36–48
44. Alihosseini F, Ju K, Lango J, Hammock BD, Sun G (2008) Antibacterial colorants: characterization of prodiginines and their applications on textile materials. Biotechnol Prog 24:742–747
45. Zheng L, Cai Y, Zhou L, Huang P, Ren X, Zuo A, Meng X, Xu M, Liao X (2017) Benzoquinone from Fusarium pigment inhibits the proliferation of estrogen receptor-positive MCF-7 cells through the NF-κB pathway via estrogen receptor signaling. Int J Mol Med 39:39–46
46. Krishnamurthy KV, Siva R, Senthil TK (2002) Natural dye-yielding plants of Shervaroy Hills of Eastern Ghats. In: Proceedings of national seminar on the conservation of the Eastern Ghats, environment protection training and research institute, Hyderabad, pp 151–153
47. Tuli HS, Chaudhary P, Beniwal V, Sharma AK (2015) Microbial pigments as natural color sources: current trends and future perspectives. J Food Sci Technol 52:4669–4678

48. Nerurkar M, Vaidyanathan J, Adivarekar R, Langdana ZB (2013) Use of a natural dye from *Serratia marcescens* subspecies Marcescens in dyeing of textile fabrics. Octa J Environ Res 1:129–135
49. Patil S, Paradeshi J, Chaudhari B (2016) Anti-melanoma and UV-B protective effect of microbial pigment produced by marine *Pseudomonas aeruginosa* GS-33. Nat Prod Res 30:2835–2839
50. Hernandez VA, Galleguillos F, Robinson S (2016) Fungal pigments from spalting fungi attenuating blue stain in *Pinus* spp. Int Biodeterior Biodegrad 107:154–157
51. Robinson SC, Weber G, Hinsch E, Vega Gutierrez SM, Pittis L, Freitas S (2014) Utilizing extracted fungal pigments for wood spalting: a comparison of induced fungal pigmentation to fungal dyeing. J Coat 2014:59073
52. Vega Gutierrez S, Vega Gutierrez P, Godine A, Pittis L, Huber M, Stanton S, Robinson S (2016) Feasibility of coloring bamboo with the application of natural and extracted fungal pigments. Coatings 6:37–47
53. Robinson SC, Hinsch E, Weber G, Freitas S (2014) Method of extraction and resolubilisation of pigments from *Chlorociboria aeruginosa* and *Scytalidium cuboideum*, two prolific spalting fungi. Color Technol 130:221–225
54. Suryanarayanan TS, Ravishankar JP, Venkatesan G, Murali TS (2004) Characterization of the melanin pigment of a cosmopolitan fungal endophyte. Mycol Res 108:974–978
55. Addy HD, Piercey MM, Currah RS (2005) Microfungal endophytes in roots. Can J Bot 83:1–13
56. Pagano MC, Dhar PP (2015) Fungal pigments: an overview. In: Fungal biomolecules: sources, applications and recent developments, pp 173–181
57. Ladygina N, Rineau F (2013) Biochar and soil biota. CRC Press, New York
58. Sonderegger M, Sauer U (2003) Evolutionary engineering of *Saccharomyces cerevisiae* for anaerobic growth on xylose. Appl Environ Microbiol 69:1990–1998
59. Nigam PS, Luke JS (2016) Food additives: production of microbial pigments and their antioxidant properties. Curr Opin Food Sci 7:93–100
60. Chen Y, Li D, Lu W, Xing J, Hui B, Han Y (2003) Screening and characterization of astaxanthin-hyperproducing mutants of *Haematococcus pluvialis*. Biotechnol Lett 25:527–529
61. Tripathi U, Venkateshwaran G, Sarada R, Ravishankar GA (2001) Studies on *Haematococcus pluvialis* for improved production of astaxanthin by mutagenesis. World J Microbiol Biotechnol 17:143–148
62. Venil CK, Aruldass CA, Dufossé L, Zakaria ZA, Ahmad WA (2014) Current perspective on bacterial pigments: emerging sustainable compounds with coloring and biological properties for the industry—an incisive evaluation. RSC Adv 4:39523–39529
63. Gunasekaran S, Poorniammal R (2008) Optimization of fermentation conditions for red pigment production from *Penicillium* sp. under submerged cultivation. Afr J Biotechnol 7:1894–1898
64. Domínguez-Espinosa RM, Webb C (2003) Submerged fermentation in wheat substrates for production of Monascus pigments. World J Microbiol Biotechnol 19:329–336
65. Pandey A (2003) Solid state fermentation. Biochem Eng J 13:81–84
66. Nimnoi P, Lumyong S (2011) Improving solid-state fermentation of *Monascus purpureus* on agricultural products for pigment production. Food Bioprocess Technol 4:1384–1390
67. Velmurugan P, Hur H, Balachandar V, Kamala-Kannan S, Lee KJ, Lee SM, Chae JC, Shea PJ, Oh BT (2011) Monascus pigment production by solid-state fermentation with corn cob substrate. J Biosci Bioeng 112:590–594
68. Babitha S, Soccol CR, Pandey A (2006) Jackfruit seed-a novel substrate for the production of Monascus pigments through solid-state fermentation. Food Technol Biotechnol 44:465–471
69. Saxena S, Raja A S M, Mutthu S (2014) Roadmap to sustainable textile and clothing. In: Textile science and clothing technology. Springer, pp 978–981
70. Gmoser R, Ferreira JA, Lennartsson PR, Taherzadeh MJ (2017) Filamentous ascomycetes fungi as a source of natural pigments. Fungal Biol Biotechnol 4:1–25
71. Venil CK, Nordin N, Zakaria ZA, Ahmad WA (2014) *Chryseobacterium artocarpi* sp. nov., isolated from the rhizosphere soil of *Artocarpus integer*. Int J Syst Evol Microbiol 64:3153–3159

72. Baysal T, Ersus S, Starmans DAJ (2000) Supercritical CO_2 extraction of β-carotene and lycopene from tomato paste waste. J Agric Food Chem 48:5507–5511
73. Hong ST, Carney JR, Gould SJ (1997) Cloning and heterologous expression of the entire gene clusters for PD 116740 from *Streptomyces* strain WP 4669 and tetrangulol and tetrangomycin from *Streptomyces rimosus* NRRL 3016. J Bacteriol 179:470–476
74. Pfeifer BA, Khosla C (2001) Biosynthesis of polyketides in heterologous hosts. Microbiol Mol Biol Rev 65:106–118
75. Mcdaniel R, Ebert-khosla S, Fu H, Hopwoodt DA, Khosla C (1994) Engineered biosynthesis of novel polyketides: influence of a downstream enzyme on the catalytic specificity of a minimal aromatic polyketide synthase. Proc Natl Acad Sci 91:11542–11546
76. Krishna JG, Basheer SM, Beena PS, Chandrasekaran M (2008) Marine bacteria as source of pigment for application as dye in textile industry. In: Proceedings of the international conference on biodiversity conservation and management, vol 1, pp 743–750
77. Shahitha S, Poornima K (2012) Enhanced production of prodigiosin production in *Serratia marcescens*. J Appl Pharm Sci 2:138–140
78. Usman HM, Farouq AA, Baki AS, Abdulkadir N, Mustapha G (2018) Production and characterization of orange pigment produced by Halophilic bacterium *Salinococcus roseus* isolated from Abattoir soil. J Microbiol Exp 6:238–243
79. Heer K, Sharma S (2017) Microbial pigments as a natural color: a review. Int J Pharm Sci Res 8:1913–1922
80. Yusof NZ (2008) Isolation and applications of Red Pigments from Serratia marcescens. Doctoral dissertation, Univ. Tekno., Malaysia
81. Sharma D, Gupta C, Aggarwal S, Nagpal N (2012) Pigment extraction from fungus for textile dyeing. NISCAIR-CSIR, India
82. Hobson DK, Wales DS (1998) Greeny dyes. J Soc Dye Color 114:42–44
83. Räisänen R (2009) Anthraquinones from the fungus Dermocybe sanguinea as textile dyes. Helsingin yliopisto
84. Hinsch EM, Weber G, Robinson SC, Chen HL (2015) Colorfastness of extracted wood-staining fungal pigments on fabrics: a new potential for textile dyes. J Text Apparel Technol Manag 9:1–11
85. Hinsch EM, Robinson SC (2016) Mechanical color reading of wood-staining fungal pigment textile dyes: an alternative method for determining colorfastness. Coatings 6:1–12
86. Weber G, Chen HL, Hinsch E, Freitas S, Robinson S (2014) Pigments extracted from the wood-staining fungi *Chlorociboria aeruginosa*, *Scytalidium cuboideum*, and *S. ganodermophthorum* show potential for use as textile dyes. Color Technol 130:445–452
87. De Santis D, Moresi M, Gallo AM, Petruccioli M (2005) Assessment of the dyeing properties of pigments from *Monascus purpureus*. J Chem Technol Biotechnol Int Res Process Environ Clean Technol 80:1072–1079
88. Velmurugan P, Chae JC, Lakshmanaperumalsamy P, Yun BS, Lee KJ, Oh BT (2009) Assessment of the dyeing properties of pigments from five fungi and anti-bacterial activity of dyed cotton fabric and leather. Color Technol 125:334–341
89. Velmurugan P, Kim MJ, Park JS, Karthikeyan K, Lakshmanaperumalsamy P, Lee P, Park YJ, Oh BT (2010) Dyeing of cotton yarn with five water soluble fungal pigments obtained from five fungi. Fibers Polym 11:598–605
90. Poorniammal R, Parthiban M, Gunasekaran S, Murugesan R, Thilagavathi G (2013) Natural dye production from *Thermomyces* sp. fungi for textile application. Indian J Fibre Text Res 38:276–279
91. Atalla MM, El-khrisy EAM, Youssef YA, Mohamed AA (2011) Production of textile reddish brown dyes by fungi. Malays J Microbiol 7:33–40

92. Shirata A (2000) Isolation of bacteria producing bluish-purple pigment and use for dyeing. Jpn Agric Res Q 34:131–140
93. Sen T, Barrow CJ, Deshmukh SK (2019) Microbial pigments in the food industry—challenges and the way forward. Front Nutr 6:7
94. Babitha S (2009) Microbial pigments. In: Biotechnology for agro-industrial residues utilisation. Springer, pp 147–162

Chapter 10
Statistical Approaches for Sustainable Textile Coloration: Current Scenario and Future Recommendations

Muhammad Aftab, Tanvir Ahmad, and Maria Javed

Abstract The current era is the age of technology where with the advancement of industrialization, the addition of novel techniques is making the product more green and sustainable. One of the causes of global pollution is textile effluent which is due to improper process management. The demand for sustainable products in applied fields needs such processing techniques that are not only cost and energy-effective but also make the atmosphere clean and pollution free. To achieve this goal, the inclusion of statistical approaches are of immense importance where response surface designs, i.e., central composite design and Box Behnken design are most commonly being used. These response surface designs explore various choices of input dyeing parameters and provide the most optimal design points among the infinite possible, leading towards clean procedures. Response surface designs also help to select the best potential parameters like nature of treatment, time, temperature, pH and other allied auxiliary levels. In this chapter discussion about the possible implementation of these existing and novel statistical techniques in the optimization of textile processes to save energy, capital and labor has been made. Hence, the findings of this chapter may guide researchers, traders to employ these methods in their processes to make the product sustainable and global friendly.

Keywords Artificial neural network · Box–Behnken design · Central composite design · Eco-friendly · Model fitting · Optimization

10.1 Introduction

The natural beauty of the world, categorized by various ranges of colors given by nature, is generally being compromised by the development of science in various fields [1, 2]. The advancement of sectors such as industrialization is aimed at improving the lifestyles of the people to bring a sense of calm and peace [3]. The

M. Aftab (✉) · T. Ahmad · M. Javed
Department of Statistics, Government College University Faisalabad, Faisalabad, Pakistan
e-mail: maftab@gcuf.edu.pk

M. Shahid et al. (eds.), *Advancements in Textile Coloration*,
https://doi.org/10.1007/978-981-96-5091-0_10

textile industry, in particular, plays a crucial role in human life by providing clothing, handicrafts, fashion, and style [4, 5]. Natural sources of colors were used to increase the beauty of many aspects of life including, house painting, body decoration, and traditional cultural occasions in the past [6, 7]. However, this advancement of industrialization led to the revolutionary discovery of synthetic dyes and pigments like Mauvein, changing the way colors are produced and used in the world [8, 9].

Nowadays, thousands of dyes are being produced and employed in various aspects of daily life [10, 11]. However, the global safety and protection agencies who are keenly observing their role found that many of these dyes after application or during their synthesis shed toxic effluents [12, 13]. Their effluents, when in contact with water, produce waterborne diseases, barren agricultural land by disturbing soil pH and the water ecosystems [14, 15]. These wastes are when released into the atmosphere also raise global heat, disturb water quality parameters and cause many hazardous and toxic aftermaths for the people [16, 17]. A great awareness about reducing the widespread use of these toxic colorants in all fields is currently being considered [18, 19]. The global protection agencies have now spread awareness about green products, their benefits and after reuse aspects. One such alternative is plant-based natural dyes, which are making a comeback in various industries, especially in the textile sector [20, 21].

10.2 Natural Dyes and Their Advantages

Natural dyes are functional bio-colorants derived from plants, minerals and animals without the use of any hazardous chemicals [22, 23]. Plants such as roots, seeds, shoots, flowers and barks etc. are commonly used sources of natural dyes [24]. These dyes contain potent isolates that have beneficial biological medicinal activities. These colorants when isolated do not possess any harmful effluents [25, 26]. Many plants have also the ability to contain such bio-colorants that are insect repellant and attractive eye-soothing shades [27, 28]. However, their isolation process using conventional methods has some limitations [29, 30]. Traditional methods of isolating these bio-colorants, such as heating, soaking, and stirring, are time-consuming and resource-intensive [31]. Now researchers are using such methods which are taking less solvent, labor, energy and cost [32]. Among these modern techniques; ultrasonic, gamma, plasma, microwave treatment are eco-friendly, uniform and rapid in action [33].

Metallic mordants have been used to improve fastness ratings for natural dyes [34, 35]. However, these metallic mordants such as salts of copper, cobalt, tin, chromium, etc. are carcinogenic and need replacement with some eco-friendly anchors [36, 37]. Now bio-mordants have been included to make color-fast shades which are employed on fiber before and after dyeing. Now after the utilization of microwave treatment and bio-mordants there is a need for standardization of extraction and optimization of effective dyeing variables at low levels [38]. Process optimization is to search for the optimum values of the dyeing parameters that increase the value of response

(K/S) and this heavily depends upon the selection of an appropriate design plan. Hence, statistical methodology along with machine learning techniques plays a vital role in chemical processes and should be given considerable attention by researchers.

10.3 Contribution of Statistics to Enhance Processes

Statistics is the scientific field that works for the collection, presentation, analysis and interpretation of experimental data [39]. In the textile industry, modern statistical techniques are being employed to properly plan experiments and data analysis. In statistics, the reliance on the results heavily depends on the reliability of the investigational data. To obtain better data, well-planned designs and their execution are required. Statistics helps researchers by providing these facilities for enhanced design planning and improved analysis techniques. These algorithms, help to reduce process cost, energy, work levels, data reliability enhancement, better data computation and guidance to use factors that affect industrial processes.

These techniques include:

- Design of experiments for proper planning and conducting of experimentation includes classical design structures such as central composite design (CCD) and Box-Behnken design (BBD).
- Computation of the reliability of the experimental data.
- Regression analysis, to study the cause and effect relationship among response variable (K/S) and set of input variables (color parameters).
- Application of analysis of variance (ANOVA) to study the statistical significance of the color parameters.
- Checking the adequacy of the employed model by using the coefficient of determination (R^2).
- Data visualization by some graphical tools such as main effect plot, interaction plot, response surface plots and contour plots.

10.4 Statistics for Lab Experimentation: A Methodology Review

Understanding statistical concepts is essential for conducting accurate and reliable lab experimentation. This helps researchers to draw meaningful conclusions from the experimental data. In the coming sub-sections, an insightful review of the statistical methodology is presented.

10.4.1 Response Surface Methodology (RSM)

RSM is a combination of modern statistical methods that plays a significant role in modeling and optimizing the relationships between multiple inputs and one or more outputs of a process. RSM involves the use of design of experiment (DOE), statistical modeling, and optimization algorithms to develop mathematical models [40]. By applying different conditions of input variables these models help to estimate the behavior of the system and find optimal input settings to determine the contribution of the inputs for the response. RSM has broader applications in disciplines such as manufacturing, engineering, agriculture, business, psychology, and healthcare research to optimize processes and improve products or services with reduced costs [41–44].

10.4.2 Central Composite Design (CCD) and Its Significance in Natural Dyes

Experiments are mainly designed for exploring response surfaces, the experimental plans are known as response surface designs (RSDs). CCD is the design that is sequential in nature and is the basis of the sequential nature assembly of the RSDs [45]. These designs have five levels; containing full factorial denoted as n_f or fractional factorial of Resolution-V runs for the k-factors, $2k$ axial or star points and finally n_c center runs. Factorial or corner points are used to find the estimates of main effects (linear) and interaction effects (bi-linear). Axial or star design points are used to estimate the quadratic effects of the model and center runs provide extra df to estimate the pure error of the model [46]. Moreover, factorial runs are denoted by -1 (low value of the given parameter) and $+1$ (high value of the given parameter). Center trials are denoted by 0 indicates the middle values of the given parameters.

In literature, researchers have commonly used CCD as an experimental design. Process optimization towards isolation of colorant from Esfand for bio-mordanted silk dyeing is investigated by CCD. For five color variables, 32 design points were experimented [22]. The wool dyeing parameters were optimized by using CCD and 29 optimal experimental trails were determined [47]. Sustainable eco-friendly extraction of yellow natural dye from haar singhar for bio-coloration of cotton fabric was studied by employing RSM. Six dyeing variables were used and 32 trials were experimented by using CCD [48]. Optimization of *Hibiscus sabdariffa* natural dye extraction was studied by classical CCD to achieve the optimized values of the color parameters [38]. RSM using CCD was utilized to observe the significant dyeing parameters for arsenic sequestration on ameliorated waste molasses nano adsorbents [49].

10.4.3 Box–Behnken Design (BBD) and Its Application in Textiles

Second-order polynomial designs generally require five levels coded as $-\alpha$, -1, 0, 1, α for every k independent factors under study. Occasionally due to economic or ease of experimentation reasons, we prefer to have a smaller number of levels, say three. An important class of three-level incomplete FDs for the estimation of the parameters in the second-order model was introduced by Box and Behnken [50]. If researchers desire three equally spaced levels for design parameters, the BBD is an efficient alternative to the CCDs.

RSM-based BBD for modeling and ANN with two hidden layers and six nodes for each; to study the photocatalysis degradation was used [51]. The evaluation of biodiesel output prediction by ANN and RSM modeling has been considered. BBD was implemented as an RSD and the ANN predictive model was developed using a feed-forward-back propagation algorithm with 14 neurons by Avinash and Murugesan [52]. Three input variables with 17 design points by using BBD were obtained and RSM and ANN were used for model fitting. One to eight hidden neurons have experimented and six hidden neurons provided minimum mean square error [53]. The roots of rumex abyssinicus plant were optimized by BBD [54]. Cotton yarn parameters were optimized by using BBD [55].

10.4.4 Estimation of Unknown Function by Polynomial Approximations

By using a suitable design, an experimental data is obtained. Now, the collected data require proper analysis. In analysis, the purpose is to estimate the unknown function i.e. the response. The inputs may be linearly or non-linearly related to the outcome variable. A polynomial of higher degree must be used, when system shows the non-linearity, such as the second-order polynomial model:

$$E(y/x) = b_0 + \sum_{i=1}^{k} b_i x_i + \sum_{i=1}^{k} b_{ii} x_i^2 + \sum_{i=1}^{k-1} \sum_{j>1}^{k} b_{ij} x_i x_j \qquad (10.1)$$

where $E\left(\frac{y}{x}\right)$ is the expected value of y for the given values of $x, x_i, \ldots, x_k$, the coded levels of the k factors, and b's are the unknown parameter estimates [56]. Model (10.1) is comprised of intercept, linear, interaction and quadratic terms and these model terms are often adequate enough to estimate the true underlying response surface. This model is most commonly used in literature. In Eq. (10.1), the linear part helps to study the main effect of the factors and interaction part, about the combined effects of the factors to be studied.

10.5 Errors in Model Fitting

After model fitting, the researchers keenly observe the difference between experimental data and their estimated values, known as error. If the observed and fitted values are very close to each other then the error value will be small (which is desired) otherwise high error value will be obtained. In experimental design, errors are of two types, i.e. pure and lack-of-fit errors [57]. Pure error is based upon the replication of the basic experiment under identical conditions and it provides information about the internal variation of the data and to estimate it, replication of the design at the center is required. Lack-of-fit error is determined by the unique trials of the experiment. Unique trails are used to estimate the model terms. It provides degrees of freedom to test the lack of fit of the model [58].

10.6 Significance of the Concept of Model Adequacy

After the selection of suitable modeling and its estimation, the next step is to investigate the reliability and adequacy of the fitted model. There is a list of methods available that studies the model performance. The coefficient of determination R^2 and adjusted R^2 are commonly used to judge the adequacy of the fitted model. The higher the value of R^2, i.e. close to 1, the better the fit will be [56].

10.7 Data Visualization and Most Commonly Used Graphs

Graphs are important to explain results in pictorial forms. In RSM, most commonly main effect, interaction effect, contour plots, and response surface plots are constructed. The detail of these graphs is presented in the coming sections.

10.7.1 Interaction (Combined) Effect Graphs

In factorial designs, the simplest graphs are interaction graphs. These graphs if show parallel lines, mean that no interaction effect is present as in Fig. 10.1a and if lines are crossed then the interaction is present as shown in Fig. 10.1b. The interaction graphs are constructed for supposed data.

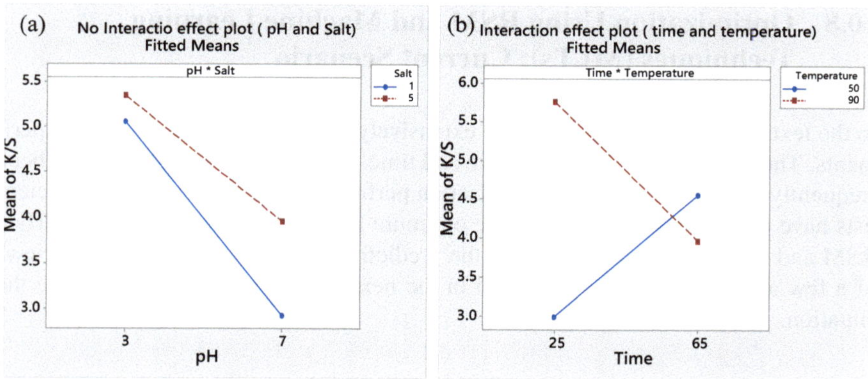

Fig. 10.1 **a** Absence of interaction. **b** Presence of interaction

10.7.2 Contour and Response Surface Plots

The contour plots revealed the relationship between 2 input variables and a dependent variable. The *X* and *Y* values are displayed along the *X* and *Y*-axes and contour lines and bands represent the Z value. The graph shows values of the *Z* variable for combinations of the *X* and *Y* variables. Response surface is also viewed in a two-dimensional plane in the form of contour plot [59].

Figure 10.2a, b presents contour and response surface plots, respectively. These are three-dimensional graphs. The plots are constructed for supposed data. In both these plots, the curved line shows the presence of interaction in the model.

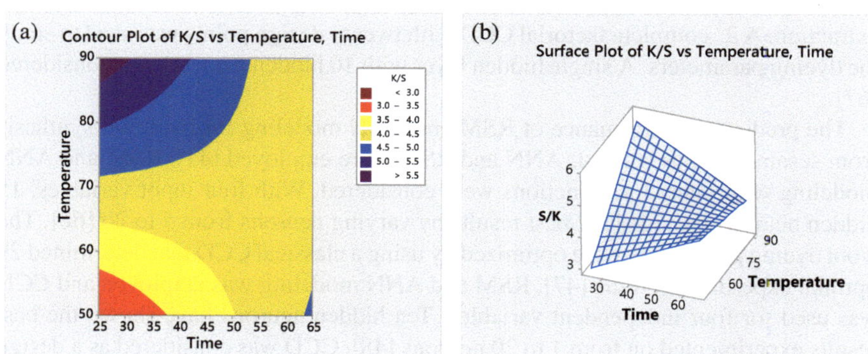

Fig. 10.2 **a** Contour plot. **b** Response surface plot

10.8 Optimization Using RSM and Machine Learning Techniques (MLTs): Current Scenario

In the textile industry, researchers are extensively using RSDs for designing experiments. These designs are cost-effective and time-saving. RSM and MLTs have been frequently used to enhance the optimization performance in textiles [60–64]. Scientists have used RSDs to search for the optimum levels for dyeing parameters. Then RSM and MLTs are used to increase the predictive capacity of the model. Reviews of a few articles have been presented in the next section to briefly summarize the situation.

10.9 RSM and MLT-Based Literature

Researchers in various fields have used the combination of RSM and MLTs. RSM is used to plan a suitable design that produces reliable data and MLTs like artificial neural networks (ANN) are used to improve the prediction capacity of the model. Some studies from the literature are included as: An improved membrane flux prediction using ANN and RSM optimization was observed, CCD was implemented to decide the network structure of ANN [65]. The prediction of quality parameters of spray-dried pomegranate juice by considering a comparative study of an ANN and RSM was performed. Levels of the three input variables were optimized by using a central composite rotatable design (CCRD) and later a feed-forward multilayered perceptron (FFMLP) trained algorithm was used to optimize the response [66]. The role of synthetic dyes using seeds of *Bixa orellana* (Annatto) was observed by RSM and ANN to explore the predictive and optimization capability of the model for dye extraction. A 2^3 complete factorial CCD with twenty design points was used to study the dyeing parameters. A single hidden layer with 10 hidden neurons was considered [67].

The prediction performance of RSM and ANN modeling for biodiesel synthesis from sesame oil was studied; ANN and RSM were employed [41]. RSM and ANN modeling with many loss functions were considered. With four input variables, 15 hidden neurons provided the best results by varying neurons from 5 to 20 [68]. The wool dyeing parameters were optimized by using a classical CCD that determined 29 optimal experimental trails [47]. RSM and ANN modeling was employed and CCD was used for four independent variables. Ten hidden neurons have shown the best results experimented on from 1 to 20 neurons [48]. CCD was considered as a design plan, and RBF-NN, ANFIS, and RSM modeling were used to optimize the process [61]. RSM and ANN modeling were employed for process optimization. CCD has been used to determine the optimal values of the dyeing variables [3]. RSM and ANN were employed to study 4 factors with five hidden neurons [42].

10.10 Research Findings from the Reviewed Literature

Many articles regarding natural dyes were included in the present work as a reference. Dyeing variables were optimized using RSM as a statistical method. Almost every case study included in this section has obtained optimal experimental trials by using CCD (five levels) and BBD (three levels). Different loss functions were studied to judge the model performance. The obtained experimental data was analyzed using ANOVA and regression analysis, as statistical techniques. Arithmetic mean as an average and standard deviation (S.D.) as a variation tool were mostly used for descriptive analysis. S.D. helps to determine the reliability of the experimental data. The lower the value of the S.D., the more precise the experimental data and validity of the analysis. ANN as an MLT was used to enhance the predictive capacity model. Conclusively, nowadays, both RSM and ANN have been extensively employed for prediction purposes.

10.11 Modern Approaches of Statistical and Machine Learning Modeling: Future Perspectives

Most Scientists have used CCD and BBD structures to design the experiment. Innovative subset designs for $K = 3, 4$ and 5 dyeing parameters are available that can be used for designing experiments. These designs include a variety of experimental trials to explore the experimental region in various ways [57]. Gilmour proposed a rich class of second-order RSDs, called subset designs, which includes many popular second-order RSDs like CCDs, BBDs, etc. [59]. The author suggested combining different subsets of factorial experiments in a single design. The work was further extended by Ahmad and Gilmour [40]. Subset designs have some important properties. Their construction is easy and it allows efficient fitting of the second-order model as well as allowing testing of lack of fit. These designs spare enough degrees of freedom for the estimation of pure error terms.

10.12 MLTs and Over-Fitting Problem with Limited Datasets

Over-fitting is a problem of model fitting, where the model trains well for the given data (training data) but not well for the unseen data (testing data). Because of this, the model starts memorizing rather than a generalization of the results. In laboratory experiments, we have small data sets due to costly experiments. Commonly researchers have studied extremely complex models with these limited data sets which introduces over-fitted problems during model fitting.

Table 10.1 Demonstration of the over-fitting problem when ANN is employed

Reference articles	Input variables	Hidden neurons	Data size	RSM unknowns	ANN unknowns	Design structure	Testing data
[49]	4	20	30	15	80	CCD	10
[36]	4	14	27	15	56	BBD	8
[70]	2	10	13	6	20	CCD	5
[71]	4	10	30	15	40	CCD	9
[53]	3	8	17	10	24	BBD	5

In Table 10.1, two of the commonly ignored problems have been highlighted. First is the order of the approximation is generally being overlooked while comparing the efficiency of both RSM and ANN models. These are the only reference articles that are cited, and it is clear that there is no comparison of RSM and ANN models as both have different approximation orders. Since RSM generally assumes second-order polynomial model approximation, the ANN model considers higher-order approximation by using more hidden neurons [69]. Secondly, with more hidden neurons, there are fewer observations in the chemical process to estimate extra unknowns. In ML models, this is the case of over-fitting again being ignored. In the over-fitting problem, the model works better for training data but not well for testing data.

10.13 Recommendations with Future Directions

From the view of the given material, it is strongly recommended that

- Innovative RSDs suggested above should also be used for designing purposes in comparison to conventional design structures.
- Proper model selection and a suitable design plan should be given considerable importance along with ANOVA.
- While comparing the prediction performance of RSM and ANN, approximation order should also be considered.
- Due to the study's small size, over-fitting mostly occurred, which should also be incorporated using an appropriate number of hidden layers and neurons. Sometimes, we have to use regularization techniques to avoid over-fitting.
- This application requires such models for natural dye isolation to make the process economical, time, and energy-saving.

Declaration The authors declare that they have no known competing financial interests or personal relationships that could have appeared to influence the work reported in this chapter. Generative AI and AI-assisted technologies have been used in the writing process to improve the readability and language of the work only. After using this tool, the authors carefully reviewed and revised the content to ensure accuracy and quality. The authors take full responsibility for the final version of this publication.

References

1. Islam MT, Jahan R, Jahan M, Howlader MS, Islam R, Islam MM, Hossen MS, Kumar A, Robin AH (2022) Sustainable textile industry: an overview. Non-Metall Mater Sci 4(2):15–32
2. Kamboj A, Jose S, Singh A (2022) Antimicrobial activity of natural dyes—a comprehensive review. J Nat Fibers 19(13):5380–5394
3. Karmakar B, Saha SP, Chakraborty R, Roy S (2023) Optimization of starch extraction from Amorphophallus paeoniifolius corms using response surface methodology (RSM) and artificial neural network (ANN) for improving yield with tenable chemical attributes. Int J Biol Macromol 237:124183
4. Makhlouf A, Belaadi A, Boumaaza M, Mansouri L, Bourchak M, Jawaid M (2022) Water absorption behavior of jute fibers reinforced HDPE biocomposites: prediction using RSM and ANN modeling. J Nat Fibers 19(16):14014–14031
5. Shah W, Faraz SM, Awan ZH (2023) Photovoltaic properties and impedance spectroscopy of dye sensitized solar cells co-sensitized by natural dyes. Physica B 654:414716
6. Farhana K, Kadirgama K, Mahamude ASF, Mica MT (2022) Energy consumption, environmental impact, and implementation of renewable energy resources in global textile industries: an overview towards circularity and sustainability. Mater Circul Econ 4(1):15
7. Irfan M, Waqas S, Arshad U, Khan JA, Legutko S, Kruszelnicka I, Skrzypczak A (2022) Response surface methodology and artificial neural network modelling of membrane rotating biological contactors for wastewater treatment. Materials 15(5):1932
8. Talib A, Adeel S, Ali A, Ahmad T, Hussaan M, Qayyum MA (2023) Sustainable isolation and application of plant extract-based natural dye for bio-dyeing of silk fabric. Coatings 13(1):112
9. Mabuza L, Sonnenberg N, Marx-Pienaar N (2023) Natural versus synthetic dyes: consumers' understanding of apparel coloration and their willingness to adopt sustainable alternatives. Resour Conserv Recycl Adv 18:200146
10. Arain ARA, Ahmad F, Khatri Z, Peerzada MH (2021) Microwave assisted henna organic dyeing of polyester fabric: a green, economical and energy proficient substitute. Nat Prod Res 35(2):327–330
11. Hilali S, Wils L, Chevalley A, Larosière CB, Delaye BL (2022) Glycerol-based NaDES as green solvents for ultrasound-assisted extraction of phycocyanin from arthrospira platensis RSM optimization and ANN modeling. Biomass Convers Biorefin 12(Suppl. 1):157–170
12. Affat SS (2021) Classifications, advantages, disadvantages, toxicity effects of natural and synthetic dyes: a review. UTJsci 8(1):130–135
13. Ragab MM, Hassabo AG (2021) Various uses of natural plants extracts for functionalization textile based materials. J Text Color Polym Sci 18(2):143–158
14. Teja AS, Srivastava A, Sathrughna JAK, Tiwari MK, Kanwade A, Yadav SC, Shirage PM (2022) Optimal processing methodology for futuristic natural dye sensitized solar cells and novel applications. Dyes Pigm 110997
15. Adeel S, Mumtaz A, Mia R, Aftab M, Hussaan M, Amin N, Khattak SP (2023) Microwave-assisted sustainable coloration of wool fabric using Rheum Emodi based natural dye. Surf Innov 1–12
16. Wang L, Zhu Q, Yu H, Xie G, Yu Z (2022) Agricultural crops of sarsaparilla root as source of natural dye for inkjet printing and bio-functional finishing of cotton fabric. Ind Crops Prod 187:115490
17. Berkani M, Bouchareb MK, Bouhelassa M, Kadmi Y (2020) Photocatalytic degradation of industrial dye in semi pilot scale prototype solar photo reactor: optimization and modeling using ANN and RSM based on Box Wilson approach. Top Catal 63:964–975
18. Habib N, Ali A, Adeel S, Aftab M, Inayat A (2023) Assessment of wild turmeric-based eco-friendly yellow natural bio-colorant for dyeing of wool fabric. Environ Sci Pollut Res 30(2):4570–4581
19. Li C, Nasir MJ (2022) Extraction of natural dyes from turmeric (Curcuma longa) and their application on polylactic acid fiber. J Des Text 1(1):21–41

20. Rehman FU, Adeel S, Haddar W, Bibi R, Azeem M, Mia R, Ahmed T (2022) Microwave-assisted exploration of yellow natural dyes for nylon fabric. Sustainability 14(9):5599
21. Aung T, Kim SJ, Eun JB (2022) A hybrid RSM-ANN-GA approach on optimisation of extraction conditions for bioactive component-rich laver (Porphyra dentata) extract. Food Chem 366:130689
22. Adeel S, Ahmad S, Habib N, Mia R, Ahmed B (2022) Coloring efficacy of Nyctanthes Arbortristis based yellow natural dye for surface-modified wool. Ind Crops Prod 188:115571
23. Bhattacharya S, Das P, Bhowal A, Saha A (2021) Thermal, chemical and ultrasonic assisted synthesis of carbonized Biochar and its application for reducing Naproxen: batch and fixed bed study and subsequent optimization with response surface methodology (RSM) and artificial neural network (ANN). Surf Interfaces 26:101378
24. Rather LJ, Zhou Q, Ali A, Haque QMR, Li Q (2020) Valorization of natural dyes extracted from Mugwort leaves (Folium artemisiae argyi) for wool fabric dyeing: optimization of extraction and dyeing processes with simultaneous coloration and biofunctionalization. ACS Sustain Chem Eng 8(7):2822–2834
25. Silva E, Alvarado-Beltrán CG, Gaxiola A, Orozco-Carmona VM, Luque PA, Castro-Beltrán A (2023) A new green procedure to obtain and photosensitize SnO_2, in one step, for solar photocatalysis using natural dyes. Ceram Int 49(11):16732–16739
26. Mahboob M, Adeel S, Barkaat S, Ahmad T, Özomay M, Mia R, Zuber M (2023) Microwave-assisted santalin extraction from Pterocarpus santalinus for mordanted woolen yarn dyeing. Sustain Chem Pharm 35:101224
27. Boutra B, Sebti A, Trari M (2022) Response surface methodology and artificial neural network for optimization and modeling the photo degradation of organic pollutants in water. Int J Environ Sci Technol 19(11):11263–11278
28. Khetib Y, Dief AHM, Alanazi AK, Rawa M, Sajadi SM, Sharifpur M (2021) Competition of ANN and RSM techniques in predicting the behavior of the CuO-liquid paraffin. Chem Eng Commun 1–13
29. Tian Y, Lu Y, Zhang Y, Hou X, Zhang Y (2022) Extraction and characterization of natural colorant from Melia azedarach bark and its utilization in dyeing and finishing of wool. Sustain Chem Pharm 27:100647
30. Kailash BR, Charles B, Ravikanth G, Setty S, Kadirvelu K (2022) Identifying the potential global distribution and conservation areas for Terminalia chebula, an important medicinal tree species under changing climate scenario. Trop Ecol 63(4):584–595
31. Doldolova K, Bener M, Lalikoglu M, Ascı YS, Arat R, Apak R (2021) Optimization and modeling of microwave-assisted extraction of curcumin and antioxidant compounds from turmeric by using natural deep eutectic solvents. Food Chem 353:129337
32. Fried R, Oprea I, Fleck K, Rudroff F (2022) Biogenic colourants in the textile industry—a promising and sustainable alternative to synthetic dyes. Green Chem 24(1):13–35
33. Adeel S, Ahmad S, Habib N, Saif MJ, Batool F, Aftab M (2023) Environmental friendly pollution free bio-dyeing of wool with haar singhar (*Coral Jasmine*) flower extract. Environ Sci Pollut Res 30(5):12453–12465
34. Rosa JM, Guerhardt F, Júnior RSER, Belan PA, Lima GA, Santana JCC, Berssaneti FT, Tambourgi EB, Vanale RB, Araújo SAD (2021) Modeling and optimization of reactive cotton dyeing using response surface methodology combined with artificial neural network and particle swarm techniques. Clean Technol Environ Policy 23:2357–2367
35. Agarwal S, Singh AP, Mathur S (2023) Removal of COD and color from textile industrial wastewater using wheat straw activated carbon: an application of response surface and artificial neural network modeling. Environ Sci Pollut Res 1–22
36. Uthayakumar H, Radhakrishnan P, Shanmugam K, Kushwaha OS (2022) Growth of MWCNTs from azadirachta indica oil for optimization of chromium (VI) removal efficiency using machine learning approach. Environ Sci Pollut Res 29(23):34841–34860
37. Cai Y, Xiao L, Ehsan MN, Jiang T, Pervez MN, Lin L, Naddeo V (2023) Green penetration dyeing of wool yarn with natural dye mixtures in D5 medium. J Market Res 25:6524–6541

38. Nwuzor IC, Adinoyi BJ, Onyesolu OCF, Oyeoka HC (2023) Hibiscus sabdariffa natural dye extraction process with central composite design for optimal extract yield. Sustain Chem Environ 100008
39. Pukelsheim F (2006) Optimal design of experiments. Society for Industrial and Applied Mathematics
40. Ahmad T, Gilmour SG, Arshad HM (2018) Comparisons of augmented pairs designs and subset designs. Commun Stat Simul Comput 49(7):1898–1921
41. Sarve A, Sonawane SS, Varma MN (2015) Ultrasound assisted biodiesel production from sesame oil using barium hydroxide as a heterogeneous catalyst: comparative assessment of prediction abilities between response surface methodology (RSM) and artificial neural network (ANN). Ultrason Sonochem 26:218–228
42. Shahmoradi GF, Razbin M, Tehrani M, Zolfipour Aghdam Vayghan L, Sadrjahani M (2024) Modeling and optimization of dyeing process of polyamide 6 and woolen fabrics with plum-tree leaves using artificial intelligence. Sci Rep 14(1):15067
43. Sharma M, Kohli D, Soni N, Mitra J (2021) Microwave assisted extraction of bioactive compounds: a review. Octa J Biosci 9(2)
44. Politis N, Colombo S, Colombo P, Rekkas G (2017) Design of experiments (DoE) in pharmaceutical development. Drug Dev Ind Pharm 43(6):889–901
45. Anderson VL, McLean RA (2018) Design of experiments: a realistic approach. CRC Press
46. Antony J (2023) Design of experiments for engineers and scientists. Elsevier
47. Haji A, Rahimi M (2022) RSM optimization of wool dyeing with Berberis thunbergii DC leaves as a new source of natural dye. J Nat Fibers 19(8):2785–2798
48. Yameen M, Adeel S, Nasreen H, Ghaffar A, Ahmad T, Inayat A (2022) Sustainable eco-friendly extraction of yellow natural dye from haar singhar (Nyctanthes arbor-tritis) for bio coloration of cotton fabric. Environ Sci Pollut Res 29(55):83810–83823
49. Baruah J, Chaliha C, Nath BK, Kalita E (2021) Enhancing arsenic sequestration on ameliorated waste molasses nanoadsorbents using response surface methodology and machine-learning frameworks. Environ Sci Pollut Res 28:11369–11383
50. Box GEP, Behnken DW (1960) Some new three level designs for the study of quantitative variables. Technometrics 2:455–475
51. Chandrika KC, Prabhu TN, Kiran RS, Krishna RH (2022) Applications of artificial neural network and Box-Behnken design for modelling malachite green dye degradation from textile effluents using TiO_2 photocatalyst. Environ Eng Res 27(1):200553
52. Avinash A, Murugesan A (2018) Prediction capabilities of mathematical models in producing a renewable fuel from waste cooking oil for sustainable energy and clean environment. Fuel 216:322–329
53. Zulfiqar M, Chowdhury S, Omar AA, Siyal AA, Sufian S (2020) Response surface methodology and artificial neural network for remediation of acid orange 7 using TiO_2P25: optimization and modeling approach. Environ Sci Pollut Res 27:34018–34036
54. Bekele N, Abewaa M, Mengistu A, Adino E, Takele T, Yilma M, Angassa K (2024) Optimization of natural dye extraction from the root of Rumex abyssinicus plant using response surface methodology. Results Chem 101577
55. Meena J, Saraswat H (2024) Optimizing dyeing parameters for indigo on cotton yarn: a comparative study of exhaust dyeing and continuous dyeing techniques. J Inst Eng India Ser E 105:33–40
56. Aftab M, Ahmad T, Adeel S, Javed M (2024) Ecofriendly process optimization of natural dye extraction from clove (Syzygium aromaticum L.) using response surface methodology and artificial neural network modeling. Sustain Chem Pharm 37:101401
57. Ahmad T, Gilmour SG (2010) Robustness of subset response surface designs to missing observations. J Stat Plan Inference 140(1):92–103
58. Myers RH, Montgomery DC, Anderson-Cook CM (2016) Response surface methodology: process and product optimization using designed experiments. Wiley
59. Gilmour SG (2006) Response surface designs for experiments in bioprocessing. Biometrics 62(2):323–331

60. Chen J, Lin Y, Liu Y (2025) Fast prediction of optimal reaction conditions and dyeing effects of natural dyes on silk fabrics by lightweight integrated learning (XGBoost) models. Color Technol 141(2):223–237
61. Deylami S, Sabzevari MH, Ghaedi M, Azqhandi MA, Marahel F (2023) Efficient photodegradation of disulfine blue dye and Tetracycline over Robust and Green g-CN/Ag_3VO_4/PAN nanofibers: experimental design, RSM, RBF-NN and ANFIS modeling. Process Saf Environ Prot 169:71–81
62. Fan T, Liu K, Zheng H (2024) Prediction and modeling of tie-dye polyester fabric's texture and color feature in supercritical carbon dioxide. J CO_2 Util 83:102801
63. Haji A, Vadood M, Öztürk M, Yigit İ, Eren S, Eren HA (2025) Prediction of colour strength in environmentally-friendly dyeing of polyester fabric with madder using supercritical carbon dioxide. Color Technol 141(1):44–53
64. Koina IM, Sarigiannis Y, Hapeshi E (2023) Green extraction techniques for the determination of active ingredients in tea: current state, challenges, and future perspectives. Separations 10(2):121
65. Ibrahim S, Abdul Wahab N (2022) Improved artificial neural network training based on response surface methodology for membrane flux prediction. Membranes 12(8):726
66. Youssefi S, Djomeh EZ, Mousavi SM (2009) Comparison of artificial neural network (ANN) and response surface methodology (RSM) in the prediction of quality parameters of spray-dried pomegranate juice. Drying Technol 27(7):910–917
67. Sinha K, Saha PD, Datta S (2012) Response surface optimization and artificial neural network modeling of microwave assisted natural dye extraction from pomegranate rind. Ind Crops Prod 37(1):408–414
68. Ahmad W, Khan A, Ali N, Khan S, Uddin S, Malik S, Ali N, Khan H, Khan H, Bilal M (2021) Photocatalytic degradation of crystal violet dye under sunlight by chitosan encapsulated ternary metal selenide microspheres. Environ Sci Pollut Res 28:8074–8087
69. Carpenter WC, Barthelemy JF (1993) A comparison of polynomial approximations and artificial neural nets as response surfaces. Struct Optim 5(3):166–174
70. Singh J, Kumar P, Eid EM, Taher MA, El-Morsy MH, Osman HE, Kumar V (2022) Phytoremediation of nitrogen and phosphorus pollutants from glass industry effluent by using water hyacinth (Eichhornia crassipes (Mart.) Solms): application of RSM and ANN techniques for experimental optimization. Environ Sci Pollut Res 30(8):20590–20600
71. El-taweel RM, Mohamed N, Alrefaey KA, Husien S, Abdel-Aziz AB, Salim AI, Radwan AG (2023) A review of coagulation explaining its definition, mechanism, coagulant types, and optimization models; RSM, and ANN. Curr Res Green Sustain Chem 6:100358

Chapter 11
Advanced Surface Modification Techniques for Sustainable Coloration of Textile Materials

Mir Abdul Gaffar, Bayazid Ahammed, Tabasirul Hoque, Sumaiya Maksura, and Md. Rashedul Islam

Abstract The textile industry uses high amounts of water, chemicals, and energy, particularly in the dyeing process. This has led to significant environmental concerns in this sector. This is why the textile sector is considering finding more efficient and sustainable ways to color textiles without deteriorating the final product's quality. One sustainable solution is the surface modification of such textiles. Surface modification of textiles alters the surface characteristics of fibers allowing the dye to bind more effectively. This approach cuts down the usage of harmful chemicals and reagents significantly. Besides just improving dyeability, surface modification techniques can produce multi-functional textiles with properties like UV protection, water repellency, and bacteria resistance. Thus, the industries' shift towards sustainable coloration is proving to be a part of the broader goal to innovate the processing of textiles.

Keywords Sustainable textile coloration · Surface modification techniques · Eco-friendly dyeing processes · Plasma treatment

11.1 Introduction

The textile industry has a serious environmental impact across its supply chain, producing more than 3.3 billion metric tonnes of greenhouse gas emissions yearly. It also demands substantial amounts of land and water, contaminates them, and adds to waste accumulation. Among the non-renewable resources, about 98 million tonnes are used for textile processing [1]. A report by UNEP (United Nations Environment Programme) indicated that the dyeing and finishing processes alone contribute around 36% of the textile industry's total impact on climate. One kilogram of cotton dyeing

M. A. Gaffar · B. Ahammed · T. Hoque · S. Maksura · Md. R. Islam (✉)
Department of Wet Process Engineering, Faculty of Textile Chemical Engineering, Bangladesh University of Textiles (BUTEX), Dhaka, Bangladesh
e-mail: rashedul.islam@wpe.butex.edu.bd

M. Shahid et al. (eds.), *Advancements in Textile Coloration*,
https://doi.org/10.1007/978-981-96-5091-0_11

uses about 125 L of water [2, 3]. Other hazardous chemicals such as detergents, dyes, and water-repellent agents are used in the process. These harmful chemicals pose substantial risks for both carcinogenic and non-carcinogenic human toxicity [4]. Each year, about 10,000 tonnes of dyes are produced globally but only 7,000 tonnes of dyes are properly utilized [5]. Surface modification of textiles is emerging as an eco-friendly alternative to these environmental challenges.

Advanced surface modification techniques can ensure precise control of dye uptake, fixation, and colorfastness. These modification techniques minimize excessive chemical usage by increasing interactions between dye molecules and textile fibers. Techniques such as plasma treatment, enzymatic processing, and nanoparticle incorporation are emerging as powerful modification methods. These methods alter the surface properties of textiles without compromising the fabric's inherent qualities. For instance, plasma treatment can functionalize fiber surfaces which makes them more receptive to dyes. The enzymatic surface modification process is a more biodegradable alternative to synthetic agents traditionally used in dyeing. The surface modification techniques can be broadly classified into two types:

- **Physical Surface Modification**: Physical modification techniques change the surface properties of textile fibers without introducing new chemical substances. Common methods include laser treatment, plasma treatment, and ultraviolet (UV) irradiation. These methods are greatly energy-saving and need no water most of the time.
- **Chemical Surface Modification**: Chemical modification techniques involve introducing specific chemical groups or coatings onto the textile surface to improve dye affinity and functionality. Enzymatic treatments, grafting, and nano-coatings are some of the well-known chemical modification approaches.

This chapter provides an in-depth overview of the fundamentals of surface modification, focusing on the influence of surface energy, morphology, and the effect of chemical functionalization on textile coloration. The discussion is followed by physical and chemical surface modification techniques, like enzymatic surface modification, plasma treatment, and sol-gel coatings. The environmental and economic benefits of such techniques, including resource efficiency and reduced lifecycle impacts are also highlighted. The chapter concludes by addressing future trends in surface modification technologies, the challenges of scaling these techniques, and the importance of continued innovation to achieve sustainable textile coloration.

11.2 Fundamentals of Surface Modification

11.2.1 Surface Energy and Wettability

Surface energy is a measurement of energy at the surface of a material compared to its bulk. This property directly influences the spreading or penetrating of dyes onto the fiber surface. This spreading and penetrating property is known as wettability. An increase in wettability ensures dye solutions can be properly attached to the textile surface. The primary goal in fiber textile treatment and coating processes is to enhance attraction between the fiber or textile and the coating ink or liquid to increase wetting. This improvement can be obtained by altering the surface properties of the fiber or by changing the coating liquids' composition. The surface tension of a dye must precisely align with that of the solid substrate to ensure proper adhesion. If the substrate's energy is too low, some areas may remain unwetted. Conversely, if the energy is too great, defects like "orange peel" or "picture framing" may occur. Complete energetic compatibility and a significant polar component on both surfaces are essential for optimal attachment [6]. To evaluate the surface energy of fibers or textiles consisting of both dispersive and polar components, it is required to have contact angle data (cos θ) for the solid fiber or textile. According to Young's equation [7]:

$$\gamma_s = \gamma_{SL} + \gamma_L \cos\theta \quad (11.1)$$

Here, γ_{SL} indicates the interfacial energy within solid and liquid, γ_s denotes solid's surface energy, and γ_L refers to the liquid's surface energy. The Young-Dupré equation is another way to describe the work of adhesion which indicates how well the dye will adhere to the textile surface.

$$W_A = \gamma_{lv}(1 + \cos\theta) \quad (11.2)$$

Here γ_{lv} is the surface tension of the dye solution. The spreading coefficient ($S_{S/L}$) determines whether a liquid spreads spontaneously on a solid surface [8]. For effective dyeing, a positive spreading coefficient indicates that the dye solution will naturally spread across the textile surface. When $S_{S/L} > 0$, the contact angle among the dye solution and the textile surface approaches 0°. A zero contact angle means that the dye solution spreads completely across the fabric (Fig. 11.1). Therefore, achieving the best possible dyeing effect without the need for extra force.

$$S_{S/L} = \gamma_S - (\gamma_L + \gamma_{SL}) \quad (11.3)$$

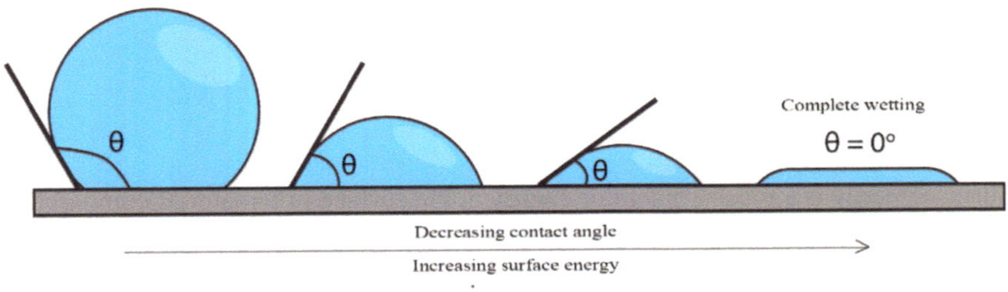

Fig. 11.1 Relationship between surface energy and contact angle [14]

11.2.2 Surface Morphology and Roughness

Surface morphology and roughness of the textile's surface properties influence the interaction with the dye solution. A rougher or porous surface enhances dye affinity whereas a smoother surface can hinder it. Roughness can enhance capillary action in highly porous or fibrous textiles. Hemi-wicking occurs when a dye solution partially wicks along the fibers of a rough surface due to capillary action. The capillary pressure (Δp) driving the dye solution into the pores can be expressed by the Young–Laplace equation:

$$\Delta p = -2\gamma H_f \tag{11.4}$$

Here, Δp represents the Laplace pressure, which is the pressure gap between exterior pressure and interior pressure, γ denotes the surface tension, and H_f is the mean curvature of dye droplets [9]. The Wenzel model modified the contact angle for a rough surface making a relation showing that the fiber roughness increases wettability.

$$\cos\theta_W = r\cos\theta_Y \tag{11.5}$$

Here, the roughness factor (r) is defined as the fraction between the actual surface area of the rough surface. θ_W is the theoretical contact angle on a rough surface and θ_Y is Young's contact angle [10]. A rougher surface (higher r) lowers the contact angle θ_W. Thus, roughness enhances the dye solution's ability to spread. However, when a fabric is porous, air pockets can form between the dye solution and the fabric's surface. This is described by the Cassie-Baxter equation:

$$\cos\theta_{\text{CB}} = f\cos\theta + (1 - f) \tag{11.6}$$

Here, θ_{CB} represents the apparent contact angle on a heterogeneous surface, f is the fraction of the dye solution in contact with the solid surface, and θ is the contact angle of the dye solution on a flat surface [11]. Increasing f enhances dye penetration and adhesion, while a lower f value can allow air pockets to form, potentially resulting in uneven dyeing.

The rate by which dye molecules penetrate fiber surfaces can be calculated using the Washburn equation [12].

$$L = \sqrt{\frac{\gamma rt \cos(\theta)}{2\eta}} \tag{11.7}$$

L represents the distance penetrated by the dye solution over time t. γ is the surface tension of the dye solution. θ denotes the contact angle and η is the viscosity of the dye solution. The relation indicates that a dye solution with lower viscosity and contact angle will penetrate the fabric more rapidly, which is desirable for achieving uniform dyeing across the textile.

11.2.3 Chemical Functionalization

Chemical functionalization refers to the insertion of distinct reactive groups onto the fiber surface to enhance dye compatibility. The addition of groups such as carboxyl, amino, or hydroxyls can improve the binding of specific dye types. This type of modification enhances dye fiber affinity, improving color yield, wash fastness, and overall dyeing efficiency. Functionalization can be achieved through processes like grafting or plasma-induced polymerization. These processes allow for selective attachment of functional groups without affecting the fabric's core properties. The efficiency of adding chemical groups to fibers through grafting can be measured by grafting yield (GY).

$$\%\text{GY} = \frac{W_g - W_o}{W_o} \times 100 \tag{11.8}$$

Here, the dry weight of the grafted fiber is denoted by W_g and the original fiber dye weight is W_O [13].

11.3 Physical Surface Modification Techniques

11.3.1 Plasma Treatment

Plasma treatment is a promising technology that offers an alternative to various traditional textile processes, significantly reducing chemical, water, and energy use. With careful selection of plasma gas and processing parameters, it enables various surface modifications, such as contaminant removal, bond breaking, crosslinking, etching, grafting, polymerization, and functionalization [15–17]. Plasma is created by heating or exposing neutral gas to electromagnetic fields forming a mixture of high-energy activated compounds capable of breaking chemical bonds and enabling eco-friendly surface modifications by using minimal gas. The surface modification effect depends on gas type and parameters like pressure and power. It modifies only the surface to a depth of less than 10 nm, leaving the bulk material unaltered [15, 18]. Plasma can be classified into cold and hot types on a working pressure basis. Since high temperatures can degrade textiles, only low-temperature plasma is suitable for textile applications [19]. Plasma can be created at either low or normal pressure. In low-pressure plasma generators, the pressure of the chamber typically ranges from 1 to 100 Pa. Plasma systems at atmospheric pressure levels can be classified into four types based on construction:

- **Corona discharge**—Two asymmetric electrodes are used which are made of a highly bent metal electrode resembling a wire or needle and paired with a flat or cylindrical electrode. The plasma forms in a gas region at a distance of 0.5 mm away from the needle tip (ionization region). The charged particles then diffuse toward the grounded electrode in the surrounding drift region [19].
- **Dielectric barrier discharge (DBD)**—It consists of at least one dielectric material coated electrode among two parallel electrodes (Fig. 11.2). DBD can occur in two forms: filamentary and homogeneous. A homogeneous DBD plasma can be achieved by fine-tuning the process parameters ensuring uniform treatment of surface. The distance between the electrodes ranges from a few millimeters, and the applicable voltage is around 20 kilo volt (AC) [20, 21].
- **Atmospheric-pressure glow discharge (APGD)**—The APGD system operates at a comparatively lower voltage (around 200 V) and high frequency (2–60 MHz) between two bare metal electrodes. The electrodes can be either curved or planar and the space between the electrodes ranges from a few millimeters. Helium is commonly used as the processing gas to avoid the formation of arc. Because of its high cost, nitrogen or argon is utilized as an alternative [20, 22].

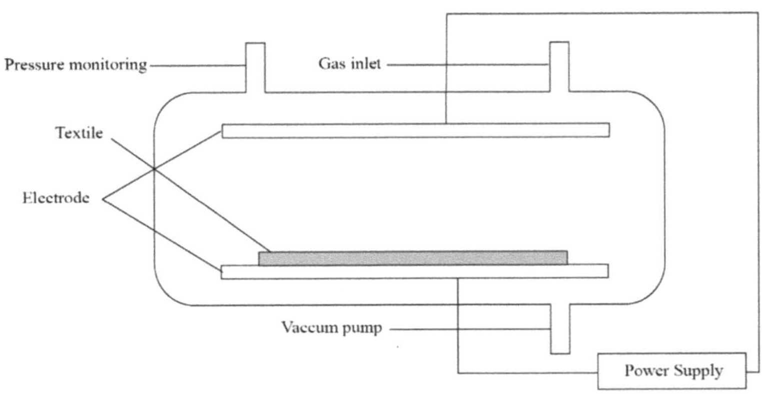

Fig. 11.2 Schematic diagram of plasma technology. Reproduced with permission [19]. Copyright (2024), Elsevier

- **Atmospheric-pressure plasma jet (APPJ)**—The APPJ plasma system consists of two concentric electrodes. When RF power (100–250 V) is applied to the inner electrode, the gas is ionized, creating plasma. This ionized gas then emits through a nozzle and is projected onto the substrate surface [19].

In the case of wool dyeing, the cuticle layer of wool fiber contains a monolayer (fatty acid) on its surface, which is hydrophobic and serves as a barrier to restrict the penetration of water, chemicals, and dyes [19]. Plasma treatment can remove this outer layer and expose the hydrophilic layer of the epicuticle [23]. Plasma treatment can also create polar functional groups and etching effects on the wool surface. As a result, wool achieves improved wettability, increased dyeing rates, enhanced dye uptake, and reduced dyeing times. Additionally, etching can increase surface roughness and reduce light reflection, leading to deeper color in the dyed product [24]. Similar to wool, DBD air plasma treatment of silk increases the density of polar functional groups, and reduces the water contact angle to 20° [25]. Additionally, argon and low-pressure oxygen plasma can cause significant etching on silk, while H_2 and N_2 treatments result in micro-pitting enhancing the printability and dyeability of silk [26, 27]. However, extended DBD treatment can lead to decreased tensile strength because of fibrillation and defects [28].

Cotton is the most commonly used natural fiber and contains many OH groups, making it hygroscopic and readily dyeable with synthetic dyes [29]. However, natural dyeing of cotton is challenging due to its mild negative charge from ionized carboxyl and hydroxyl groups. This charge creates electrostatic repulsion with negatively charged natural dyes resulting in the reduction of dye uptake and lowering color strength. Synthetic polymers, such as polyester, are highly crystalline and hydrophobic. Acrylic fibers are known for their low moisture absorption, high flammability, and low thermal stability. These fibers can benefit from plasma treatment to enhance dye uptake. The primary functions of plasma treatment for surface modifications include:

- **Cleaning**: Removes contaminants, allowing better interaction between dye molecules and fiber surfaces.
- **Activation**: Enhancing reactivity and bonding capabilities by increasing surface energy.
- **Grafting**: Introduces functional groups that strengthen chemical interactions between the dye and fiber.
- **Etching**: Creates micro-roughness that increases surface area. As a result, more effective dye retention.
- **Polymerization**: Forms thin films that modify surface characteristics, enhancing dye compatibility and absorption.
- **Functionalization**: Adds specific chemical functionalities to improve dye affinity. This results in better uptake due to increased interaction.

11.3.2 Laser Treatment

Laser treatment uses a concentrated beam of light (laser) to modify the color, surface or texture of fabric without the use of conventional techniques or chemicals. It demonstrates remarkable efficiency in passive cutting and can also be used for modifying textile surfaces. Lasers have been employed as alternatives to wet processing methods to replicate traditional textile surface design effects. Devoré and stone-washed denim are such examples [30]. The process is entirely dry, rapid and precise while ensuring excellent reproducibility and repeatability [31]. Surface modification of materials has been achieved through the utilisation of a variety of commercial lasers since the late 1990s [32].

CO_2 Laser: CO_2 lasers are widely used, less harmful, and powerful. However, thermal damage may result if power is not adjusted appropriately. This risk can be reduced by employing pulsed-mode CO_2 lasers that provide superior control over continuous wave lasers [32].

Excimer Laser: Excimer lasers are a form of ultraviolet gas lasers. It creates laser through an electric discharge by using an excimer as the lasing medium. Excimers are distinctive molecules that exist solely with one atom in a higher energy state (excited state). The release of this energy as a photon leads to the disintegration of the molecule, a decrease in the frequency of the lower energy state, and promoting population inversion. These are some of the essential requirements for laser action. Excimer lasers use chemically inert noble gas compounds and form bonds solely in a higher energy state. These lasers predominantly function at UV wavelengths [33].

In terms of morphology, laser treatment generates ripple-like textures and increased surface roughness, which results in a larger surface area that improves dye adhesion [34]. Upon exposure to laser irradiation exceeding the ablation threshold, the surface can develop well-oriented grooves or ripple structures characterised by micrometer-scale dimensions. This treatment leads to unique modifications in the surface structure of highly absorbing polymers, resulting in alterations to their physical and chemical characteristics. The alterations to the surface greatly influence the properties of textile polymers, possibly improving their functionality and appropriateness for a range of uses [30]. These combined effects lead to dyeing results that are more durable and effective. Bartlett [35] explored laser irradiation effects on dye-wetted cloth, noting a slight enhancement in dye uptake. Morgan [36] furthered investigating dye reactions at the laser's site of contact, refining laser dye-fixation techniques, and exploring innovative applications. This resulted in the creation of "peri-dyeing," a method in which the term "peri" indicates adjacent or around. A laser is used in peri-dyeing to precisely fix dye locally on fabric by irradiating the areas where the dye is applied. This allows dye reactions to occur at or near the laser's point of interaction with the textile.

A method for the three-dimensional shaping of synthetic textiles through the application of infrared laser irradiation has been developed. This approach utilises the photothermal characteristics of lasers to produce precise surface architectures and designs on materials such as PET and polyamide. This approach integrates three-dimensional moulding with laser dyeing, yielding results akin to shibori dyeing (a japanese method for resist dyeing) while offering improved precision and consistency. This innovative method, utilising lasers, eliminates the need for physical moulds or intricate setups, enabling adaptable design modifications through digital techniques. Furthermore, the utilisation of synthetic materials can enhance sustainability by facilitating more efficient recycling and material recovery processes [36].

11.3.3 Ultraviolet (UV) Irradiation

UV irradiation is a useful technique in the textile industry for achieving enhanced coloration by modifying the surface and chemistry of fibers. When cellulosic fabrics, such as cotton, are exposed to UV light at a wavelength of 184.9 nm, oxygen is absorbed, resulting in the generation of ozone. The high-energy radiation initiates oxidative reactions that break chemical bonds and generate new reactive groups, such as carbonyls, hydroxyls, and carboxylic acids. These new polar groups increase the fabric's affinity for dyes, creating more bonding sites for dye molecules.

Cellulose undergoes significant chemical modifications upon exposure to UV irradiation. UV irradiation generates radicals which reacts with oxygen and form reactive groups such as –COOH. These functional groups enhance dye-fiber interactions by providing more bonding sites for ionic or hydrogen bonding with dye molecules. This modification occurs through the abstraction of hydrogen atoms from carbon positions such as C1, C2, C3, or C4. Hydroxyl group abstraction can also occur at higher irradiation intensities. The C6 position is particularly susceptible here due to its lower steric hindrance. In cases of elevated UV irradiation, the CH_2OH group at C6 may be removed. This leads to the creation of reactive radical compounds (e.g., CHO or H radicals) [37]. The continuous oxidation of these radicals leads to the creation of –COOH groups, due to which dye fiber interaction is increased [38].

UV irradiation has similar effects on protein-based fibers like wool. Wool contains disulfide groups that act as cross-linking barriers, preventing fiber swelling. The disulfide groups (S–S) in keratin fibers undergo surface oxidation when treated with UV radiation at 253.7 nm for 40–60 min. A reduction in disulfide bonds can be observed when tested with X-ray photoelectron spectroscopy (XPS) [39].

In addition, UV irradiation creates micro-cracks or spaces in between fiber linings (Fig. 11.3) providing additional surface area for dye penetration. The degree of surface modification is influenced by the following factors:

- **Wavelength of UV Light**: Different wavelengths can penetrate and affect surface properties to varying extents.
- **Duration of Exposure**: Longer exposure time mostly enhances surface modification but can lead to degradation if exposed for too long.
- **Oxygen Availability**: Oxygen presence can accelerate surface oxidation.

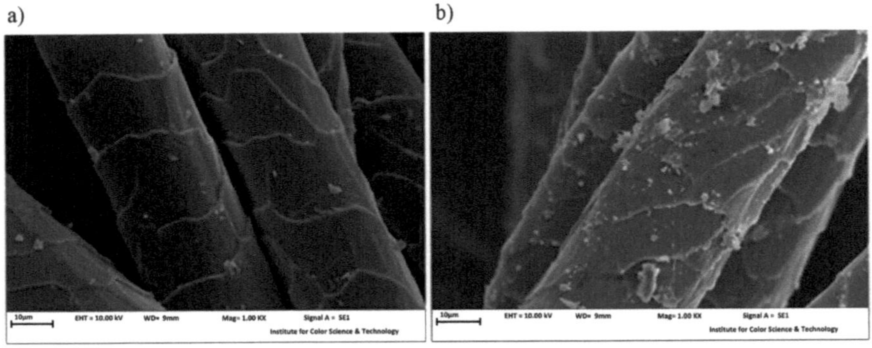

Fig. 11.3 Scanning Electron Microscope (SEM) images depicting **(a)** wool in its untreated state and **(b)** wool subjected to 24 h of UV treatment [40]

- **Distance from UV Source**: A shorter distance from the source increases intensity resulting in enhanced modification effects.
- **Fiber's Chemical Composition**: The chemical nature of the fiber can determine its susceptibility to UV-induced changes.

11.3.4 Electron Beam Irradiation

Electron beam (EB) irradiation is an eco-friendly, energy-efficient, and highly effective technique that has become increasingly popular for textile surface modification. Electron beam irradiation has been applied to enhance materials, including antibacterial cotton, metallic substrates, and carbon nanotubes [41]. The EBI process involves a stream of high-energy electrons produced by an accelerator to alter the physical and chemical characteristics of materials. EBI is regarded as a pure energy carrier because electrons are fundamental components of atoms. The process can be divided into three distinct phases: physical, physicochemical, and chemical when accelerated electrons interact with polymers.

In the physical phase, the energetic (primary) electrons transfer their energy incrementally to the targeted substrate. This leads to the generation of functional species such as secondary electrons, ions, and excited states. This is followed by the physicochemical phase, where these transient species are converted into stable polymer radicals. Then the subsequent chemical phase begins as the polymer radicals initiate various chemical reactions within the polymer matrix. The degree of such modification depends on the polymer's composition and treatment parameters maintained during the EBI process [42, 43].

When polypropylene fibers are modified with sulfonic acid groups through EBI, they achieve optimal dyeing results when dyed with cationic dyes. Polypropylene polymer undergoes decomposition and crosslinking dye to the transition and formation of radicals. As a result, the substrate chemically reacts with oxygen in the air, creating new functionally active groups such as –COOH, –OH, –COC, and –COO. Such functional groups also increase the dye fiber interaction leading to better uptake [44].

11.3.5 Gamma Irradiation

Gamma radiation is a form of penetrative, electromagnetic ionizing radiation. It is generated by the radioactive decay of atomic nuclei. The most notable effect of gamma irradiation is electron excitation which leads to ionization. This ionization can cause bond breaking, radical formation, generation of excited states and finally leading to the formation of new chemical bonds [45].

Gamma radiation transitions the atomic energy level leading to ionizing radiation. This radiation affects the surface of textiles by causing functional groups at the surface to excite and ionize which ultimately leads to the formation of radicals and changes in both physical and chemical properties [46]. Gamma-ray irradiation can convert the –OH groups on cotton surface to –COOH. This significantly improves the dye-fiber interaction when the fabric is dyed with red calico leaves extract [27]. Rehman et al. [47] explored the dyeing behaviors of cotton with natural flavonoid dye. The dye was extracted from irradiated onion shells. The optimal irradiation was 4 kGy for extracting natural quercetin, with dyeing conditions at 60 °C yielding maximum color shade and exhaustion.

11.3.6 Physical Vapor Deposition

Physical vapor deposition (PVD) involves the vaporization of material from a solid state followed by the deposition of material onto the textile surface atom by atom or molecule by molecule. This technique is proven to be effective in creating protective coatings resistant to corrosion and wear [48].

Vacuum evaporation, ion implantation, and sputtering are some of the PVD techniques. Each method offers unique advantages for creating durable and functional textiles.

- **Vacuum evaporation**: This technique involves material evaporation under vacuum conditions and condensation onto a substrate to form thin, high-quality coatings [49]. However, it comes with several limitations such as low bonding strength and poor handling of high melting point materials [48].
- **Ion implantation**: In this method, high-energy ions are created in solid material changing its physical and chemical properties. For example, it can create new functional groups by modifying the surface of electrospun PVA nanofibers [50]. Although effective, this method can damage the crystal structure of the target material.

- **Sputtering**: It is a key PVD technique that uses ion bombardment to eject target atoms and condense them on a textile surface [51]. Magnetron sputtering is an advanced modification of this method. It enhances deposition rates and efficiency by trapping electrons near the target surface with magnetic fields [52].

11.4 Chemical Surface Modification Techniques

11.4.1 Enzymatic Surface Modification

Enzymes are large proteins with active groups that interact with substrates to produce products by reducing the activation energy of reactions. These biocatalysts are derived from bacterial metabolites of living organisms [53]. The enzymatic surface modification treatment is used at different levels of textile processing bringing chemical and physical changes to the fabric surface along with the introduction of functional groups. Low water and energy consumption, improved product quality, high specificity, and eco-friendliness are some of the many benefits of enzymatic treatment [54]. The enzyme catalysis rate is dependent on factors like temperature, pH, and salt concentration, which affect the structure of protein. Some enzymes also require non-protein-type catalytic compounds called cofactors [55]. In general, there are three steps in Enzyme-catalyzed reactions [53]:

- The substrate (S) binding with the enzyme (E) to form an enzyme–substrate complex (ES).
- The enzyme–substrate complex (ES) transforms into an enzyme-product complex (EP).
- The product (P) is released from the enzyme-product complex (EP). The released enzyme (E) is now capable of beginning another catalytic cycle (Fig. 11.4) (Table 11.1).

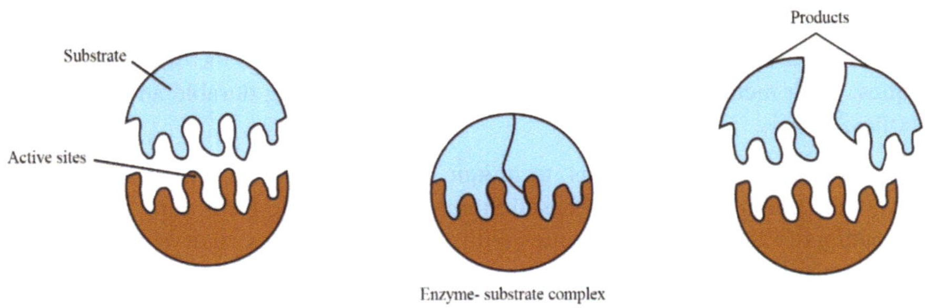

Fig. 11.4 Mechanism of enzymatic treatment [56]

Table 11.1 Enzymatic surface modification application on different textiles

Enzyme name	Substrate	Dye used/ natural dye source	w/V %	Time (min)	Temperature (°C)	Color strength (K/S)	Refs
Lipase (*Pseudomonas fluorescens*)	Polyester/ cotton	C.I. Reactive Dye Red S-8 G	2.75	50	80	4.2	[57]
Protease (Alcalase 2.4 L)	Polyamide 6.6	C.I. Reactive Blue 19	2	50	60	0.36	[58]
Cellulase (*T. reesei*)	Cotton	Copper phthalocyanine (CuPc) pigment	4	120	50	6	[59]
Proteolytic enzyme (Esperase 4.0 T)	Jute	C.I. Reactive Orange 18	2	60	80	10.34	[60]
Endo-1,4-β-D-xylanase (Dyadic Xylanase 2XP)	Jute	C.I. Reactive Orange 18	2	60	80	10.5	[60]
Lipase	Polyester	C.I. Disperse Blue 284	0.5	40	60	3.685	[61]
Cutinase	Polyester	C.I. Reactive Red 195	5	60	90	0.15	[62]

11.4.2 Polymer Grafting

Grafting of textile surfaces is a technique where specific functional groups or polymer chains are chemically bonded to the textile fibers, enhancing the fabric's ability to hold dyes and other treatments. This modification improves the coloration efficiency by creating additional irreversible reactive sites on the fabric. This results in a strong, uniform, and durable color due to a higher degree of interaction. Grafting techniques can be categorized as 'grafting from' or 'grafting to'.

In the 'grafting to' approach, reactive monomers bond with the main polymer chain to produce the grafted polymer (Fig. 11.5). 'Grafting from' involves chemical treatment of the substrate to create immobilized initiators leading to polymerization [63]. Table 11.2 briefly highlights some chemical grafting applications that resulted in improved color strength for the final dyed product.

11.4.3 Electrochemical Surface Modification

Electrochemical surface modification is commonly used to remove pollutants, degrade dyes, and treat wastewater. This eco-friendly method uses electrical current to initiate redox reactions at the electrode interface [75].

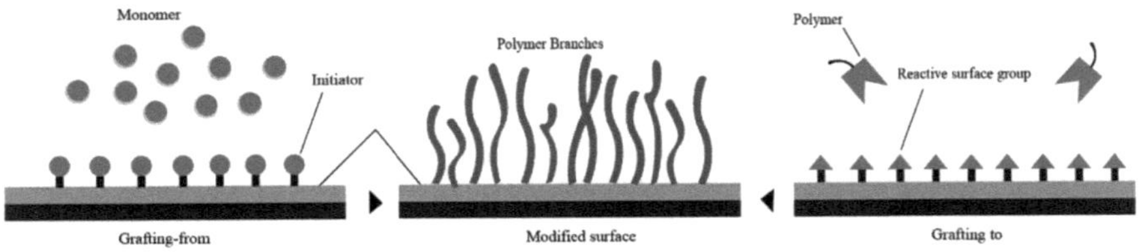

Fig. 11.5 Schematic representation of "grafting-from" and "grafting-to" techniques for surface modification [64]

Table 11.2 Effect of various modifiers on dye uptake (K/S value) for different textile substrates before and after polymer grafting

Substrate	Modifier	Dye used	K/S value		Refs.
			Before grafting	After grafting	
Silk	Chitosan	Acid Black NB.B	0–1	11.87	[65]
Silk	PAMAM dendrimer	*Seaweed Sargassum* sp. extract	0.15	0.24	[66]
Cotton	PAMAM dendrimer	*Seaweed Sargassum* sp. extract	0.06	0.20	[66]
Cotton	Sodium hydroxide	1,4-Naphthoquinone dyes	1.27	7.05	[67]
Wool	Acrylic acid (AA)	Bismethacrylated Quinizarin dye	1.4	4.7	[68]
Jute	Acrylic acid (AA)	Coracryl Red C4G Supra	2.679	4.648	[69]
Flax	Bovine serum albumin (BSA) and α-casein, catechin, quercetin	Catechin and quercetin	1.05	12.22	[70]
Hemp	Chitosan/silicone oil	Remazol Brillant Blue R vinyl sulphone	3.01	4.38	[71]
Polyester	Zinc acetate, copper acetate, NaOH	C.I. Basic Red 24	4.01	20.50	[72]
Nylon 6	Benzaldehyde-sodium bisulfite	C.I. Disperse Orange 45	3.24	2.1	[73]
Nylon 6	2-Ethylmethacrylate phosphoric acid (EMPA)	Basic dye: C.I. Red 18	0.488	0.469	[74]
PET/V (polyester/ viscose)	Sodium hydroxide (NaOH), Cu and Zn acetates	C.I. Basic Red 24 and C.I. Basic Blue 9	3.22	6.25	[72]

Electrochemical surface modification greatly improves coloration by optimizing how dyes are absorbed and enhancing the durability of colors. At first, electrochemical reduction transforms dyes into their soluble leuco forms by supplying electrons to dye particles. The dye particles can deeply penetrate into fibers in this soluble form [76]. After being absorbed, the dye molecules go through a re-oxidation process. This oxidation process turns the dye into its original, insoluble forms and securely entraps the dye particles into the fiber structure. This mechanism improves color durability, wash and light fastness, and a brighter shade.

Alongside reduction, electrochemical oxidation alters the fiber surface by producing reactive species like hydroxyl radicals ($OH^{\bullet}$) and metal oxides (MO) at the anode. These reactive species assist in breaking down contaminants on the fiber surface while also reinforcing the bonds between the dye and the fiber. This creates a stable layer that boosts dye adhesion, greatly improving color durability [77]. This strong connection makes sure that the dye stays securely attached to the fiber, even in tough situations [78].

Moreover, during electrochemical treatment, the creation of reactive moieties like hydroxyl (–OH) and carboxyl (–COOH) on the fiber layer enhances the fibers' reactivity and hydrophilicity. This change improves how the dye interacts, as the polar functional groups draw in dye molecules, making it easier for the dye to be absorbed. The greater hydrophilicity makes sure that dye solutions soak in better, leading to consistent and rich coloration [76]. Additionally, the electrochemical process leads to surface roughening, resulting in a micro- and nano-scale texture on the fiber. The rough surface really boosts the effective area for dye adsorption, enhancing both mechanical entrapment and chemical bonding. The improved interaction between the dye and fiber results in stronger adhesion. When this is combined with the previously mentioned mechanisms, it enhances color fastness and durability. Additionally, the electrochemical treatment can create an extra bonding matrix for dyes by introducing polyphenylene or azo layers on the fiber surface. This new binding interface keeps dye particles securely in place and lowers color fading [78].

11.4.4 Sol–Gel Coating

Successful modification of almost every type of fiber is possible by using different sols. The core advantage of sol–gel coating is its ability to integrate multiple functionalities in just one coating process [79].

The sol–gel coating is categorized into three main steps based on chemical reactions: hydrolysis, coating, and curing. Hydrolysis is the initial step of the coating process. An alkoxysilane precursor is used here, such as tetraethyl orthosilicate (TEOS). TEOS is broken down to form silica particles by hydrolysis. These silica particles have a low to medium degree of crosslinking in the solution and can be used to create thin silica layers when applied as a coating. Dyes can be added to the solution with the TEOS before the formation of the silica particles, or to the pre-prepared silica sol (Fig. 11.6). Regardless of when the dye is introduced, the dye molecules will

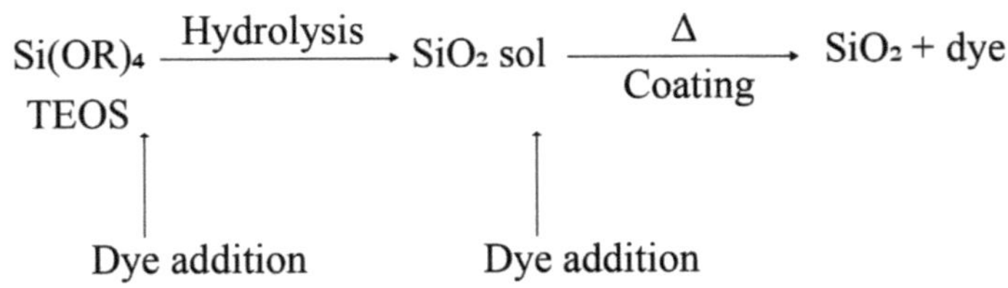

Fig. 11.6 Application of sol–gel coating in the dyeing process [80]

become immobilized (trapped or fixed) within the silica coating as it forms during the application process. The dye becomes entrapped into the silica network as the silica coating dries and hardens. The fixated color remains stable and resists fading even after repeated washing. The effectiveness of this dye immobilization relies on the dye type and the silica matrix composition [80].

Aside from TEOS, other common precursors for the sol–gel technique are glycidyloxypropyltrimethoxysilane (GPTMS), Triethoxypropylsuccinic anhydride (TESP-SA), and aminopropyltriethoxysilane (APTES). However, when cotton fabric is treated with TESP-SA along with anionic dyes, dye adsorption is reduced. This occurs because TESP-SA deposition increases the fabric's negative charge. As the dye particles are also negatively charged, electrostatic repulsion arises between the cotton and the dye, limiting dye uptake. This reduction in dye adsorption does not occur in fabrics treated with GPTMS, APTES, and TEOS [81]. The fixing mechanism for dyes by using GPTMS gel involves two key effects: the caging effect and chemical interactions. The caging effect refers to the physical entrapment of dye molecules within the three-dimensional silicate network formed during hydrolysis and condensation. The small pores created during these processes securely trap the dyes and prevent leaching during washing (Fig. 11.7). Higher curing temperature reduces pore size, enhancing this entrapment and improving dye fastness. Chemical interactions also play a crucial role, particularly for dyes with polar functional groups (such as $-NH_2$ and –OH), which can establish hydrogen and covalent linkages with the silicone hydroxides on the silica surface. These interactions further anchor the dye within the gel matrix. This results in stronger adhesion and improved dye retention when exposed to moisture or washing [82].

11.4.5 Chemical Vapor Deposition

Surface modification and functionalization via chemical vapor deposition (CVD) involve depositing or coating functional materials directly onto a substrate surface by initiating chemical reactions. The process begins with a gas flow over the substrate. The gas is mixed with volatile precursors and reactants and used as a carrier medium to initiate reactions. Such interactions create a coating on the surface.

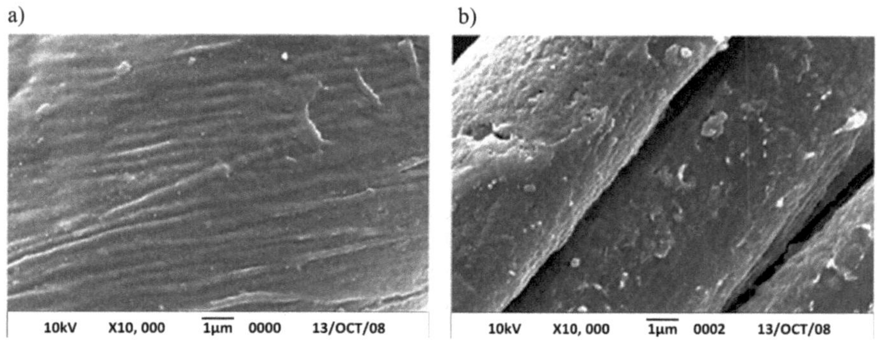

Fig. 11.7 SEM images of **(a)** cellulose fiber and **(b)** sol–gel coated cellulose fiber. Reproduced with permission [83]. Copyright (2024), Elsevier

One of the main advantages of CVD is its ability to create uniform and precisely controlled thin layers at a constant deposition rate. Due to adjustable microscopic pore structures, these layers can be engineered to be impermeable or highly selective. These functional layers can be used as a highly reactive layer for dyes enabling further exhaustion and minimizing post-dyeing chemical treatments. However, the technique comes with certain drawbacks. CVD demands sophisticated setups, high temperatures, and vacuum conditions to ensure uniform deposition. There is also a risk of potential toxicity with some precursors [84].

11.4.6 Layer-by-Layer Assembly

The layer-by-layer (LbL) assembly involves the ion interaction between a charged material and substrate. The process begins with adsorption of a charged material onto an oppositely charged substrate. This process is further followed by alternate deposition of oppositely charged layers [85]. After film formation, removing the substrate can result in a substrate with distinct morphologies and functionalities [86].

This process is an ion exchange process. Here polymer-counter ion pairs are changed by polymer–polymer ion pairs [87]. The LbL assembly process has evolved into several methods such as dip assembly, spin-assisted assembly, microfluidic systems, and 3D printing [88].

11.5 Applications of Modified Surfaces in Sustainable Coloration

11.5.1 Functional Textile Coloration

Surface modification of textiles can produce new chemical reactive species on the surface. This addition of chemical groups provides multifunctional textiles, such as antibacterial, UV protection, water repellency, etc. For example, surface modification of wool by treating it with polyamide, and fluorocarbon polymers can provide water repellency, shrink resistance, and good handling properties [89]. Cellulosic fibers such as viscose, cotton, linen, and lyocell can achieve anti-bacterial properties via N_2-based plasma treatment. The surface treatment creates active reactive sites by developing amide (–NH_2) groups. The treated textile shows antibacterial activity against both G +ve (*S. aureus*) and G −ve (*E. coli*) pathogens and lasts even after 15 wash cycles [90]. Silver nano-particle treatment to cellulosic material has also been shown to provide protection against these pathogens [91]. ZnO, TiO_2 nanoparticles have UV protective properties when treated with cotton [92]. Sójka-Ledakowicz et al. [93] applied aminosilane-modified nano-TiO_2 and nano-TiO_2 particles onto PET by padding, sol–gel coating, and spraying. The treated fabric showed excellent UV protection capability and also demonstrated high effectiveness in degrading red wine stains. ZnO has a broader UV absorption spectrum compared to TiO_2 and offers additional functionalities, such as antifungal and antistatic properties [94, 95].

11.5.2 Innovations in Patterning and Digital Printing

Digital printing is an innovative printing method that handles the entire process from pattern design to final printing. To achieve clear prints, especially with reactive dyes, the fabric is required to have some form of pre-treatment [96]. This is because reactive dyes have a fixation rate ranging from 50–80% which results in wastewater [97]. Surface modification technologies like plasma treatment can create grooves on the fiber surface increasing the total roughness. These rough textures increase the fabric's capacity to hold water and inks resulting in deeper penetration of color [98]. Corona and flame plasma systems can raise the energy of the surface by creating functional groups containing oxygen. Flame plasma treatment can achieve the highest oxidation level and surface energy due to the extra oxygen in the combustion process. This results in enhanced print quality, excellent tone, and visual quality. Corona treatment also raises the surface energy but can cause an "electret phenomenon" leading to reduced visual quality and defects like print mottling. Atmospheric plasma treatment gives more uniformity even with lower oxidation levels and surface energy in comparison to corona treatment [99]. CO_2 laser treatment can alter the printability of textiles by creating engraving on the surface. Thus having a potential for integrating laser and digital inkjet printing techniques to create various textures and adjust color based on apparel and textile design [100].

11.6 Environmental and Economic Benefits of Surface Modification Techniques

11.6.1 Reduction in Water, Chemical, and Energy Usage

Surface modification techniques in textile dyeing significantly reduce water, chemical, and energy consumption ensuring sustainability. Traditional dyeing demands huge amounts of water to transport dyes and chemicals, high energy for heat, and various chemicals to secure dye fixation and colorfastness. Surface modification techniques enhance dye uptake and adhesion by changing the textile surface at a molecular or nanoscale level. As a result, minimizing the need for excess resources. Table 11.3 outlines the benefits offered by various surface modification techniques.

11.6.2 Lifecycle Impact and Sustainability Metrics

The evaluation of environmental and economic advantages of surface modification techniques for sustainable dyeing demands an assessment of lifecycle impact and sustainability metrics. This section examines the lifecycle assessment (LCA) of these techniques based on costs, technical quality, and environmental pollution reduction.

11.6.2.1 Cost Efficiency

Surface modification techniques can greatly impact the overall cost structure of dyeing processes by reducing energy, water, and chemical usage. Techniques like plasma treatment offer the potential for utility cost reduction as the process operates at lower temperatures and requires less water [113]. However, initial investment for these equipment can be high for some manufacturers [114]. The modification techniques reduce the amount of dye and chemicals required by enhancing dye uptake efficiency [102]. Reduction of chemical usage thus lowers the expenditure on these inputs.

11.6.2.2 Technical Quality

To determine industry-scale adoption surface modification techniques, the technical quality of the final dyed product plays an important role. Technical parameters like

Table 11.3 Environmental and economic benefits of different surface modification techniques

Modification technique	Advantages	References
Physical surface modification techniques		
Plasma treatment	– Low application temperature, leading to energy savings – Minimal or no need for water and solvents – Elimination of drying steps post-treatment, reducing energy consumption – Significant reduction in dye and auxiliary usage – Shorter processing times – Notable enhancements in wool print quality	[18, 101, 102]
Laser treatment	– Does not require any chemical treatment – Requires less energy – Can create controlled irregularities at the surface – Different types of patterns can be according to fiber type	[103, 104]
UV irradiation	– Generation of free radicals at the surface – Increases dye uptake by surface oxidation – Requires low energy and a straightforward setup	[40, 105–107]
Chemical surface modification techniques		
Enzymatic treatment	– Introduces multiple functional groups to the surface – Minimal energy and water usage – Eco-friendly process – Highly specific application	[54, 108]
Electrochemical dyeing	– Recycling of dyeing mediators and liquors, lowering water and chemical usage – No toxic sulfates or sulfites in effluents, protecting aquatic life – Reduced chemical concentration requirements – Significantly decreased wastewater discharge, reducing energy at treatment plants – Elimination of odors and issues from sulfur-based reducing agents	[102, 109, 110]
Sol–gel coating	– Improves color and wash fastness – Simple processing – No pollution is created – Silica particles can be chemically modified with other compounds	[80, 83, 111, 112]

colorfastness, uniformity, and durability of dyes need to be ensured. For example, plasma-treated dyed fabric shows superior colorfastness and dye uniformity than untreated ones [115]. These improvements reduce the need for re-dyeing and enhance textile quality. Chemical modification approaches like cross-linking or nanocoatings enhance the durability of treated fabrics, extend product lifespans, and maintain dye integrity over time. This durability indirectly benefits cost and environmental sustainability by reducing the rate of textile replacement.

11.6.2.3 Environmental Pollution Reduction

Environmental pollution is an essential metric while assessing the sustainability of surface modification techniques. This is due to dyeing processes contributing significant water and chemical pollution. Improved dye uptake from surface modifications results in reduced dye effluent, minimizing water pollution and the load on wastewater treatment facilities [116]. Techniques like supercritical CO_2 dyeing are becoming highly favorable from an environmental standpoint because of its near-zero liquid discharge [117]. These processes also limit the need for toxic chemicals and minimize their presence in effluents. For example, plasma treatment can lower COD/ BOD levels in effluents and enhance chemical exhaustion from the dye bath [114, 118]. The energy efficiency of many surface modification techniques contributes to lower carbon emissions. Dyeing at reduced temperatures or with alternative energy sources (e.g., plasma treatment) helps decrease reliance on fossil fuels and reduces the overall carbon footprint.

11.7 Future Trends and Challenges

11.7.1 Innovations in Surface Modification Technologies

11.7.1.1 Bio-Based Modifiers

Bio-based modifiers are emerging as sustainable alternatives to traditional synthetic chemicals in dyeing. For example, green tea and curcuma extracts applied to cotton fabrics treated with ammonia plasma showed increased dye yields [119]. Ammonia plasma increased curcumin dye yield due to nitrogen-functional groups on cotton. Green tea extract exhibited an even stronger effect. Chitosan-modified cotton showed higher dyeability when pomegranate rinds were used as the dye [120]. Bio-mordants such as catechu, myrobalan, and pomegranate show higher color depth compared to metallic mordants (ferrous sulfate and stannous chloride) when applied to polyester [121].

11.7.1.2 Green Nanotechnology

Green nanotechnology is also another innovative solution for sustainable dyeing. Nanoparticles such as silver oxide can be derived from green synthesis methods (e.g., using plant extracts) [122]. These nanoparticles can be integrated into surface coatings to enhance the durability of dyes on fabrics, reducing the frequency of dyeing and thus cutting down on resource consumption. Another example is nanocellulose which is extracted from plants or even waste materials. It is used to form thin, functional coatings on textile fibers. Rai et al. [123] conducted an eco-friendly indigo-dyeing method using nanocellulosic materials to reduce water usage by up to 25 times and achieving over 90% dye fixation compared to 70–80% in conventional methods.

11.7.2 Challenges in Scaling and Adoption

Scaling and adopting advanced surface modification techniques for sustainable dyeing can face several challenges. Cost and economic viability act as immediate barriers as many of these surface modification techniques demand high initial investments in specialized machinery and skilled labor. For instance, plasma and nanotechnology-based treatments use expensive equipment and expertise, making it difficult for smaller manufacturers to invest in additional processing. Although these techniques may promise long-term savings, the limited short-term financial returns might hinder proper adoption. Techniques like laser treatment need precise control over parameters. Even minor deviations in process parameters such as temperature, and pressure can deteriorate the overall quality of the textile. This demands a sophisticated infrastructure that may not be accessible to all production facilities. Moreover, textile fibers are not compatible with all surface modification techniques. Additional processing is required for modification which adds further complexity and cost. Ensuring uniform results across large-scale production poses another challenge as most of the surface modification techniques are not market-ready. These modification methods must adhere to strict environmental regulations as chemical modification processing can bring out more ecological pollution to an already hazardous process.

11.8 Conclusion

Advanced surface modification techniques have a huge potential for sustainable textile coloration. Techniques ranging from plasma and laser treatments to chemical grafting and nano-coatings offer noticeable improvements in dye uptake, deeper shade, and durability, making the dyeing process efficient and eco-friendly. Physical and chemical modifications contribute not only to color depth improvement and adhesion but also reduce the environmental impact by minimizing water, chemical, and energy consumption. Furthermore, innovations in sol–gel innovations are now becoming demanding as they use low-water dyeing methods. However, challenges remain in widespread application for these technologies as they require high investment and constant regulatory support. Continuous research into utilizing the potential of surface modification techniques should be prioritized worldwide to pave the way for a greener and more sustainable coloration process.

References

1. Pizzicato B, Pacifico S, Cayuela D, Mijas G, Riba-Moliner M (2023) Advancements in sustainable natural dyes for textile applications: a review. Molecules 28(16), Article 16. https://doi.org/10.3390/molecules28165954
2. United Nations Environment Programme, Sustainability and Circularity in the Textile Value Chain: Global Stocktaking. https://wedocs.unep.org/20.500.11822/34184
3. Turley DB, Horne M, Blackburn RS, Stott E, Leybourn SR, Copeland JE, Harwood J (2011) The role and business case for existing and emerging fibres in sustainable clothing. Food and Rural Affairs (DEFRA) Department for Environment. https://library.unccd.int/Details/books/1364
4. Sandin G, Roos S, Peters G (2019) Environmental assessment of Swedish clothing consumption—six garments, sustainable futures. https://core.ac.uk/download/pdf/270109142.pdf
5. Bauer C, Jacques P, Kalt A (2001) Photooxidation of an azo dye induced by visible light incident on the surface of TiO_2. J Photochem Photobiol A 140(1):87–92. https://doi.org/10.1016/S1010-6030(01)00391-4
6. Minch R (2018) Methods to characterize textile/fibre surface for different treatments like printing, dyeing, coating etc. https://www.ifatcc.org/wp-content/uploads/2018/01/O25.pdf
7. Owens DK, Wendt RC (1969) Estimation of the surface free energy of polymers. J Appl Polym Sci 13(8):1741–1747. https://doi.org/10.1002/app.1969.070130815
8. Berg JC (2002) Chapter 1—Semi-empirical strategies for predicting adhesion. In: Dillard DA, Pocius AV, Chaudhury M (eds) Adhesion science and engineering. Elsevier Science B.V., pp 1–73. https://doi.org/10.1016/B978-044451140-9/50001-9
9. Liu H, Cao G (2016) Effectiveness of the Young-Laplace equation at nanoscale. Sci Rep 6(1):23936. https://doi.org/10.1038/srep23936
10. Patankar NA (2004) Mimicking the lotus effect: influence of double roughness structures and slender pillars. Langmuir 20(19):8209–8213. https://doi.org/10.1021/la048629t
11. Cortese B, Caschera D, Padeletti G, Ingo GM, Gigli G (2013) A brief review of surface-functionalized cotton fabrics. Surf Innov 1(3):140–156. https://doi.org/10.1680/si.13.00008
12. Guancheng J (2018) Chapter 2—Evaluation methods and influencing factors of gas wettability. In J. Guancheng (Ed) Gas wettability of reservoir rock surfaces with porous media. Gulf Professional Publishing, pp 29–84
13. Zahran MK (2016) A novel initiating system for wool grafting. Arab J Chem 9(3):408–420. https://doi.org/10.1016/j.arabjc.2013.07.027
14. Spooner E (n.d.) Surface energy: formula & definition. Ossila. https://www.ossila.com/pages/a-guide-to-surface-energy. Accessed 11 Nov 2024
15. Dave H, Ledwani L, Nema SK (2019) 8—Nonthermal plasma: a promising green technology to improve environmental performance of textile industries. In: Shahid-ul-Islam, Butola BS (eds) The impact and prospects of green chemistry for textile technology. Woodhead Publishing, pp 199–249. https://doi.org/10.1016/B978-0-08-102491-1.00008-3
16. Haji A (2013) Eco-friendly dyeing and antibacterial treatment of cotton. Cellul Chem Technol 47:303–308
17. Haji A, Semnani Rahbar R, Mousavi Shoushtari A (2014) Improved microwave shielding behavior of carbon nanotube-coated PET fabric using plasma technology. Appl Surf Sci 311:593–601. https://doi.org/10.1016/j.apsusc.2014.05.113
18. Kan C, Yuen CM (2007) Plasma technology in wool. Text Prog. https://doi.org/10.1080/00405160701628839
19. Haji A, Naebe M (2020) Cleaner dyeing of textiles using plasma treatment and natural dyes: a review. J Clean Prod 265:121866. https://doi.org/10.1016/j.jclepro.2020.121866
20. Herbert T (2007) 4—Atmospheric-pressure cold plasma processing technology. In: Shishoo R (ed) Plasma technologies for textiles. Woodhead Publishing, pp 79–128. https://doi.org/10.1533/9781845692575.1.79
21. Samanta K, Jassal M, Agrawal AK (2006) Atmospheric pressure glow discharge plasma and its applications in textile. Indian J Fibre Text Res 31:83–98

22. McCoustra MRS, Mather RR (2018) Plasma modification of textiles: understanding the mechanisms involved. Text Prog 50(4):185–229. https://doi.org/10.1080/00405167.2019.1637115
23. Naebe M, Cookson PG, Rippon J, Brady RP, Wang X, Brack N, van Riessen G (2010) Effects of plasma treatment of wool on the uptake of sulfonated dyes with different hydrophobic properties. Text Res J 80(4):312–324. https://doi.org/10.1177/0040517509338308
24. Wakida T, Cho S, Choi S, Tokino S, Lee M (1998) Effect of low temperature plasma treatment on color of wool and Nylon 6 fabrics dyed with natural dyes. Text Res J 68(11):848–853. https://doi.org/10.1177/004051759806801110
25. Zhang C, Guo F, Li H, Wang Y, Zhang Z (2019) Study on the physical-morphological and chemical properties of silk fabric surface modified with multiple ambient gas plasma for inkjet printing. Appl Surf Sci 490:157–164. https://doi.org/10.1016/j.apsusc.2019.06.053
26. Ghalab S, Raslan WM, El-Khatib EM, El-Halwagy AA (2011) Improving of silk printability by atmospheric air plasma treatment. Res J Text Appar 15(3):115–123. https://doi.org/10.1108/RJTA-15-03-2011-B013
27. Khan AA, Iqbal N, Adeel S, Azeem M, Batool F, Bhatti IA (2014) Extraction of natural dye from red calico leaves: gamma ray assisted improvements in colour strength and fastness properties. Dyes Pigm 103:50–54. https://doi.org/10.1016/j.dyepig.2013.11.024
28. Rani KV, Chandwani N, Kikani P, Nema SK, Sarma AK, Sarma B (2018) Optimization and surface modification of silk fabric using DBD air plasma for improving wicking properties. J Text Inst 109(3):368–375. https://doi.org/10.1080/00405000.2017.1347230
29. Zhou C-E, Kan C (2013) Review of antibacterial finishing processes with plasma for cotton. Res J Text Appar 17(4):12–24. https://doi.org/10.1108/RJTA-17-04-2013-B002
30. Kane F, Shen J, Morgan L, Prajapati C, Tyrer J, Smith E (2020) Innovative technologies for sustainable textile coloration, patterning, and surface effects. In: Muthu SS, Gardetti MA (eds) Sustainability in the textile and apparel industries. Springer International Publishing, pp 99–127. https://doi.org/10.1007/978-3-030-38545-3_4
31. Hung ON, Song LJ, Chan CK, Kan CW, Yuen CWM (2011) Using artificial neural network to predict colour properties of laser-treated 100% cotton fabric. Fibers Polym 12(8):1069–1076. https://doi.org/10.1007/s12221-011-1069-1
32. Esteves F, Alonso H (2007) Effect of CO_2 laser radiation on surface and dyeing properties of synthetic fibres. Res J Text Appar 11(3):42–47. https://doi.org/10.1108/RJTA-11-03-2007-B006
33. Excimer Laser (2024) In: Wikipedia. https://en.wikipedia.org/w/index.php?title=Excimer_laser&oldid=1256359050
34. Štěpánková M, Wiener J, Dembický J (2010) Impact of laser thermal stress on cotton fabric. Fibres Text East Eur 18(3:80):70–73
35. Bartlett SN (2006) Lasers and textiles: an exploration into laser dye–fibre interaction and the process of technology transfer. Doctoral dissertation, Loughborough University. https://repository.lboro.ac.uk/articles/thesis/Lasers_and_textiles_an_exploration_into_laser_dye_fibre_interaction_and_the_process_of_technology_transfer/9336272/1
36. Morgan L (2016) Laser textile design: the development of laser dyeing and laser moulding processes to support sustainable design and manufacture. Doctoral dissertation, Loughborough University. https://repository.lboro.ac.uk/articles/thesis/Laser_textile_design_the_development_of_laser_dyeing_and_laser_moulding_processes_to_support_sustainable_design_and_manufacture/9333209/1
37. Zuber M, Zia KM, Bhatti IA, Ali Z, Arshad MU, Saif MJ (2012) Modification of cellulosic fibers by UV-irradiation. Part II: after treatments effects. Int J Biol Macromol 51(5):743–748. https://doi.org/10.1016/j.ijbiomac.2012.07.001
38. Bhatti IA, Zia KM, Ali Z, Zuber M, Fazal-ur-Rehman (2012) Modification of cellulosic fibers to enhance their dyeability using UV-irradiation. Carbohydr Polym 89(3):783–787. https://doi.org/10.1016/j.carbpol.2012.04.010
39. Xin J, Zhu R, Hua J, Shen J (2006) Surface modification and low temperature dyeing properties of wool treated by UV radiation. Color Technol 118:169–173. https://doi.org/10.1111/j.1478-4408.2002.tb00095.x

40. Sadeghi-Kiakhani M, Safapour S, Sabzi F, Tehrani-Bagha AR (2020) Effect of ultra violet (UV) irradiation as an environmentally friendly pre-treatment on dyeing characteristic and colorimetric analysis of wool. Fibers Polym 21(1):179–187. https://doi.org/10.1007/s12221-020-9154-y
41. Zhang S, Ding F, Wang Y, Ren X, Huang T-S (2020) Antibacterial and hydrophilic modification of PET fabrics by electron beam irradiation process. Fibers Polym 21(5):1023–1031. https://doi.org/10.1007/s12221-020-9765-3
42. Abou Elmaaty T, Okubayashi S, Elsisi H, Abouelenin S (2022) Electron beam irradiation treatment of textiles materials: a review. J Polym Res 29(4):117. https://doi.org/10.1007/s10965-022-02952-4
43. Stein JM (1975) The effect of adrenaline and of alpha- and beta-adrenergic blocking agents on ATP concentration and on incorporation of 32Pi into ATP in rat fat cells. Biochem Pharmacol 24(18):1659–1662. https://doi.org/10.1016/0006-2952(75)90002-7
44. Kim H, Bae J-S (2009) Modification of polypropylene fibers by electron beam irradiation. I. Evaluation of dyeing properties using cationic dyes. Fibers Polym 10(3):320–324. https://doi.org/10.1007/s12221-009-0320-5
45. Gulfishan M, Mir RA, Hussain SA, Khan S, Shabir F, Khan E (2022) Gamma ray irradiation technology for textile surface treatment/modification. In: Yusuf M, Shahid M (eds) Emerging technologies for textile coloration, 1st edn. CRC Press, pp 157–170. https://doi.org/10.1201/9781003140467-9
46. Zhang Y, Shahid-ul-Islam, Rather LJ, Li Q (2022) Recent advances in the surface modification strategies to improve functional finishing of cotton with natural colourants—a review. J Clean Prod 335:130313. https://doi.org/10.1016/j.jclepro.2021.130313
47. Rehman F, Adeel S, Shahid M, Bhatti IA, Nasir F, Akhtar N, Ahmad Z (2013) Dyeing of γ-irradiated cotton with natural flavonoid dye extracted from irradiated onion shells (*Allium cepa*) powder. Radiat Phys Chem 92:71–75. https://doi.org/10.1016/j.radphyschem.2013.07.002
48. Wei Q, Xu Y, Wang Y (2009) 3—Textile surface functionalization by physical vapor deposition (PVD). In: Wei Q (ed) Surface modification of textiles. Woodhead Publishing, pp 58–90. https://doi.org/10.1533/9781845696689.58
49. Jankowski A, Hayes J (2004) The evaporative deposition of aluminum coatings and shapes with grain size control. Thin Solid Films 447–448:568–574. https://doi.org/10.1016/j.tsf.2003.07.018
50. Ho Wong KK, Zinke-Allmang M, Wan W (2006) N+ surface doping on nanoscale polymer fabrics via ion implantation. Nucl Instrum Methods Phys Res Sect B 249(1–2):362–365. https://doi.org/10.1016/j.nimb.2006.04.029
51. Tanaka K, Kasahara M, Usui H (2000). Preparation of ITO electrode on the organic layer by sputtering. Electron Commun Jpn (Part II: Electron) 83(4):23–30. https://doi.org/10.1002/(SICI)1520-6432(200004)83:4<23::AID-ECJB3>3.0.CO;2-F
52. Ohring M (1992) The materials science of thin films. Academic Press
53. Essa D (2024) 04. Recent techniques of surface modification for textile fabrics. J Meas Sci Appl (JMSA) 4(1):66–87. https://doi.org/10.21608/jmsa.2024.360350
54. Cui L, Fan X, Wang P, Wang Q, Fu G (2011) Casein and transglutaminase-mediated modification of wool surface. Eng Life Sci 11(2):201–206. https://doi.org/10.1002/elsc.201000110
55. Palmer T, Bonner PL (2007) Enzymes: biochemistry, biotechnology, clinical chemistry. Elsevier
56. Korkmaz G, Kılınç M, Kılınç N, Kut YD, Korkmaz G, Kılınç M, Kılınç N, Kut YD (2023) The role of surface modification methods for sustainable textiles. In: Roadmap to sustainable textiles. IntechOpen. https://doi.org/10.5772/intechopen.112792
57. Tegegne W, Haile A, Feleke T (2024) One-bath one-step dyeing of polyester/cotton blend fabrics through lipase enzyme surface modification. J Nat Fibers 21(1):2399661. https://doi.org/10.1080/15440478.2024.2399661

58. Kanelli M, Vasilakos S, Ladas S, Symianakis E, Christakopoulos P, Topakas E (2017) Surface modification of polyamide 6.6 fibers by enzymatic hydrolysis. Process Biochem 59:97–103. https://doi.org/10.1016/j.procbio.2016.06.022
59. Hao L, Wang R, Wang L, Fang K, Liu J, Men Y (2016) The influences of enzymatic processing on physico-chemical and pigment dyeing characteristics of cotton fabrics. Cellulose 23(1):929–940. https://doi.org/10.1007/s10570-015-0804-y
60. Hassan MM, Saifullah K (2019) Effect of enzymatic bio-scouring on the dyeability, physicochemical, and mechanical properties of jute fabrics. Fibers Polym 20(3):578–587. https://doi.org/10.1007/s12221-019-1108-x
61. Kumar JA, Kumar MS (2020) A study on improving dyeability of polyester fabric using lipase enzyme. Autex Res J 20(3):243–249. https://doi.org/10.2478/aut-2019-0030
62. Toprak-Cavdur T, Anis P (2020) The effect of enzymatic modification on the dyeability of polyester fabric with reactive dye. AATCC J Res 7:41–47. https://doi.org/10.14504/ajr.7.6.6
63. Bhattacharya A, Rawlins JW, Ray P (2008) Polymer grafting and crosslinking. Wiley
64. Nadi A, Boukhriss A, Bentis A, Jabrane E, Gmouh S (2018) Evolution in the surface modification of textiles: a review. Text Prog 50(2):67–108. https://doi.org/10.1080/00405167.2018.1533659
65. Davarpanah S, Mahmoodi NM, Arami M, Bahrami H, Mazaheri F (2009) Environmentally friendly surface modification of silk fiber: chitosan grafting and dyeing. Appl Surf Sci 255(7):4171–4176. https://doi.org/10.1016/j.apsusc.2008.11.001
66. Kadir MIA, Ahmad MR, Jabbar HA (2023) Dyeability of surface-modified cotton and silk fabrics with PAMAM dendrimer using Sargassum sp. seaweed extract. Trends Sci 20(9), Article 9. https://doi.org/10.48048/tis.2023.5807
67. Zhang L, Li H, Zhu J, Yan J (2020) Structure and dyeing properties of cotton fabric modified by protein from waste feathers. In: IOP conference series: materials science and engineering, vol 774, no 1, p 012043. https://doi.org/10.1088/1757-899X/774/1/012043
68. Dong Y, Zheng L, Wei J, Jang J (2016) UV-grafting coloration of cotton and wool fabrics using bismethacrylated quinizarin dye. Fibers Polym 17(10):1598–1604. https://doi.org/10.1007/s12221-016-6481-0
69. Teli MD, Sheikh J, Rajamohan A (2015) Cationic dyeable antibacterial jute obtained through grafting. Cellulose Chem Technol 49:529–535
70. Kim S, Cavaco-Paulo A (2012) Laccase-catalysed protein–flavonoid conjugates for flax fibre modification. Appl Microbiol Biotechnol 93(2):585–600. https://doi.org/10.1007/s00253-011-3524-8
71. Zhang H, Zhang LM (2010) Improving the dyeing properties and softness of hemp fabric using chitosan and epoxy modified silicone oil. J Text Inst 101(9):849–857. https://doi.org/10.1080/00405000902935015
72. Ibrahim N, Eid B, Youssef M, Ameen H, Salah A (2013) Surface modification and smart functionalization of polyester-containing fabrics. J Ind Text 42(4):353–375. https://doi.org/10.1177/1528083712440899
73. Samaha SH, Morsy MS, Essa DM, Tera FM (2004) Acrylonitrile grafting onto Nylon-6 fabric. II. Thermal, mechanical, and dyeing characterization. Polym-Plast Technol Eng 43(4):1213–1227. https://doi.org/10.1081/PPT-200030079
74. Mosleh S (2000) Chemical grafting of 2-ethyl methacrylate phosphoric acid onto nylon 6 fabric. J Appl Polym Sci 77(6):1357–1361. https://doi.org/10.1002/1097-4628(20000808)77:6%3c1357::AID-APP22%3e3.0.CO;2-R
75. Martínez-Huitle CA, Brillas E (2009) Decontamination of wastewaters containing synthetic organic dyes by electrochemical methods: a general review. Appl Catal B 87(3–4):105–145. https://doi.org/10.1016/j.apcatb.2008.09.017
76. Ahmed NSE, El-Shishtawy RM (2010) The use of new technologies in coloration of textile fibers. J Mater Sci 45(5):1143–1153. https://doi.org/10.1007/s10853-009-4111-6
77. Sala M, Gutiérrez-Bouzán MC (2012) Electrochemical techniques in textile processes and wastewater treatment. Int J Photoenergy 2012:1–12. https://doi.org/10.1155/2012/629103

78. Doppelt P, Hallais G, Pinson J, Podvorica F, Verneyre S (2007) Surface modification of conducting substrates. Existence of Azo bonds in the structure of organic layers obtained from diazonium salts. Chem Mater 19(18):4570–4575. https://doi.org/10.1021/cm0700551
79. Al Kashouty M, Elsayad H, Salem T, Elhadad S, Twaffiek S (2020) An overview: textile surface modification by using sol-gel technology. Egypt J Chem. https://doi.org/10.21608/ejchem.2020.24441.2464
80. Ak̨·it AC, Onar N (2008) Leaching and fastness behavior of cotton fabrics dyed with different type of dyes using sol-gel process. J Appl Polym Sci 109(1):97–105. https://doi.org/10.1002/app.27284
81. Schramm C, Rinderer B (2011) Dyeing and DP treatment of sol-gel pre-treated cotton fabrics. Fibers Polym 12(2):226–232. https://doi.org/10.1007/s12221-011-0226-x
82. Juan D, Li Z, Shuilin C (2005) Wash fastness of dyed fabric treated by the sol–gel process. Color Technol 121(1):29–36. https://doi.org/10.1111/j.1478-4408.2005.tb00245.x
83. Hou A, Shi Y, Yu Y (2009) Preparation of the cellulose/silica hybrid containing cationic group by sol–gel crosslinking process and its dyeing properties. Carbohyd Polym 77(2):201–205. https://doi.org/10.1016/j.carbpol.2008.12.022
84. Mostafavi AH, Mishra AK, Gallucci F, Kim JH, Ulbricht M, Coclite AM, Hosseini SS (2023) Advances in surface modification and functionalization for tailoring the characteristics of thin films and membranes via chemical vapor deposition techniques. J Appl Polym Sci 140(15):e53720. https://doi.org/10.1002/app.53720
85. Ariga K, Hill JP, Ji Q (2007) Layer-by-layer assembly as a versatile bottom-up nanofabrication technique for exploratory research and realistic application. Phys Chem Chem Phys 9(19):2319. https://doi.org/10.1039/b700410a
86. Wang Y, Angelatos AS, Caruso F (2008) Template synthesis of nanostructured materials via layer-by-layer assembly. Chem Mater 20(3):848–858. https://doi.org/10.1021/cm7024813
87. Lu P, Ding B (2009) 10—Nano-modification of textile surfaces using layer-by-layer deposition methods. In: Wei Q (ed) Surface modification of textiles. Woodhead Publishing, pp 214–237. https://doi.org/10.1533/9781845696689.214
88. Zhang Z, Zeng J, Groll J, Matsusaki M (2022) Layer-by-layer assembly methods and their biomedical applications. Biomater Sci 10(15):4077–4094. https://doi.org/10.1039/D2BM00497F
89. Hassan MM, Leighs SJ (2017) Effect of surface treatments on physicomechanical, stain-resist, and UV protection properties of wool fabrics. Appl Surf Sci 419:348–356. https://doi.org/10.1016/j.apsusc.2017.05.046
90. Ibrahim NA, Eid BM, Abdel-Aziz MS (2017) Effect of plasma superficial treatments on antibacterial functionalization and coloration of cellulosic fabrics. Appl Surf Sci 392:1126–1133. https://doi.org/10.1016/j.apsusc.2016.09.141
91. Abdel-Aziz MS, Eid BM, Ibrahim NA (2014) Biosynthesized silver nanoparticles for antibacterial treatment of cellulosic fabrics using O_2-plasma. AATCC J Res 1(1):6–12. https://doi.org/10.14504/ajr.1.1.2
92. Tsuzuki T, Wang X (2010) Nanoparticle coatings for UV protective textiles. Res J Text Appar 14:9–20. https://doi.org/10.1108/RJTA-14-02-2010-B010
93. Sójka-Ledakowicz J, Lewartowska J, Kudzin M, Leonowicz M, Jesionowski T, Siwińska-Stefańska K, Krysztafkiewicz A (2009) Functionalization of textile materials by alkoxysilane-grafted titanium dioxide. J Mater Sci 44:3852–3860. https://doi.org/10.1007/s10853-009-3522-8
94. Sawai J, Yoshikawa T (2004) Quantitative evaluation of antifungal activity of metallic oxide powders (MgO, CaO and ZnO) by an indirect conductimetric assay. J Appl Microbiol 96(4):803–809. https://doi.org/10.1111/j.1365-2672.2004.02234.x
95. Zhang F, Yang J (2009) Preparation of Nano-ZnO and its application to the textile on antistatic finishing. Int J Chem 1(1), Article 1. https://doi.org/10.5539/ijc.v1n1p18
96. Krump H, Šimor M, Hudec I, Jaššo M, Luyt AS (2005) Adhesion strength study between plasma treated polyester fibres and a rubber matrix. Appl Surf Sci 240(1):268–274. https://doi.org/10.1016/j.apsusc.2004.06.109

97. Khatri A, Peerzada MH, Mohsin M, White M (2015) A review on developments in dyeing cotton fabrics with reactive dyes for reducing effluent pollution. J Clean Prod 87:50–57. https://doi.org/10.1016/j.jclepro.2014.09.017
98. Zhang C, Fang K (2009) Surface modification of polyester fabrics for inkjet printing with atmospheric-pressure air/Ar plasma. Surf Coat Technol 203(14):2058–2063. https://doi.org/10.1016/j.surfcoat.2009.02.006
99. Tuominen M, Lahti J, Lavonen J, Penttinen T, Räsänen JP, Kuusipalo J (2010) The influence of flame, corona and atmospheric plasma treatments on surface properties and digital print quality of extrusion coated paper. J Adhes Sci Technol 24(3):471–492. https://doi.org/10.1163/016942409X12561252292224
100. Jung U, Lamar T (2022) The effects of CO_2 laser treatment on a digital velvet printing. J Text Inst 113(7):1291–1301. https://doi.org/10.1080/00405000.2021.1926119
101. Atav R (2013) The use of new technologies in dyeing of proteinous fibers. In: Eco-friendly textile dyeing and finishing. IntechOpen. https://doi.org/10.5772/53912
102. Schoenberger H, Schäfer T (2003) Best available techniques for the textile industry. 84:321–324
103. Gunay M (2013) Eco-friendly textile dyeing and finishing. InTech Open
104. Tunakova V, Hrubosova Z, Tunak M, Kasparova M, Mullerova J (2018) Laser surface modification of electrically conductive fabrics: material performance improvement and design effects. Opt Laser Technol 98:178–189. https://doi.org/10.1016/j.optlastec.2017.07.017
105. Gao SL, Häßler R, Mäder E, Bahners T, Opwis K, Schollmeyer E (2005) Photochemical surface modification of PET by excimer UV lamp irradiation. Appl Phys B 81(5):681–690. https://doi.org/10.1007/s00340-005-1928-9
106. Iqbal J, Adeel S (2008) Effect of UV radiation on dyeing of cotton fabric with extracts of henna leaves. Indian J Fibre Text Res 33:157–162
107. Millington K (2000) Comparison of the effects of gamma and ultraviolet radiation on wool keratin. Color Technol 116:266–272. https://doi.org/10.1111/j.1478-4408.2000.tb00045.x
108. Singh M, Vajpayee M, Ledwani L (2021) Eco-friendly surface modification of natural fibres to improve dye uptake using natural dyes and application of natural dyes in fabric finishing: a review. Mater Today Proc 43:2868–2871. https://doi.org/10.1016/j.matpr.2021.01.078
109. Kulandainathan MA, Patil K, Muthukumaran A, Chavan RB (2007) Review of the process development aspects of electrochemical dyeing: its impact and commercial applications. Color Technol 123(3):143–151. https://doi.org/10.1111/j.1478-4408.2007.00082.x
110. Roessler A, Crettenand D, Dossenbach O, Marte W, Rys P (2002) Direct electrochemical reduction of indigo. Electrochim Acta 47(12):1989–1995. https://doi.org/10.1016/S0013-4686(02)00028-2
111. Gaponenko NV, Mudryi AV, Sergeev OV, Borisenko VE, Stepanova EA, Baran AS, Rat'ko AI, Pivin JC, McGilp JF (1998) Erbium luminescence in sol–gel derived oxide glass films. Spectrochim Acta Part A Mol Biomol Spectrosc 54(13):2177–2182. https://doi.org/10.1016/S1386-1425(98)00135-8
112. Xiao X, Chen F, Wei Q, Wu N (2009) Surface modification of polyester nonwoven fabrics by Al_2O_3 sol–gel coating. J Coat Technol Res 6(4):537–541. https://doi.org/10.1007/s11998-008-9157-x
113. Morent R, De Geyter N, Verschuren J, De Clerck K, Kiekens P, Leys C (2008) Non-thermal plasma treatment of textiles. Surf Coat Technol 202(14):3427–3449. https://doi.org/10.1016/j.surfcoat.2007.12.027
114. Radetic M, Jovancic P, Puac N, Petrovic ZL (2007) Environmental impact of plasma application to textiles. J Phys Conf Ser 71(1):012017. https://doi.org/10.1088/1742-6596/71/1/012017
115. Samanta KK, Basak S, Chattopadhyay SK (2014) Environment-friendly textile processing using plasma and UV treatment. In: Muthu SS (ed) Roadmap to sustainable textiles and clothing: eco-friendly raw materials, technologies, and processing methods. Springer, pp 161–201. https://doi.org/10.1007/978-981-287-065-0_6

116. Božič M, Kokol V (2008) Ecological alternatives to the reduction and oxidation processes in dyeing with vat and sulphur dyes. Dyes Pigm 76:299–309. https://doi.org/10.1016/j.dyepig.2006.05.041
117. Goñi ML, Gañán NA, & Martini RE (2021) Supercritical CO_2-assisted dyeing and functionalization of polymeric materials: a review of recent advances (2015–2020). J CO_2 Utiliz 54:101760. https://doi.org/10.1016/j.jcou.2021.101760
118. Jelil RA (2015) A review of low-temperature plasma treatment of textile materials. J Mater Sci 50(18):5913–5943. https://doi.org/10.1007/s10853-015-9152-4
119. Gorjanc M, Mozetič M, Vesel A, Zaplotnik R (2018) Natural dyeing and UV protection of plasma treated cotton. Eur Phys J D 72(3):41. https://doi.org/10.1140/epjd/e2017-80680-9
120. Haji A (2017) Improved natural dyeing of cotton by plasma treatment and chitosan coating. Optimization by response surface methodology. Cellul Chem Technol 51:975–982
121. Sen A, Bhowal A, Datta S (2019) Application of natural dye on polytrimethylene terephthalate fiber. Res J Text Appar 23(1):71–90. https://doi.org/10.1108/RJTA-06-2018-0041
122. Yu Z, He H, Liu J, Li Y, Lin X, Zhang C, Li M (2020) Simultaneous dyeing and deposition of silver nanoparticles on cotton fabric through in situ green synthesis with black rice extract. Cellulose 27(3):1829–1843. https://doi.org/10.1007/s10570-019-02910-2
123. Rai S, Saremi R, Sharma S, Minko S (2021) Environment-friendly nanocellulose-indigo dyeing of textiles. Green Chem 23(20):7937–7944. https://doi.org/10.1039/D1GC02043A

Chapter 12
Advancements in Textile Printing from Conventional to Modern Practices

Habibur Rahman Anik and Mahin Akter

Abstract With the exponential demand for colorful and aesthetic clothing, textile printing has revolutionized the industry in so many ways. It has continuously developed with integration of different innovative techniques and machinery. A significant shift happened from the conventional printing process to the modern systems toward sustainability, automation, and customization with the variation of the products. It has significantly transformed traditional printing methods. Advanced printing technology can make better-quality designs with high precision and durability, reduced resource consumption, and minimal environmental impact. Also, the chemicals used for the printing are mostly environment-friendly in terms of pollution, energy-intensive production, and health risks. Printing technology has advanced printing techniques like inkjet, heat transfer, and screen printing, showing improved efficiency, reproducibility, and large-scale production. This chapter discussed the recent innovations in textile printing with the most focus on the transition to digital and modern-day practices in natural and synthetic textiles. It also highlighted these innovations' limitations and future aspects that must be addressed.

Keywords Textiles · Printing · Sustainable textiles · Fibers · Dyes

12.1 Introduction

Textile printing is the process of making a simple fabric into functional and aesthetic textiles or wearable products [1]. It also serves a variety of choices to the wearers as well as the industries to bring variations from fashion to many functional applications in sectors like healthcare, construction, defense, transportation, military, and

H. R. Anik (✉)
Department of Chemistry and Chemical and Biomedical Engineering, University of New Haven, 300 Boston Post Road, 06516, West Haven, CT, USA
e-mail: habib15anik@gmail.com; hanik1@unh.newhaven.edu

M. Akter
Department of Textile Engineering Management, Bangladesh University of Textiles, Dhaka, Bangladesh

M. Shahid et al. (eds.), *Advancements in Textile Coloration*,
https://doi.org/10.1007/978-981-96-5091-0_12

aerospace [2]. With its enriched history in the past, it all started with hand-crafted printing over the textile surface, i.e., block and screen printing, which has eventually evolved into advanced industrial processes meeting increasing demand all over the world [3]. Therefore, textile printing has progressed a long way in recent years. With some significant transformations blended with the touch of nanomaterials and advanced technology. Textile printing is not just putting some colors on the textile substrates but involves subsequent processes coming with many unique advantages, opportunities, and challenges [4]. Since the beginning of human civilization, humans have been fabricating clothing, home textiles, and different artifacts to enhance the aesthetics of their surroundings. It all started with playing with colors derived from nature. They tried to make some strains over the fabrics or textile substrates with the plant extracts, and gradually, they came up with mordants, dyes, and other chemicals. Ochres and charcoal were pretty common for coloration [5].

Printing methods in the past, i.e., rotary and flatbed screen printing, were appropriate for large-scale textile production [6]. However, it requires an extensive setup and an alarming usage of water and dyes, which eventually wastes groundwater. Therefore, industries in modern days have been looking for sustainable and easy alternatives to deal with the abovementioned shortcomings [7]. For example, digital printing eliminates the necessity for stencils, and it is possible to put the color and design over the fabrics directly [8]. It helps to have a better resolution of multicolor patterns, and it can minimize materials usage. Although it is mostly convenient for smaller batch production, it is still advantageous in crafting different interior designs and fashion arenas, where customization has greatly benefited [9]. Another advanced innovation over time is the sublimation printing method to have the permanent transfer of the dye molecules happen through the synthetic fabrics using optimum heat and pressure [10]. It helps not only with the durability but also with the quality of the printed materials, which stay top-notch. Printing textiles using different inks and coating may produce specialized functional textiles [11, 12]. For example, using graphene with different printing inks would help fabricate wearable textiles [13].

Printing is one of the most diverse coloring techniques used on natural and man-made fibers for introducing color and design to the fabric, where dyestuff is transferred to the substrates using methods like block or stencil printing [14]. Analytically speaking, printing is the process of mixing a textile substrate (often cloth), one or more colorants, and a design concept using a method that precisely applies the colorants [15, 16]. Block printing and stenciling were among the earliest methods used to embellish cloths thousands of years ago, marking the beginning of textile printing [8]. Over time, the industry has changed because of technological advancements and the invention of synthetic dyes, which produce intricately patterned fabrics in huge quantities with faster turnaround times and longer lasting effects [17]. Depending on the fabric type and the intended pattern, various printing techniques are used, including block printing, burn-out printing, and rotary screen printing [18]. With the growth of the textile business in a place where textile printing has become more sophisticated, adopting sustainability concepts and environmental concerns to reduce the ecological impact of dyeing and printing processes [19]. The emergence of eco-friendly printing methods like digital printing reflects growing awareness of the need for greener solutions in textile manufacturing [20]. This chapter highlights the ongoing developments impacting the industry today while examining the newest materials, techniques, and textile printing applications in creating functional textiles.

There are concerns about the adverse environmental impacts of the textile processing in the printing section, as it discharges and uses a large amount of chemicals and water. Nowadays, researchers are working on the printing process with less water usage and chemical consumption [21]. Therefore, the prime concern is to reduce the water footprint to ensure global sustainability with textile products. This chapter discussed the evolution of textile printing, looking at the transition from traditional practices to modern-day printing technologies. It explores different aspects of textile printing while keeping pace with the demands of functional textile applications and mentioning the challenges of upscaling efficiency, sustainability, and functionality. Highlighting the latest developments and their implications, this chapter would play a great role in communicating with the researchers working in the field of textile printing in shaping the future of the sustainable textile industry.

12.2 Process and Essential Aspects of Textile Printing

12.2.1 Process

Textile printing is generally the controlled process of imparting color on the textile substrates, even at a micro level containing certain patterns or images to display [22]. Unlike the uniform coloration observed in the dyeing process, textile printing essentially follows selective deposition of color on confined areas to create desired designs [8, 23]. The selection of colors or dyes in textile printing is influenced by various factors, most notably the chemical composition of the fiber, to ensure the desired level of colorfastness [24]. While dyeing immerses fabrics entirely in a dye solution to achieve uniform coloration, printing employs thickened dye pastes to localize the color application. This key distinction underscores the nuanced techniques involved in textile coloring processes [25]. In textile dyeing, excess dye is typically removed from the bath after fabric immersion. In contrast, printing requires the preparation of a dye paste by blending dyes with auxiliary materials and a thickening agent to achieve the required viscosity [26]. This paste is then applied to the fabric using techniques such as roller or screen printing, allowing for precise, sharp designs without color bleeding [27]. Thickening agents, often polymer-based, play a pivotal role in maintaining the integrity of the printed pattern by confining the dye to the designated design areas [28]. Advancements in textile printing arrived with different sections of the printing process particularly in machinery, thickening agents, auxiliaries, and fixation methods, have significantly enhanced its role in modern textile production [29]. The aesthetic appeal and versatility of the designs have been some of the extra touch of printed textiles contributing to the widespread adoption around the world.

Various techniques and styles are employed in textile printing, including screen printing, automated machine printing, block printing, photographic printing, spray, stencil printing, and resist and discharge printing [30, 31]. Modern adaptations like rotary screen printing and roller printing facilitate continuous operation with higher productivity. Hand screen printing has still been a simpler method rooted in the stenciling tradition. These diverse techniques set the evolving landscape of textile printing promoting more adaptability to artistic and industrial needs.

12.2.1.1 Pretreatment

Pretreatment is the step to make the fabric prepared for effective printing. Different types of fabrics go through different treatment steps, and that requires different printing methods in the end [32]. For example, scouring can remove natural waxes, dirt, oils, and impurities from the fabrics. Also, the fabrics can be washed in the presence of detergent and hot water and other alkaline solutions so that the fabric becomes soft and more receptive to dyes [33]. Another pretreatment process, desizing removes the sizing agents used at the time of weaving to make the fabrics stiffer, and it helps in dye penetration [34]. Bleaching in pretreatment normally whitens the fabric as it removes other colors or unexpected stains with the help of hydrogen peroxide or sodium hypochlorite. Mercerization is a process where the cotton fabric is treated with NaOH, and it facilitates clustering and increases the strength along with enhanced dye receptivity. Considering different printing methods, fixing agents or mordants can also be used before or after printing to improve dye adhesion and color fastness. Also, the heat setting can improve the interaction between the fabrics and the dyes with the help of high temperatures [35].

The main reason for the pretreatment is to prepare the fabric surface for better penetration of dye or pigment adhesion. By removing impurities, sizing agents, and other unwanted stuff, it ensures that the colorants would create a bond effectively with the fibers for maintaining the quality of printing [36]. Adequate pretreatment ensures consistent results across different fabric batches with a smaller number of defects such as uneven coloring, bleeding or streaks. With the advancement of technology, using plasma for treatment has been a really sustainable approach causing less harm to the environment [37]. However, it is not as effective as the wet process as plasma can reach up to a certain layer of the fabric. Therefore, it becomes not-so-effective in large-scale production. Vacuum plasma technology has been used in large-scale production with reduced pressure and control over the treatment parameters [38]. There is yet a lot to advance in this part of textile manufacturing.

12.2.1.2 Transfer

Transfer is the process of putting the design over the surface using a media in textile printing. Transfer paper is one of the mediums that plays an important role in the performance of printing. When the paste is coated on the surface of the transfer paper, the printing quality mostly goes on that, i.e., the K/S values or the durability [39]. Different types of paper may affect dye sublimation and dye transfer ability in different ways. For instance, when polyethylene film is placed with a glazed-coated paper, it gives better results in terms of the K/S values [40]. The uncoated papers with porous structures mostly led to less efficient dye transfer compared to coated papers. The ink deposited over the uncoated paper tends to be dispersed into the pores, and it causes more sublimation of the dye rather than effective transfer to the substrate. However, now in most facilities, inkjet printers are already in action to print designs and images with sublimable ink on transfer media [15]. Moreover, this method ends up providing high-quality image transfer with some options for multi-type printing. However, it is still not that good for large-scale printing. Hence, screen printing or offset printing is mostly used for the large-scale production of printed products [41]. The thermoplastic polymeric paste may improve the printing performance using it as a pre-coat on the transfer paper before starting the printing process for the silk fabrics [42]. The polyethylene transfer films are generally better than most of the transfer films in terms of crock and wash fastness [39]. However, the deposition of different nanoparticles on the transfer papers can be the horizon for future research to work on.

12.2.1.3 Fixation

Setting up the molecules as absorbed and chemically bonded to the fabric surface is the process known as fixation in textile printing [43]. It mostly depends on the interaction between dye and printing molecules and the substrates supported by the strong covalent bonds and interaction with the fibers. However, the mechanism of transferring the dyes to the surface and making them settle down is a bit tricky to understand [15]. The printed fabric is initially exposed to high-temperature steaming to heat both the fabric and the print paste film. Then the heat from the steam is transferred there [44]. First, the print paste film does the absorption of the condensed steam to activate other auxiliaries like urea to form the eutectic mixture, which eventually initiates the dye aggregation [45]. The water absorption facilitates the dissemination of the paste to transfer and diffuse over the fabric. The distribution of the dye particles, type of the fibers, force over the fabrics, and the interaction between the fabric and the dyes influence the color of the fixation. A stronger affinity between the dyes and the thickening agents is preferred to ensure higher dye retention in the paste. It may eventually impact the color yield of the fabric negatively. On the other hand, the dyes with lower affinity for the thickening agents result in having more effective transfer on the fabric [46]. The dispersion of the paste film should be heterogeneous in the dye transfer process. The thickening agents have some inherent Van der Waals forces causing stronger adhesive properties that hinder the dye release leading to lower color yields [47]. The gummy thickening agents have higher substitution levels with strong adhesive forces, and they have greater resistance to dye absorption. With the lower molar replacements, the adhesives may lose a portion of their adhesive power [48]. Hence, the dye and water can penetrate easily into the fabrics in a more effective way. The adhesive or thickening agents with low solid contents and molecular weight are generally good for more heterogeneity and lower substitution levels [49]. Moreover, it exhibits a minimal adhesion and elasticity resulting in better dye transfer and improved color yield.

12.2.1.4 Post-treatment

Once the printing steps are done, the fabric is not yet ready for further steps. It requires some other processing to ensure the quality of the substances. Curing, soaking, steaming, washing, soaking, drying, rinsing, and treatment with different fixing agents are some of the post-treatments that need to be done to have the printed fabrics of good quality [50]. In most cases, the post-treatments facilitate improving the adhesion, fixation, or penetration of the dyes on the fabric substances. Curing helps in adhering the inks uniformly on the fabrics and steaming helps in ensuring proper fixation and penetration of inks into the fabric [51]. Therefore, post-treatment is necessary for improving colorfastness, visual appearance, strength, durability, longevity, and comfort. They help the printed textiles meet the required standards with both functional performance and aesthetic appeal.

12.2.2 Materials

12.2.2.1 Fibers and Fabrics

Dyes and printing materials mainly interact with the textile fibers to create desirable unique patterns, designs, and effects on the textile surface. Textile fabrics are mainly made of natural or textile fibers [52]. Natural fibers are generally derived from natural sources like plants, animals, vegetables, and sometimes from minerals. Most of the plant fibers have cellulose units in the composition, while the wool fiber like wool is made of protein keratin. Cotton is one of the most common cellulosic fibers, fluffy in structure that maintains excellent dimensional stability, making it more chemically stable [53, 54]. Synthetic fibers like polyester are produced through chemical processes using condensation reaction between ethylene glycol and terephthalic acid, and are widely used for their durability and strength. Blended fibers, mostly natural with synthetic fibers, also exhibit many benefits, i.e., cost-effectiveness and other specific properties. Polyester/cotton and polyester/wool are some of the common blends exhibiting some excellent characteristics like moisture-wicking, shape stability, high tensile strength, and superior wear characteristics in textile fabrics [55, 56].

The performance of the textile printing mostly depends on the substrates. For example, using different polymeric agents on natural fiber fabrics enhances the retention of dyes for transfer printing. Polymers create a film-like barrier between the dye and the fabrics to improve the dyeing ability as it enhances adherence to the fibers

[47]. Treating wool fibers with benzoic anhydride increases the wash fastness of the fabrics significantly. Using complex-forming dyes requires some significant pretreatment to have improved printing quality. The chromium-complex dispersion dyes help in having good K/S values with excellent fastness after some steps of appropriate pretreatment [57].

12.2.2.2 Paste

The composition of the printing pastes has been a crucial factor for printing quality. It may vary with different printing methods, fabric type, or desired design. The primary colorants can be dyes or pigments to insert the design on the textiles. There are some thickeners, i.e., starch, sodium alginate, guar gum, and carboxymethyl cellulose (CMC) are used to maintain the appropriate viscosity for printing [58]. Binders, like acrylics and polyvinyl acetate (PVA) can fix the colorants on the fabrics. They improve adhesion and color fastness during curing and drying. The anionic or cationic fixing agents are used to prevent color fading. There are some modifiers that can adjust the brightness of the color, texture, and even the durability of the printing [59]. Some other chemicals like dispersing agents, wetting agents, pH modifiers, and crosslinking agents may help in achieving specific effects [60].

Print paste, commonly known as print ink, is basically a non-Newtonian fluid that has a shear-dependent flow behavior. With pseudoplastic performance, the viscosity falls under high shear stress. Hence, the resistance to flow is lower at higher shear rates, then the paste gets stuck over the surface. In the rheogram of measuring shear rate and shear stress, it would show an inclined profile for this case [61, 62]. Some of the printing pastes are thixotropic, and the flow resistance decreases over time under a constant shear. This rheological behavior decides how easy it can be to put the color onto the surface. Dyes in the paste mainly puts the color and the color variation in the printing process. The dyes selection happens mostly based on the fiber characteristics. Therefore, the dye must be compatible with the fabric as well as other chemicals that are supposed to be used further. The high affinity of the dyes to the fabrics ensures the quality of the printing as it is the dye that creates the bond with the fibers [14]. Reactive dyes are mostly used for cellulosic fibers, while disperse dyes favor synthetic fibers like nylon or polyester, and they are mostly good for heat transfer printing [63]. Acid dyes are suitable for protein fibers like silk or wool. All these things are decided based on the fixation and chemical composition of the fibers [64]. Direct dyes are mostly used for both cellulosic and protein fibers, and the basic dyes are reserved for acrylic and polyester fibers [65]. Vat dyes are quite complex to process yet they provide better color fastness. For denim and cotton fabric, sulfur dyes are used for lower lightfastness, then solvent dyes go for synthetic fabrics like nylon [66]. However, extensive research has been going on exploring dyes from different natural resources, and the wash and color fastness from them are not so satisfying [47].

12.2.2.3 Thickener

Thickeners are used to increase the viscosity of the printing paste. The viscous substances enhance the gumminess and flexibility of printing pastes. Hence, it can ensure precise application to fabric surfaces with distinct pattern outlines. Moreover, it can prevent issues like color bleeding or scattering. Long ago, people used polysaccharides as the primary polymeric thickening agents for fabric printing. As it was one of the natural thickeners, it had a poor interaction with other materials when multiple colorants were introduced [67]. Then researchers started to think of developing various polysaccharides and their derivatives, synthetic polymers, and emulsion-based thickening agents. The thickeners are mostly selected based on dyes, the printing methods, and the structure and characteristics of the fabric. The thickeners mainly help colorants, solvents, and other additives keep in close interaction with the fabric. The physical and chemical characteristics of the thickeners are important in terms of preventing dye separation with clear and well-defined prints. Moreover, the thickeners must be highly resistant to thermal activity. They need to be easy to remove once the fixation is done. The biodegradability and environmental concerns come when it comes to disposal [68, 69].

12.2.3 Other Parameters

12.2.3.1 Time

It takes a certain time to complete the printing process steps like transferring a design, pattern, or effect onto the selected fabric surface. Sometimes, the printing time and duration of the printing process can impact the quality and durability of the printing over the surface. The K/S values and the rate of dye transfer increase may rise from 5 to 30 s for an extension of the printing time up to 40 s [8, 22]. The printing time may vary depending on the composition of the dye and fibers. Less printing time may end up having the dyestuff not settled on the textile substrate. However, more printing time may also result in melting the dye or design over the surface damaging the product in the end. The printing time can be compromised by the temperature at a very short margin. Yet, the entire transfer depends on other many factors like the transfer medium, additives, dyes, temperature, and other factors controlling the printing time. Mostly, printing at 180–240 °C for 30–40 s is optimal in terms of safe printing transfer [70]. Also, it requires good quality films or medium to ensure better printing without causing any harm.

12.2.3.2 Temperature

It is necessary to maintain an optimum temperature for the printing process to ensure the successful transfer of dye over the fabric surface. The temperature is a very delicate parameter that has to be controlled very carefully [39]. Using a thermoplastic sheet like polyethylene and keeping the temperature ranging from 140 to 200 °C in the 30 s, the dye transfer significantly improved along with a better K/S value. It continued to rise slightly as the printing temperature increased up to 220 °C. However, the K/S value drops when cotton or wool fabrics are at temperatures more than 200 °C without the polyethylene film [71]. As cotton shows a better K/S value than wool, it proves that the composition of the materials is also important in determining the color strength. Appropriate temperature must be optimized according to the materials and the substrates that are used in the printing process. However, it is also important to know about the heat resistance of the printing after the post-treatment, otherwise, it may end up degrading at a very low temperature [72].

12.2.3.3 Pressure

It is necessary to maintain the necessary pressure in the printing process. It influences the interaction between the transfer medium, dye, and the surface of the material. The pressure affects the substrate, and the amount of dye penetrates into the fabric. Excessive pressure may cause harm to the tensile strength of the fabric while low pressure can also make the process slow [39]. Pressure ensures the necessary transportation of dyes into the fiber core with enhanced dye penetration and transfer efficiency. With the heat press, it became possible to maintain the pressure and optimal coloring over the surface according to the requirements of the design. The pressure level varies with the type of thickeners, dyes, and fabrics. The lower the pressure, the higher the chance of getting the inks passed out wasted in the printing process [73]. It ends up having a hazy print lacking clarity and wasting materials with severe environmental contamination. It is not only about the physical pressure, the pressure during the dye release at which the dye vapor starts to diffuse through the paper layers is also an important factor. Moreover, the dye supply at the beginning of the printing process should be suppressed and kept minimal to ensure the sharpness of the image, it will prevent the picture from blurring and keep the dye concentrated with better definition and precision [74].

12.3 Materials Used in Printing

12.3.1 Nanomaterials

Printing has long been used to enhance the color and design of textiles [75]. However, it has gained increasing traction in recent years due to the usage of nanomaterials and their diverse range of applications. The unique surface properties of nanomaterials, combined with their cost-effectiveness, have made them highly sought after worldwide [76]. Because printing provides a simpler method for integrating nanomaterials into textiles, researchers are increasingly turning to this approach to synthesize nanoparticles within textile materials. Nanomaterials can be classified according to their dimensions: (i) zero-dimensional, (ii) one-dimensional (1D), (iii) two-dimensional (2D), and (iv) three-dimensional (3D) nanomaterials (Fig. 12.1; Table 12.1).

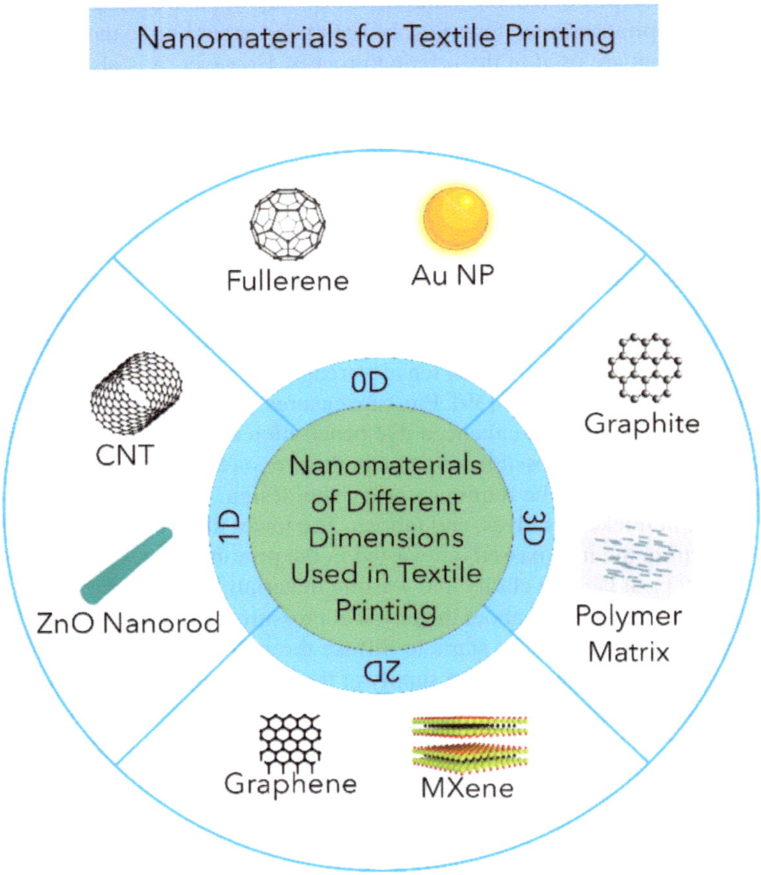

Fig. 12.1 Different nanomaterials on dimensions with example 0D: fullerene, gold nanoparticle (Au NP); 1D: carbon nanotubes (CNTs), zinc oxide nanorod (ZnO nanorod); 2-D: graphene, MXene; 3-D: graphite or 'stacked graphene', polymer matrix nanocomposite (PMNC) [12]

Table 12.1 Studies on the printing techniques of nanomaterials incorporation with textiles

Nanomaterials	Incorporation technique	Other material(s) used	Uses	Refs.
Graphene ink	Screen printing, gravure printing, inkjet printing	Liquid-based solvents	Smart textiles, like wearable sensors, energy-harvesting devices, field-effect transistors, and antennas	[77]
Hybrid graphene-based ink (e.g. graphene/silver nanoparticle (AgNP), graphene/poly(3,4-ethylenedioxythiophene)	Inkjet printing, spray coating	Water		[77, 78]
Silver nanoparticle	Inkjet printing	Silver salt	Antimicrobial cotton fabrics	[79]
Graphene nanoplatelets, carbon nanotubes	Screen Printing	Polymeric resin	Electrically conductive textiles	[80]
Carbon-based nanomaterials, metallic nanomaterials (carbon nanotubes)	Screen printing, gravure printing, inkjet printing, 3D printing, stencil printing	Oxygen-containing graphene oxide	Wearable health monitoring	[81]
Single-walled carbon nanotubes, multi-walled carbon nanotubes, graphene, silver nanowire, AgNP, Reduced graphene oxide (rGO)	Screen printing, gravure printing, inkjet printing	–	Textile sensors	[82]
Silica nanoparticles	Silkscreen printing	Butylacrylate/acrylonitrile (BA/AN) copolymer	UV protection and self-cleaning	[83]
Carbon nanotube	Screen printing	Deionized water	Textile antennas for wearable textiles	[84]

12.3.2 Dyes and Chemicals with Natural Materials

Aromatic compounds known as dye chemicals absorb light with wavelengths ranging from 350 to 700 nm. They consist of chromophores and auxochromes, which have conjugated double bonds and electron-donating substituents [39]. Figure 12.2 illustrates how dyes are grouped based on their compositions, respectively.

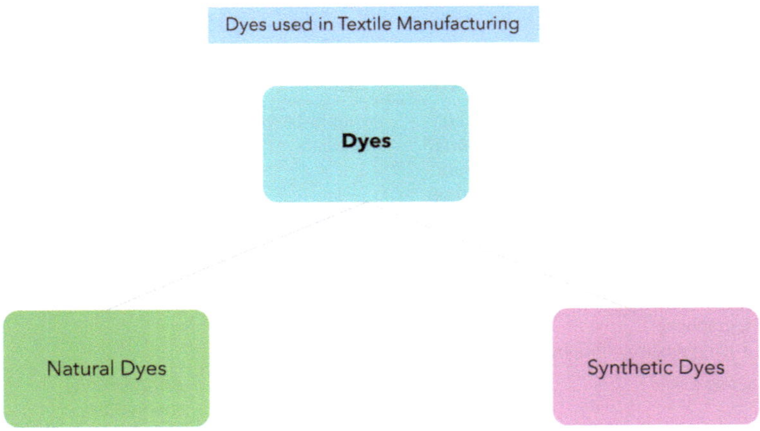

Fig. 12.2 Classification of dyes used in textiles [85]

12.3.3 Natural Dyes

In response to the global challenges posed by climate change, sustainability has been a critical focus in the textile printing industry. As demand for environmentally friendly products rises, it is essential for textile printing to adopt innovative technologies that prioritize natural materials over synthetic dyes and pigments [86]. Natural dyes, which are eco-friendly and biodegradable, offer additional benefits such as antibacterial, insect-repellent, and UV protection properties. Moreover, they are cost-effective, sustainable, and safe for both human health and the environment [47, 87].

Natural dyes used in textile printing belong to various classes, including indigoid, anthraquinone, alpha-naphthoquinone, flavonoids, carotenoids, di-hydropyrans, and anthocyanidins. These dyes are derived from a range of sources, such as Indigo (blue), Woad (blue), Tyrian (purple), Juglone (brown), Madder (orange, red), Lac (brown, red), Cochineal (red), Kermes (purple, red), Henna (brown, yellow), Manjit (brown, red), Weld (orange, yellow), Crocin (yellow to orange), Bixin (yellow), Awobanin (blue), Brazilwood (black), Sappanwood (black), and Carajurin (orange) [88]. A few natural dyes that are being studied now by the research to be used in different printing techniques are described herein. The orange extract from the rhizome of Japanese knotweed was used as a natural dye in the formulation of screen printing inks [89]. Catechu, a brown natural dye from the heartwood of *A. catechu* (Leguminosae family), is readily available and commonly used in natural dyeing. Studies on its use in fine powder form for reactive printing on cotton fabric showed the highest K/S value, with fair to good fastness properties [90]. In another study, cotton fabric was printed using extracts from five natural dyes: pomegranate peel, nutshells, alkanet roots, orange tree leaves, and Dyer's chamomile. The fastness properties of the prints were evaluated with pomegranate peel-based prints, which showed the best fastness across all tested standards. Overall, the tested fabrics showed satisfactory fastness results [91].

12.3.4 Synthetic Dyes

In 1856, dyeing changed when synthetic dyes emerged, replacing natural sources like cochineal, which had been costly and labor-intensive to process [92]. According to Sabnis et al., azo dyes presently account for the highest amount of dyestuff production, and their significance is only expected to increase. The high molar extinction of azo compounds, their great potential for structural diversity, and the ease of the coupling reaction are all aspects that contribute to their success and suitability for a variety of applications [52]. Acid dyes have an affinity for the fundamental properties of fibers, including polyamides, because they contain one or more acidic groups (SO_3H, COOH). Acid dyes are commonly used on wool, cotton, and nylon and make up between 30 and 40% of all dyes used. Usually, they are applied at a pH that is acidic. Acid Orange IV, Acid Blue 349, Acid Red 337, and Acid Yellow 36 are a few examples of acid dyes [52]. Figure 12.3 describes the classification of different dyes for printing.

Reactive dyes have better wet fastness with a wide range of hues. Their popularity is due to the reactive groups' ability to bind to fibers, their stability, and favorable processing conditions. Disperse dyes have grown alongside the rise of polyethylene terephthalate (PET) and new fibers like polytrimethylene terephthalate (PTT) and polylactic acid (PLA). They are used for dyeing nylon, where they cover barré defects but are limited to lighter shades due to poor fastness. Disperse dyes also dye acrylic fibers with excellent light fastness, though their use is restricted to pale shades because of limited build-up properties [94]. Commonly used for dyeing cellulose, they are reduced to their sodium-derived leuco form using sodium sulfide and applied to fibers in a manner similar to vat dyes [95, 96].

12.3.5 Thickener

Thickening agents are crucial in the printing process. These high-molecular-weight chemicals form a viscous, sticky paste with water, adding viscosity, preventing premature chemical reactions, and helping the print paste adhere to textiles [97]. Polysaccharides derived from plants, seeds, seaweeds, and microbes, such as guar gum, alginate, and plant gums, are frequently used as natural thickeners. Guar gum and sodium alginate are commonly used in reactive dye printing. Sodium alginate works best with reactive and acid dyes, which can coagulate when natural pigments are added. Sodium alginate is a common thickener used in reactive dye printing. When printing with bifunctional reactive dyes, guar gum and alginate mixtures have been shown to provide excellent color yield and fastness; however, certain thickeners,

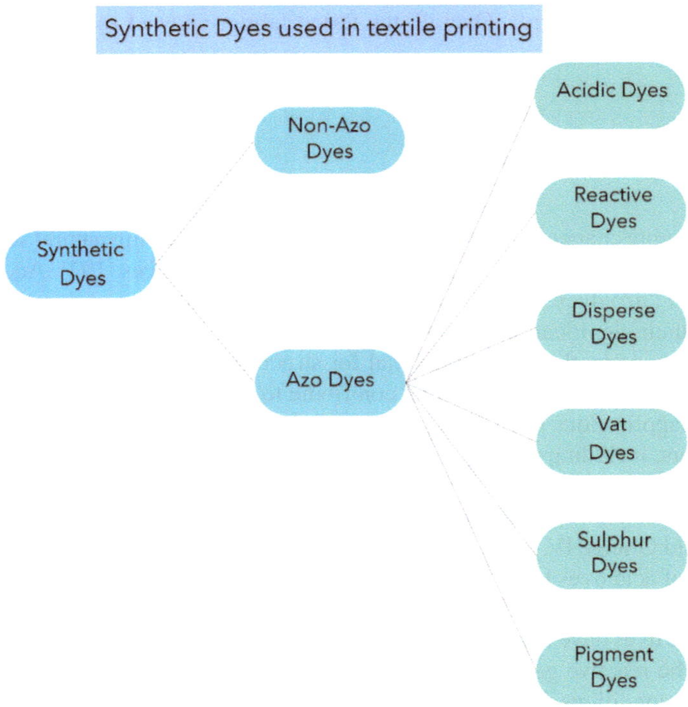

Fig. 12.3 Classification of synthetic dyes used in textile printing [93]

like goat horn and Arabic gums, may cause colors to become drab. Inkjet printing, printing, and coloring synthetic fibers, cotton, silk, wool, and polypropylene are some of the recent textile applications for chitosan [98].

Synthetic thickeners used in textile printing include acrylic acid, butyl acrylate, methacrylic acid, PAA/polyethylene glycol (PAA/PEG), ethyl acrylate, ethylene, and maleic acid. Polyacrylic acid (PAA), and PAA/polymethacrylic acid (PAA/PMA) are commonly used with reactive dyes for printing cotton fabrics. Styrene acrylate is also used for screen printing cotton and polyester fabrics with pigment colors. Polyacrylic acid is applied to print regenerated cellulosic fibers. Kerosene emulsion thickeners have been replaced by copolymers of acrylic or methacrylic acid, acrylamide, maleic anhydride, and crosslinked polymers with agents like ethyleneglycol dimethacrylate, divinylbenzene, N,N-methylene bisacrylamide, and methylol acrylamide for cotton printing [99].

12.4 Technologies in Textile Printing

12.4.1 Block Printing

Block printing is considerably the oldest manual printing method coming from the ancient art of textile and home-textiles decoration. It is still alive today, particularly for creating unique, one-of-a-kind designs on fabric in the garment industry [3]. This method involves using wooden blocks, originating from Europe during the fourteenth century, particularly in France, Italy, and Germany. Over time, the skilled printmakers replaced mineral pigments with dyes for these prints. This sort of printing is quite available in cliché wallpapers of older times in Europe, which is quite expensive for reproductions due to its complicated brocade [100]. However, block printing is the simplest method of all in terms of printing. Block printing starts with the selection of the printing paste (natural or synthetic dyes), carved wooden block, the fabric, and the printing setup. The paste is generally applied over the blocks, and then the blocks are pressed accurately at a position of the fabrics. There is a wooden sticker to hit it on the correct point to generate a mirror image over the fabric surface [101]. The blocks are mostly prepared with a carved surface made from terracotta or clay using the xylography machine. If it requires multiple colors to be deposited, the first color for the design is printed first, then it is dried. Then the other colors are deposited in a succeeding manner. The precision of the printing is mostly dependent upon setting the block at the correct place on the fabric [101]. Historically, the blocks were carved from wood using an engraving tool known as a xylography machine, although some were made from clay or terracotta. When printing with multiple colors, the fabric is typically printed with the first color, allowed to dry, and then printed again with the next color. This process is repeated until all the colors are applied. Then the block is repeated several times until the artisan finds the desired image on the fabric surface. The uniquely patterned fabrics are quite popular in the marketplace [102].

12.4.2 Screen Printing

Screen printing is a versatile textile printing method originating from Japan using printing basic patterns and writings on walls. The technique in Japan back in the past, was used to create intricate fabric designs, using stencils bound with human hair or silk threads. To the history, screen printing became popular beyond Japan in the nineteenth century, and people started using silk gauze as the template materials in Lyon in 1850 [19]. This development has been popularly known as template printing. The timeframe between 1930 and 1954, with the increasing use of knitted fabrics and synthetic fibers, screen printing also using hand stencil printing, became one of the most preferred methods for printing. Template printing has been considered one of the roots of screen printing that starts with stretching a fabric support cloth over a frame to create a stable base for the design [103]. Following that, a photo-sensitive coating is applied to the fabric, and the fabric is kept in a dark room to be dry. The desired design is prepared through the process of cutting patterns out of black paper on the areas with unwanted light exposure. The template is then brought to the light-exposed room. And then, after a brief wash in warm water for 3–4 min, the image is found there. After the preparation of the template, with the help of rubber squeegee the printing starts on the surface of the template having many open spaces there. And then, the design gets transferred onto the fabric [20]. Although the printing process is mostly done manually, the researchers are still working on having an automated system for having the template designed in an easy and automated way and the printing process to be done with an automated squeegee. The frame is generally placed on the fabric followed by the paste to be put on the screen. The squeegee is normally pushed and pulled over the screen to get the design on the fabric. Screen printing can generate a precise and vibrant design on a variety of fabrics, and it offers small and large-scale production [104].

Different kinds of screens are used in the screen-printing process. Flat and rotary screens are common for screen-printing. Flat screen printing requires a printing table, screen frame, squeegee, printing paste, and the desired fabric [18]. In the flat-screen printing, the size of the screen frame limits the size of the printed design, hence, the size of the design is constrained by the dimensions of the screen. The fabric is typically spread flat on a long table, and there is a printing blanket underneath. This blanket is normally made of cotton/polyester fabric, which is coated in neoprene rubber that makes it waterproof and helps ensure the fabric does not move during the printing process. After each application of printing, the printed fabric moves to the next frame. Then it is dried and prepared for fixation [105]. The blanket and fabric are separated once the printing is over for that term. Having the desired design fixed on the fabric, it is then washed to ensure that there is no color or gum on the fabric. Rotary screen printing is also a popular and efficient printing method used for large-scale and high-speed production. It operates almost on a similar principle as flat screen printing, but with a key difference is cylindrical metal foil screens made of stainless steel or nickel are used here instead of the flat screens. These screens are in rotating cylinders for continuous printing as the fabrics are always there underneath. In this printing process, the fabric is passed under a series of rotating cylindrical screens, then there is printing paste storage adjacent to the interior of each screen, and then the screen is forced through the mesh onto the fabric. For different colors, each cylinder roller is associated with a different color [106]. The fabric moves through the machine as the printing paste is injected with the speed of the screen and injecting the color into the fabric, and it creates the design. There is also a washing unit for cleaning the belts, screens, and other components. Zehra et al. [107] worked on pre-consumed colored cotton wastes using them in screen printing. The waste was incorporated into a commercial printing paste for cotton fabrics. They found a way to utilize the screen-printing process for utilizing the waste cotton fabrics. Rainert et al. [108] investigated improving the performance of reactive dyes used in screen-printing on cellulose fabrics by using the 3-chloro-2-hydroxypropyl trimethylammonium chloride. Reactive dyes have a low affinity for cellulose which leads to a problem in fixation. They analyzed the potential effects of factors like paste, pH, color intensity, viscosity, zeta potential, print penetration, and dye concentration in wash baths. They found the optimal combination for the concentrations of 60 g/kg for CHPTAC, 100 g/kg for urea, and 30 g/kg for sodium carbonate.

12.4.3 Heat Transfer Printing

Heat transfer printing is one of the modern techniques in the textile industry. This method transfers designs over the fabric with the help of pressure and heat. It sets a design over a heat-resistant medium like a transfer paper coated with a special heat-sensitive ink or dyes. The design is prepared to set face-down on the fabric [39]. There is a heat press machine to apply controlled heat and pressure for transferring the design. This method can produce sharp designs with high resolution without compromising the quality. It is mostly effective for synthetic fibers like nylon or polyester as they have a lower melting point. It is easy for the dyes to sublimate and fix over the fabrics [73]. However, natural fibers require certain treatments to get the desired performance. It is recommended to keep the temperature between 175 and 205 °C while printing. This method is good for improving precision and reducing waste, and it does not require any screen or template for a complete-color design. Durability has been an issue with this method as sometimes there appear some cracks and color fades over time [109].

Ding et al. [71] worked on a thermal transfer printing method for fabricating high-performance textiles with circuits of silver flakes with thermoplastic polyurethane (TPU) composites that enabled fabric electrodes to serve the properties like original wearability, washability, and stretchability of textiles. The electrodes showed high conductivity of 5.48×10^4 S/cm and high adhesion of more than 1750 N/m. The interesting part is it had a good abrasion and washing resistance with a high patterning resolution of almost 40 μm. They also prepared a textile-based radio frequency identification (RFID) tag that could do remote identification as well as a large-sized heater for wearable thermotherapy. Kwon et al. [110] worked on thermal transfer printing for the inkless fabrication of a graphene-based smart textile with a micro-supercapacitor. They used an adhesive film as a heat transfer paper for buffer adhesive layer. Laser-induced graphene (LIG) on the polyimide films has been transferred onto the adhesive film area of the textile substrates. The inkless transfer strategy has been a low cost and scalable method of making electrode on a smart textile material.

12.4.4 Ink-Jet Printing

Inkjet printing is a non-contact and highly precise printing method. There is an inkjet head that sprays tiny droplets of dye over the fabric. This can create high-quality and detailed designs and the ability to print in a wide range of colors and patterns. The inkjet printing has a digital control that does not require any physical contact between the printhead and the fabric [111]. Moreover, it offers faster and more efficient printing on the surface. The print system software generally interprets data from a digital image file, hence, that can control the parameters like, the specific color, ink volume, and placement of the microdroplets on the fabric [112]. Figure 12.4 demonstrates how ink-jet printing functions. This level of control ensures vibrant and precise prints reproduce a wide variety of colors with excellent print fastness. Inkjet printers use less material and can be used for different fabric types and patterns. The process reduces water and energy consumption compared to traditional methods. Furthermore, as there is no physical contact between the printhead and the fabric, it reduces the risk of fabric damage or distortion [113]. Thakker et al. [114] tried to investigate the impact of plasma surface treatment of inkjet printing on wool and cotton fabrics, and they used herbal inks for that. The plasma treatment can enhance the durability of the fabrics printed by the inkjet. They also highlighted the importance of plasma treatment for making the designs more sustainable. Inkjet printing typically uses dye-based or pigment-based inks. The dye-based inks provide vibrant colors and are mostly used for cotton or silk fabrics. On the other hand, the pigment-based inks show better durability, such as enhanced wash and lightfastness. Moreover, they require less chemical treatment, hence, it is mostly environmentally friendly. This kind of printing is mostly useful for customized and small-scale printing with more flexibility and lower setup cost. It is still striving to gain adaptability for large-scale printing.

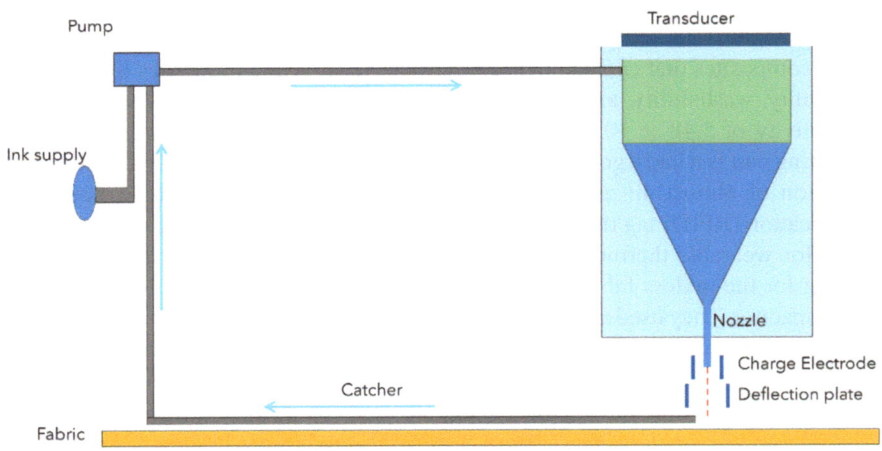

Fig. 12.4 Schematic of the ink-jet printing [113]

12.5 Functional Properties via Printing

The textile industry has rapidly expanded, offering versatile products for various applications through surface modifications [115]. These changes can be achieved using techniques ranging from traditional treatments such as lamination technique and mattress protection. Although these techniques are frequently employed to produce different functional properties, they have several drawbacks, such as the requirement for expensive extra machinery and processes, a lack of flexibility in creating various functional textile materials, and negative characteristics caused by the membrane material [116]. Due to the drawbacks of lamination and coating, research is shifting toward using printing techniques to impart functional properties to fabrics, some of which are briefly explained in this section.

12.5.1 Antimicrobial

Textiles, with their large surface area and moisture retention, provide an ideal environment for microorganisms like bacteria and fungi, which can rapidly multiply depending on moisture, nutrients, and temperature. This microbial growth leads to undesirable effects, including unpleasant odors, reduced mechanical strength, stains, discoloration, and a higher risk of user contamination [117]. As consumers become more aware of the health risks, demand for antimicrobial textiles has surged in recent years. These textiles are designed to kill or inhibit microbial growth [118], printing methods are becoming more and more common for giving textiles antibacterial qualities. For example, Tarek et al. [119] investigated an eco-friendly procedure to combine thermochromic pigment printing with antibacterial finishing in a single print paste formulation. This approach produced high-quality cotton and cotton/polyester prints with significant antibacterial activity against both G+ and G− bacteria. The results showed that incorporating Ag-NPs into the thermochromic pigment paste enhanced color strength, fastness properties, and antibacterial functionality (Fig. 12.5).

Ražić et al. [120] examined the effects of oxygen and argon low-pressure plasma pre-treatment on antibacterial properties and its interaction with digital inkjet printing. The study, conducted on cotton and lyocell knitwear, found that printed antimicrobial and plasma-treated fabrics exhibited excellent antimicrobial effects, with lyocell samples showing superior efficacy against *Staphylococcus aureus.* A different study created novel acid-disperse dyes based on thiadiazol derivatives and assessed how well they worked as antimicrobial colorants in polyester fabric digital transfer printing inks [121]. Strong contact with DNA Gyrase was demonstrated by the results, which included molecular docking studies, underscoring the inks' robust antibacterial properties. Soleimani et al. [79] developed a promising green production technique for in situ inkjet printing with antibacterial properties. They used a green water-based ink containing silver nitrate to impart antibacterial activity on

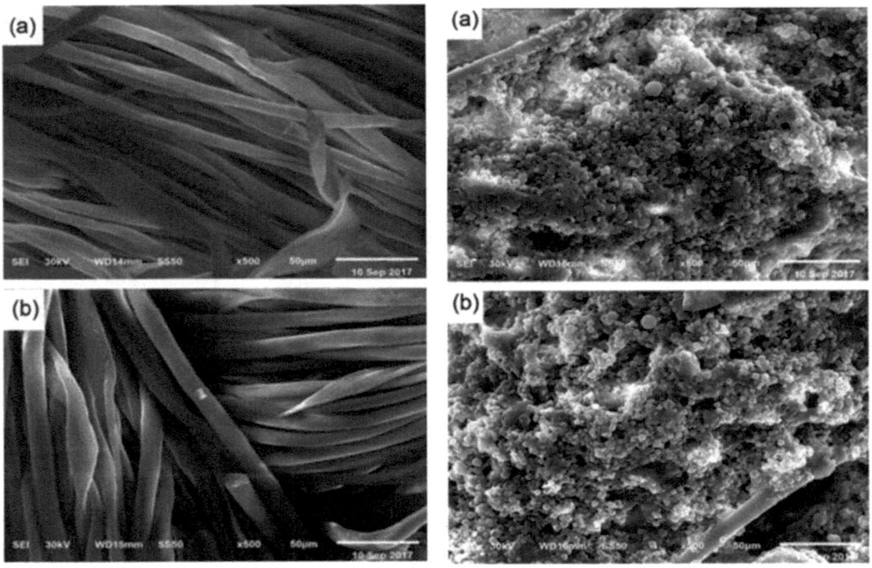

Fig. 12.5 **a, b** Display the SEM images of unprinted cotton and cotton/polyester samples, respectively, with generally smooth surfaces. However, SEM images of printed cotton show that the surface morphology has undergone significant modification because of Ag-NPs, and there were various characteristics of the fibers' surface, like voids and grooves. Reproduced with permission from Ref. [119], Copyright Springer Nature, 2018

pretreated cotton fabric, incorporating a bio-reducing agent based on carbohydrate polymers, such as sodium alginate and aloe vera extracts, as thickeners.

12.5.2 Water Repellency

Water repellency in treated fabrics refers to their ability to resist liquid penetration under static conditions, where only capillary forces and the weight of the drop are involved. It is defined as the surface's resistance to wetting by liquid water, achieved by lowering the textile's surface energy. With the contact between the liquid drop and the textile changing according to cohesive forces, this minimizes the free energy at the fiber surface, reducing wettability [122]. Water repellency, or even superhydrophobicity, can be achieved by applying low surface energy compounds (e.g., fluorocarbons) and introducing surface roughness through techniques like dip-coating, padding, knife coating, or spray coating. While fluorocarbons (CxFy) are known for their excellent water and oil repellency, their bio-accumulative nature and persistence in the environment make them unsuitable for sustainable development and environmentally friendly textile practices [123]. Considering these factors, Yu et at. [124] developed an environmentally friendly solution using inkjet printing to apply water-repellent ink onto polyester and polyamide 66 for functional textiles in sports and workwear. The resulting material, created with a waterborne, fluorocarbon-free ink containing polysiloxane micro-emulsion, featured a microstructure where polysiloxane wrapped around the filaments, enhancing the textile's surface properties, wash fastness, and abrasion resistance.

12.5.3 UV Protection

Increased exposure to UV radiation in recent years due to ozone layer depletion can result in adverse health outcomes such as photodermatosis, erythema, accelerated skin aging, and even skin cancer. As a result, the development of textiles with UV protection functionality has been a focus of extensive research [125]. The possibility of textile printing to effectively add UV-blocking chemicals to textiles has been brought to light by recent studies. The study by Song et al. [126] demonstrated the development of UV-protective cotton fabric using graphene oxide/waterborne polyurethane (GO/WPU) composites applied via a simple screen printing process. The GO/WPU composite paste was prepared by mixing GO dispersion with WPU, and thickeners and binders were added to improve the paste's rheological properties for printing. The composite paste (GO/WPU 0.9%) was applied to cotton and polyester fabrics, followed by reduction of GO to reduced graphene oxide (RGO). The printed patterns on cotton fabric were clearer compared to polyester fabric, and SEM images showed that GO and WPU were evenly distributed on the fabric surface without significant aggregation (Fig. 12.6).

Another recent study has explored the use of printing techniques to enhance UV protection in cellulosic fabrics by applying titania-coated phthalocyanine pigments stabilized by hyperbranched polyesters (HBPs) of varying generations (HBPs16

Fig. 12.6 GO/WPU 0.9% composite paste after GO reduced to RGO. Reproduced with permission from Ref. [126], Copyright Elsevier, 2019

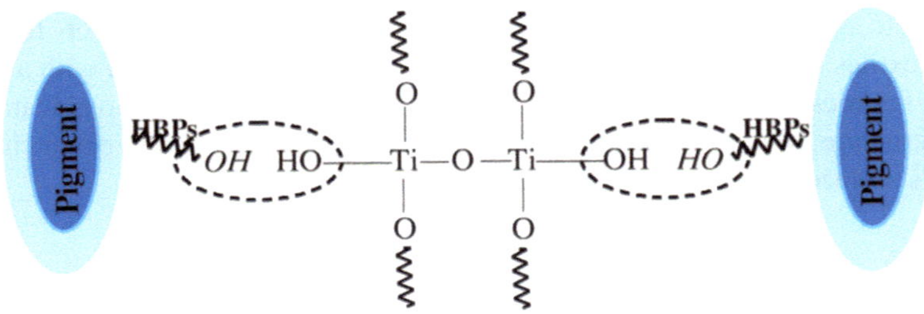

Fig. 12.7 Mechanism of interaction between HPBs encapsulated pigments and the hyperbranched polyester. Reproduced with permission from Ref. [127], Copyright Elsevier, 2019

and HBPs64) [127]. In this approach, the printing paste formulations were carefully prepared by gradually adding a synthetic thickener to water, followed by the incorporation of binders for conventional organic pigments and titania-coated HBPs-encapsulated pigments. An acid catalyst was used for conventional pigments, and finally, either conventional or titania-coated HBPs-encapsulated pigments were added to the mixture. The prepared pastes were then applied to cellulosic fabrics using a flat silk screen printing technique. The results demonstrated that prints made with titania-coated HBPs-encapsulated pigments exhibited significantly higher UV protection compared to traditional pigments. This improvement in UV shielding was attributed to the presence of functional hydroxyl end groups in the hyperbranched polyesters. This approach offers a promising, sustainable alternative to conventional UV protection methods, leveraging the versatility and precision of the printing technique (Fig. 12.7).

12.5.4 Flame Retardancy

The need for textiles with reduced flammability has been recognized throughout human history due to their wide use across various sectors, including home textiles, transportation, and clothing. As such, flame retardancy is essential, driven by both associated risks and regulatory demands [128, 129]. To address this, numerous research efforts have focused on developing textiles with enhanced flame-retardant properties. For instance, an interlayer-confined synthesis strategy was employed to direct the growth of two-dimensional α-zirconium phosphate (ZrP) within reduced graphene oxide (RGO) interlayers, resulting in the formation of hierarchical ZrP–RGO hybrids [130] (Fig. 12.8). The slurry mixture of ZrP–RGO and ammonium polyphosphate (APP) was then automatically screen-printed onto cotton fabrics, creating an intumescent flame-retardant coating. To screen print, wet ZrP–RGO segments (40 g) were uniformly blended with 2.5 and 5 g APP powders under high-speed stirring (1500 r/min) to obtain ZrP–RGO/2.5APP and ZrP–RGO/5.0APP slurries, respectively. The cotton/ZrP–RGO/5.0APP fabric demonstrates excellent self-extinguishing performance and also exhibits superior ultraviolet-blocking properties.

Another study focused on developing a new nanomaterial synthesis process for carbon dots (CDs) for use as "ink" for printing highly fluorescent patterns and as eco-friendly flame retardants for highly flammable materials, including polymeric

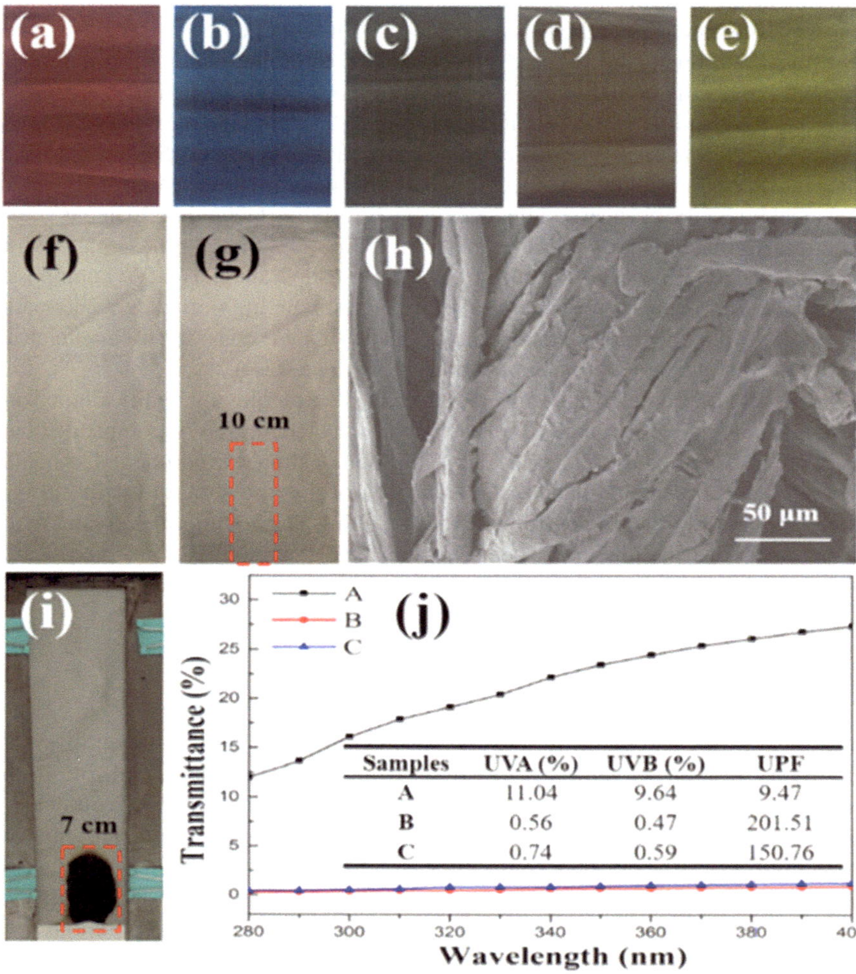

Fig. 12.8 **a–e** Colorful COT/ZrP–RGO/5.0APP fabrics. Comparison of COT/ZrP–RGO/5.0APP **f** before and **g** after 300 abrasion cycles. The abrasion area is marked with a red rectangle. **h** SEM images of COT/ZrP–RGO/5.0APP surface after 300 abrasion cycles. **i** Vertical flame testing images of COT/ZrP–RGO/5.0APP after 300 abrasion cycles. **j** UV transmittance spectra of control cotton (A), COT/ZrP–RGO/5.0APP fabrics before (B) and after (C) 300 cycles of abrasion. Reproduced with permission from Ref. [130] Copyright Springer Nature, 2021

foams and fibers. The CDs were used in an inkjet printer, producing stable fluorescent patterns [131].

12.5.5 Thermal Management

Clothing has historically served both practical and symbolic roles, providing protection from the elements and ensuring modesty. While its physical attributes—such as softness and breathability—are important, its role in thermal comfort is crucial. Thermal comfort is an individual's temperature perception, influenced by environmental factors and personal conditions. The most widely used measurement is the seven-point thermal sensation scale from the American Society of Heating, Refrigerating, and Air-conditioning Engineers (ASHRAE), where neutral sensation is considered ideal for comfort. Hu and colleagues used 3D printing techniques to manufacture a thermally conductive textile (thermoelectric films—TE film) for personal cooling by using highly aligned boron nitride (BN)/poly (vinyl alcohol) (PVA) composite fibers. The film makes it possible to directly incorporate high-efficiency nanoparticles into various intricate systems where TE devices are considered beneficial for energy harvesting and cooling applications [132] (Fig. 12.9).

Gao et al. [133] created a thermal regulation textile employing highly aligned and thermally conductive boron nitride (BN)/poly(vinyl alcohol) (PVA) composite fibers that were also created using 3D printing. They demonstrated experimentally that the fiber synthesis procedure enables the formation of a dense structure with uniform dispersion and high alignment of BNNSs, resulting in good thermal conductivity, uniform heat distribution, and exceptional mechanical strength (355 MPa).

12.6 Perspective and Challenges

The global textile and garment industry, which includes textiles, clothing, footwear, and luxury fashion, is valued at nearly $3 trillion and plays a crucial role in the GDP and economies of many countries [134]. However, it is a resource-intensive sector, with both natural and man-made textile fibers having significant environmental impacts. The industry contributes around 10% of the world's total carbon footprint [135], and rapid industrialization and population growth have worsened global water pollution. Industrial textile processing, which involves operations such as knitting, dyeing, printing, and finishing, generates large volumes of effluent, with printing contributing significantly to high biological and chemical oxygen demand (BOD and COD) [136].

Many dyes used in printing and their degradation products are toxic, carcinogenic, or mutagenic, and most synthetic dyes are non-biodegradable, meaning they do not degrade during conventional effluent treatment processes. Instead, they precipitate and separate as solid sludge, which is sent to landfills where it can contaminate soil

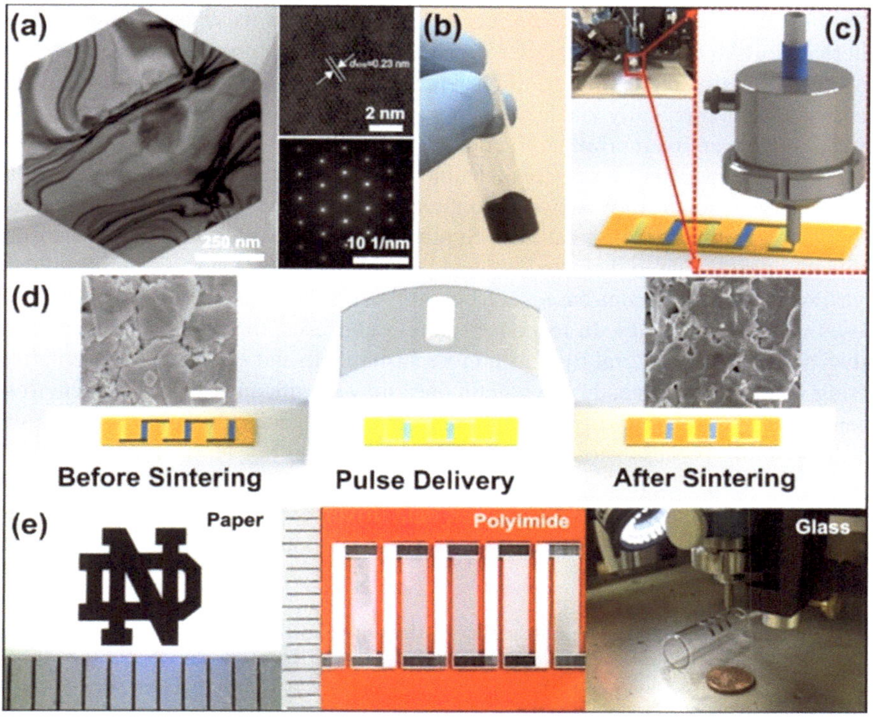

Fig. 12.9 3D conformal aerosol jet printing and intense pulsed light sintering of TE films: **a** TEM and HRTEM images of $Bi_2Te_{2.7}Se_{0.3}$ nanoplates, **b** high-stability synthesized ink printable for months, **c** aerosol jet printer deposition head, **d** schematic of the intense pulsed light sintering process with SEM images of films pre- and post-sintering (scale bar: 500 nm), and **e** as-printed films with custom patterns on diverse 2D and 3D substrates. Reproduced with permission from Ref. [132] Copyright John Wiley and Sons, 2019

and groundwater. The large amounts of Glauber's salt used to exhaust dyes contribute to high total dissolved solids (TDS) content, rendering the effluent unfit for human, industrial, or agricultural use. Additionally, many commercial dyes contain heavy metals like copper and chromium, which can increase the heavy metal content in textiles and effluent, violating eco-label standards. Dyeing auxiliaries, which are often non-biodegradable, increase the BOD and COD of the effluent. Also, traditional thickening agents have environmental challenges in terms of water pollution when they are disposed of improperly, and it causes problems with sedimentation in water streams and the water quality. Toxic carriers used for dyeing polyester can harm workers and complicate effluent treatment. Decomposition products from reducing agents like sodium hydrosulphite and sodium sulphide can raise the sulphide content in effluent, creating disposal issues [137]. Urea is used in textile printing to enhance dye solubility and moisture during steam fixation. However, its high nitrogen content (46%) can promote algae growth, increasing BOD and COD in water and leading to ecological issues and water pollution [137]. To solve these issues, sustainable alternatives like natural biodegradable products have been proposed as eco-friendly and cost-effective for textile printing, which has to be non-toxic, environmentally benign, and economically viable, offering a promising solution to the environmental impact of conventional printing practices.

The textile industry alone is responsible for a substantial portion of water pollution, as large amounts of wastewater containing dye residues and other chemicals are discharged without proper treatment. Synthetic dyes in textiles have raised environmental concerns due to their toxicity and persistence in the environment. Synthetic dyes, which are petroleum-based, can contaminate water and soil, and their disposal poses serious challenges. In response to these concerns, there has been increasing advocacy for using natural dyes as a more sustainable and eco-friendly alternative. Using natural dyes instead of synthetic ones may significantly reduce the environmental footprint, minimizing pollution in rivers, lakes, and seas. The machines used in the printing process should be designed so they do not leave much water and carbon footprint. The processes can be more automated and energy efficient with its design and efficient planning. Moreover, data-driven analyses can pave the way for better and more sustainable solutions to the existing issues regarding printing in the textile industry.

Declaration The authors declare that they have no known competing financial interests or personal relationships that could have appeared to influence the work reported in this chapter. Generative AI and AI-assisted technologies have been used in the writing process to improve the readability and language of the work only. After using this tool, the authors carefully reviewed and revised the content to ensure accuracy and quality. The authors take full responsibility for the final version of this publication.

References

1. Ujiie H (2006) Digital printing of textiles. Woodhead Publishing
2. Strand EA et al (2010) Old textiles—new possibilities. Eur J Archaeol 13(2):149–173
3. Ragab MM, Othman H, Hassabo A (2022) An overview of printing textile techniques. Egypt J Chem 65(8):749–761
4. Forza C, Romano P, Vinelli A (2000) Information technology for managing the textile apparel chain: current use, shortcomings and development directions. Int J Log Res Appl 3(3):227–243
5. Mahapatra N (2024) Textile printing. CRC Press
6. Chatterjee K, Ghosh TK (2020) 3D printing of textiles: potential roadmap to printing with fibers. Adv Mater 32(4):1902086
7. Knecht E (1952) The principles and practice of textile printing. Рипол Классик
8. Tyler DJ (2005) Textile digital printing technologies. Text Prog 37(4):1–65
9. Shang SM (2013) 13—Process control in dyeing of textiles. In: Majumdar A et al (eds) Process Control in Textile Manufacturing. Woodhead Publishing, pp 300–338
10. Haji A, Naebe M (2020) Cleaner dyeing of textiles using plasma treatment and natural dyes: a review. J Clean Prod 265:121866
11. Zhang Y et al (2023) Functional textiles with smart properties: their fabrications and sustainable applications. Adv Func Mater 33(33):2301607

12. Syduzzaman M et al (2023) Nanotechnology for high-performance textiles: a promising frontier for innovation. ChemNanoMat 9(9):e202300205
13. Karim N et al (2017) All inkjet-printed graphene-based conductive patterns for wearable e-textile applications. J Mater Chem C 5(44):11640–11648
14. Yıldırım FF, Yavas A, Avinc O (2020) Printing with sustainable natural dyes and pigments. In: Muthu SS, Gardetti MA (eds) Sustainability in the textile and apparel industries: production process sustainability. Springer International Publishing, Cham, pp 1–35
15. Abdelrahman M et al (2020) Review in textile printing technology. Egypt J Chem 63(9):3465–3479
16. Nambela L, Haule LV, Mgani Q (2020) A review on source, chemistry, green synthesis and application of textile colorants. J Clean Prod 246:119036
17. Choudhury AKR (2022) Principles of textile printing. CRC Press
18. Kozicki M, Sąsiadek E (2012) UV-assisted screen-printing of flat textiles. Color Technol 128(4):251–260
19. Locher I, Tröster G (2007) Screen-printed textile transmission lines. Text Res J 77(11):837–842
20. Azizul Hoque SM et al (2024) Environmental sustainability analysis of rotary-screen printing and digital textile printing. AATCC J Res 11(6):464–473
21. Grand N (1995) Environmental considerations for textile printing products. J Soc Dyers Colour 111(1–2):7–10
22. Kašiković N, Vladić G, Novaković D (2016) Textile printing-past, present, future. Gazette of Chemists, Technologists & Environmentalists of Republic of Srpska/Glasnik Hemicara, Tehnologa i Ekologa Republike Srpske
23. Dawson TL (1994) The use of digital systems in textile printing. Textile Printing, p 301
24. Yaman N, Ozdogan E, Seventekin N (2012) Improvement fastnesses and color strength of pigment printed textile fabric. J Eng Fibers Fabr 7(2):40–46
25. Ding Y, Freeman HS, Parrillo-Chapman L (2017) Color gamut analysis and color fastness evaluation for textile inkjet printing application on polyester. J Imaging Sci Technol 61(5). https://doi.org/10.2352/J.ImagingSci.Technol.2017.61.5.050503
26. Hamdy D, Hassabo AG, Othman H (2021) Recent use of natural thickeners in the printing process. J Text Color Polym Sci 18(2):75–81
27. Lin H-W et al (2008) The rheological behaviors of screen-printing pastes. J Mater Process Technol 197(1):284–291
28. Kan CW, Yuen CWM, Tsoi WY (2011) Using atmospheric pressure plasma for enhancing the deposition of printing paste on cotton fabric for digital ink-jet printing. Cellulose 18(3):827–839
29. Ibrahim NA et al (2006) New thickening agents for reactive printing of cellulosic fabrics. J Appl Polym Sci 101(6):4430–4439
30. Ragab MM, Hassabo AG, Othman H (2021) Synthetic thickener in textile printing. J Text Color Polym Sci 18(1):65–74
31. Ibrahim* NA et al (2005) Environmentally sound pigment printing using synthetic thickening agents. Polym-Plast Technol Eng 44(1): 111–132
32. Fan Q et al (2003) Fabric pretreatment and digital textile print quality. J Imaging Sci Technol 47(5):400–407
33. Ding Y et al (2019) Pretreatment effects on pigment-based textile inkjet printing—colour gamut and crockfastness properties. Color Technol 135(1):77–86
34. Sarkodie B et al (2023) Desizability and biodegradability of textile warp sizing materials and their mechanism: a review. J Polym Environ 31(8):3317–3337
35. Holme I (2016) 9—Coloration of technical textiles. In: Horrocks AR, Anand SC (eds) Handbook of technical textiles, 2nd edn. Woodhead Publishing, pp 231–284
36. Hussain T (2016) Textile engineering and textile processing. In: Nawab Y (ed) An introduction. De Gruyter Oldenbourg, pp 83–110
37. Hamdy D, Othman H, Hassabo AG (2022) A recent uses of plasma in the textile printing. J Text Color Polym Sci 19(1):1–10

38. Park Y, Koo K (2014) The eco-friendly surface modification of textiles for deep digital textile printing by in-line atmospheric non-thermal plasma treatment. Fibers Polym 15(8):1701–1707
39. Sarkodie B et al (2018) Status and development of transfer printing in textiles—a review. AATCC J Res 5(2):1–18
40. Shakoor A et al (2017) Pigment and disperse printing of needlepunched polyethylene terephthalate nonwovens. Dyes Pigm 136:865–872
41. Li B et al (2023) Facile and binder-free fabrication of deep colors on cotton fabrics with hand-feel enhancement via screen printing. Colloids Surf A 665:131178
42. Othman H et al (2024) Printing techniques of silk fabric. J Text Color Polym Sci 21(1):1–10
43. Marie MM, Salem AA, El Zairy EMR (2011) A novel printing method to enhance the fixation of reactive dyes on wool–polyamide fabrics. J Text Inst 102(9):790–800
44. Ujiie H (2015) Chapter 20—Fabric finishing: printing textiles. In: Sinclair R (ed) Textiles and fashion. Woodhead Publishing, pp 507–529
45. Yang J et al (2024) Constructing polyfunctional reactive dyes for printing of cotton fabrics with ultra-high fixation and ultra-low staining on the white ground. Cellulose
46. El Hamaky YH et al (2007) Printing cotton fabrics with reactive dyes of high reactivity from an acidic printing paste. Color Technol 123(6):365–373
47. Repon MR et al (2024) Natural dyes in textile printing: parameters, methods, and performance. Environ Sci Pollut Res 31(35):47552–47583
48. Rainert KT et al (2024) Textile printing: an integrated view of processes, properties, and future prospects. Color Technol 140(4):556–570
49. Lin L et al (2021) Combination of wet fixation and drying treatments to improve dye fixation onto spray-dyed cotton fabric. Sci Rep 11(1):15403
50. Majeed H, Iftikhar T, Mukhtar U (2024) Novel approach to water-efficient bulk industrial textile printing production of cotton fabric. Int J Biol Macromol 262:130064
51. Neral B, Šostar-Turk S, Vončina B (2006) Properties of UV-cured pigment prints on textile fabric. Dyes Pigm 68(2):143–150
52. Benkhaya S, M'rabet S, El Harfi A (2020) A review on classifications, recent synthesis and applications of textile dyes. Inorg Chem Commun 115:107891
53. Rowell RM (2008) 1—Natural fibres: types and properties. In: Pickering KL (ed) Properties and performance of natural-fibre composites. Woodhead Publishing, pp 3–66
54. Thomas S et al (2011) Natural fibres: structure, properties and applications. In: Kalia S, Kaith BS, Kaur I (eds) Cellulose fibers: bio- and nano-polymer composites: green chemistry and technology. Springer Berlin Heidelberg, Berlin, Heidelberg, pp 3–42
55. Salvador Cesa F, Turra A, Baruque-Ramos J (2017) Synthetic fibers as microplastics in the marine environment: a review from textile perspective with a focus on domestic washings. Sci Tot Environ 598:1116–1129
56. Carney Almroth BM et al (2018) Quantifying shedding of synthetic fibers from textiles; a source of microplastics released into the environment. Environ Sci Pollut Res 25(2):1191–1199
57. Shi F et al (2021) Eco-friendly pretreatment to the coloration enhancement of reactive dye digital inkjet printing on wool fabrics. ACS Sustain Chem Eng 9(30):10361–10369
58. Thibaut C et al (2019) A fibrous cellulose paste formulation to manufacture structural parts using 3D printing by extrusion. Carbohyd Polym 212:119–128
59. Agrawal R et al (2019) Paste-based 3D printing of metallic materials: effect of binders and precursor sizes. Mater Res Expr 6(10):106561
60. Tian D, Song Y, Jiang L (2013) Patterning of controllable surface wettability for printing techniques. Chem Soc Rev 42(12):5184–5209
61. Saffour Z, Viallier P, Dupuis D (2006) Rheology of gel-like materials in textile printing. Rheol Acta 45(4):479–485
62. Fijan R, Šostar-Turk S, Lapasin R (2007) Rheological study of interactions between non-ionic surfactants and polysaccharide thickeners used in textile printing. Carbohyd Polym 68(4):708–717

63. Lewis DM (2014) Developments in the chemistry of reactive dyes and their application processes. Color Technol 130(6):382–412
64. Billah SMR, Christie RM, Shamey R (2012) Direct coloration of textiles with photochromic dyes. Part 3: dyeing of wool with photochromic acid dyes. Color Technol 128(6):488–492
65. Selvius DeRoo C, Armitage RA (2011) Direct identification of dyes in textiles by direct analysis in real time-time of flight mass spectrometry. Anal Chem 83(18):6924–6928
66. Michie AGH, Thornton R (1953) The printing efficiency of vat dyes. J Soc Dyers Colour 69(13):629–647
67. Ebrahim SA, Hassabo AG, Othman H (2021) Natural thickener in textile printing (a mini review). J Text Color Polym Sci 18(1):55–64
68. Hassabo AG, El-Aty A, Othman H (2022) A critique on synthetic thickeners in textile printing. J Text Color Polym Sci 19(1):99–109
69. Abdelrahman MS et al (2020) Polymerization products of lactic acid as synthetic thickening agents for textile printing. J Mol Struct 1203:127421
70. Maheshwari N, Abd-Ellah M, Goldthorpe IA (2019) Transfer printing of silver nanowire conductive ink for e-textile applications. Flexible Print Electron 4(2):025005
71. Ding C et al (2022) Durability study of thermal transfer printed textile electrodes for wearable electronic applications. ACS Appl Mater Interfaces 14(25):29144–29155
72. Hwang JP, Kim S, Park CK (2015) Development of a color matching algorithm for digital transfer textile printing using an artificial neural network and multiple regression. Text Res J 85(10):1076–1082
73. Gorondy EJ (1977) Analysis of the dye transfer mechanism in heat transfer printing. Text Res J 47(9):604–615
74. Khalil E et al (2023) Innovative textile printing technology. In: Rahman MM, Mashud M, Rahman MM (eds) Advanced technology in textiles: fibre to apparel. Springer Nature Singapore, Singapore, pp 139–159
75. Agrawal TK, Campagne C, Koehl L (2019) Development and characterisation of secured traceability tag for textile products by printing process. Int J Adv Manuf Technol 101(9):2907–2922
76. Saleem H, Zaidi SJ (2020) Sustainable use of nanomaterials in textiles and their environmental impact. Materials 13(22):5134
77. Htwe YZN, Mariatti M (2022) Printed graphene and hybrid conductive inks for flexible, stretchable, and wearable electronics: progress, opportunities, and challenges. J Sci Adv Mater Dev 7(2):100435
78. Htwe YZN, Abdullah MK, Mariatti M (2022) Water-based graphene/AgNPs hybrid conductive inks for flexible electronic applications. J Mater Res Technol 16:59–73
79. Soleimani-Gorgani A, Avinc O, Alborz R (2023) Sustainable antibacterial cotton fabrics with in situ formed silver nanoparticles by bio-inkjet printing. J Clean Prod 386:135796
80. Sanivada UK et al (2022) Joule-heating effect of thin films with carbon-based nanomaterials. Materials 15(12):4323
81. Hasan MM, Hossain MM (2021) Nanomaterials-patterned flexible electrodes for wearable health monitoring: a review. J Mater Sci 56(27):14900–14942
82. Islam MR et al (2024) Advances in printed electronic textiles. Adv Sci 11(6):2304140
83. Ahmed HM et al (2019) Mini-emulsified copolymer/silica nanocomposite as effective binder and self-cleaning for textiles coating. Prog Org Coat 129:52–58
84. Ibanez Labiano I et al (2021) Screen printing carbon nanotubes textiles antennas for smart wearables. Sensors 21(14):4934
85. Yadav S et al (2023) A brief review on natural dyes, pigments: recent advances and future perspectives. Results Chem 5:100733
86. Li Q et al (2018) Dry transfer printing of silk and cotton with reactive dyes and mixed polysaccharide thickeners. Color Technol 134(3):222–229
87. Ahmed T et al (2024) Chapter 8—New trends in printing applications of natural dyes and pigments. In: Ul Islam S (ed) Renewable dyes and pigments. Elsevier, pp 139–163

88. Yusuf M, Shabbir M, Mohammad F (2017) Natural colorants: historical, processing and sustainable prospects. Nat Prod Bioprospect 7(1):123–145
89. Klančnik M (2021) Screen printing with natural dye extract from Japanese knotweed rhizome. Fibers Polym 22(9):2498–2506
90. Samanta AK, Agarwal P (2009) Application of natural dyes on textiles. Ind J Fibre Text Res 34:384–399
91. Savvidis G et al (2014) Ink-jet printing of cotton with natural dyes. Color Technol 130(3):200–204
92. Abel A (2012) 24—The history of dyes and pigments: from natural dyes to high performance pigments. In: Best J (ed) Colour design, 2nd edn. Woodhead Publishing, pp 557–587
93. Chua CK, Leong KF (2014) 3D Printing and additive manufacturing: principles and applications (with companion media pack)-of rapid prototyping. World Scientific Publishing Company
94. Gulrajani ML (2011) 10—Disperse dyes. In: Clark M (ed) Handbook of textile and industrial dyeing. Woodhead Publishing, pp 365–394
95. Maulik SR et al (2022) Vat dye and its evolution in dyeing. In: Textile dyes and pigments, pp 87–105
96. Griffiths J (1990) Introduction: the evolution of present-day dye technology. In: Waring DR, Hallas G (eds) The chemistry and application of dyes. Springer US, Boston, MA, pp 1–16
97. Ebrahim SA et al (2023) Eco-friendly natural thickener (pectin) extracted from fruit peels for valuable utilization in textile printing as a thickening agent. Textiles 3:26–49. https://doi.org/10.3390/textiles3010003
98. Chattopadhyay SN, Pan NC (2019) Ecofriendly printing of jute fabric with natural dyes and thickener. J Nat Fibers 16(8):1077–1088
99. Bajaj P, Goyal M, Chavan RB (1993) Synthetic thickeners in textile printing: a critique. J Macromol Sci Part C 33(3):321–348
100. Kamble R, Sahu A, Pandit S (2022) Occupational ergonomic assessment of hand pain symptoms among Bagh hand block print artisans of the handicraft textile industry in Madhya Pradesh, India. Int J Occup Saf Ergon 28(4):2324–2332
101. Fardhani AYS (2024) Utilization of vegetable texture as an alternative tool for printing plates in block printing techniques and textile motif design. In: Sustainability in creative industries. Springer Nature Switzerland, Cham
102. Dawson TL, Hawkyard CJ (2000) A new millennium of textile printing. Rev Prog Color Relat Top 30(1):7–20
103. Rathinamoorthy R, Raja Balasaraswathi S (2023) Characterization of microfibers originated from the textile screen printing industry. Sci Tot Environ 874:162550
104. Marra F et al (2021) Production and characterization of graphene nanoplatelet-based ink for smart textile strain sensors via screen printing technique. Mater Des 198:109306
105. Gorjanc M, Gerl A, Kert M (2022) Screen printing of pH-responsive dye to textile. Polymers 14(3):447
106. Mikuž M, Turk SŠ, Tavčer PF (2010) Properties of ink-jet printed, ultraviolet-cured pigment prints in comparison with screen-printed, thermo-cured pigment prints. Color Technol 126(5):249–255
107. Yildiz Z et al (2024) Sustainable fabric printing by using pre-consumed cellulosic textile wastes: the effect of waste particle content. J Clean Prod 448:141635
108. Rainert KT et al (2024) Optimization of the application of cationic agent in the screen printing paste with reactive dye on cotton substrate. Cellulose 31(3):1909–1926
109. Jurič I et al (2015) The influence of heat treatment on print mottle of screen printed textile knitted fabrics. Appl Therm Eng 90:215–220
110. Kwon S et al (2021) Scalable fabrication of inkless, transfer-printed graphene-based textile microsupercapacitors with high rate capabilities. J Power Sources 481:228939
111. Shahariar H et al (2019) Inkjet printing of reactive silver ink on textiles. ACS Appl Mater Interfaces 11(6):6208–6216

112. Tawiah B, Howard EK, Asinyo BK (2016) The chemistry of inkjet inks for digital textile printing-review. BEST 4:61–78
113. Cie C (2015) Ink jet textile printing. Elsevier
114. Thakker AM, Sun D, Bucknall D (2022) Inkjet printing of plasma surface-modified wool and cotton fabrics with plant-based inks. Environ Sci Pollut Res 29(45):68357–68375
115. Boukhriss A et al (2015) Sol–gel based water repellent coatings for textiles. Cellulose 22(2):1415–1425
116. Omerogullari Basyigit Z, Coskun H (2024) Enhancing antibacterial and water-repellent properties for the production of high-performance fabrics in home textiles. Fibers Polym 25(5):1789–1804
117. Shahid ul I, Shahid M, Mohammad F (2013) Green chemistry approaches to develop antimicrobial textiles based on sustainable biopolymers—a review. Ind Eng Chem Res 52(15):5245–5260
118. Morais DS, Guedes RM, Lopes MA (2016) Antimicrobial approaches for textiles: from research to market. Materials 9(6):498
119. Elmaaty TA et al (2018) One step thermochromic pigment printing and Ag NPs antibacterial functional finishing of cotton and cotton/PET fabrics. Fibers Polym 19(11):2317–2323
120. Ražić SE et al (2020) Plasma pre-treatment and digital ink jet technology: a tool for improvement of antimicrobial properties and colour performance of cellulose knitwear. Mater Today Proc 31:S247–S257
121. Ali AA et al (2023) Preparation and evaluation of antimicrobial thiadiazol azo disperse dyes as colored materials in digital transfer printing ink for printing onto polyester fabric. Pigm Resin Technol 52(1):19–32
122. Ziya Özek H (2018) 7—Silicone-based water repellents. In: Williams J (ed) Waterproof and water repellent textiles and clothing. Woodhead Publishing, pp 153–189
123. Wang Z et al (2013) Fluorinated alternatives to long-chain perfluoroalkyl carboxylic acids (PFCAs), perfluoroalkane sulfonic acids (PFSAs) and their potential precursors. Environ Int 60:242–248
124. Yu J, Seipel S, Nierstrasz VA (2018) Digital inkjet functionalization of water-repellent textile for smart textile application. J Mater Sci 53(18):13216–13229
125. Alebeid OK, Zhao T (2017) Review on: developing UV protection for cotton fabric. J Text Inst 108(12):2027–2039
126. Song W et al (2019) Graphene oxide/waterborne polyurethane composites for fine pattern fabrication and ultrastrong ultraviolet protection cotton fabric via screen printing. Appl Surf Sci 463:403–411
127. Hakeim OA, Abdelghaffar F, Haroun AA (2019) UV-protection of cellulosic fabric prints using hyperbranched polyester-stabilized titania coating. Prog Org Coat 136:105295
128. Rosace G et al (2015) Flame retardant finishing for textiles. In: Visakh PM, Arao Y (eds) Flame retardants: polymer blends, composites and nanocomposites. Springer International Publishing, Cham, pp 209–246
129. Bourbigot S (2008) 2—Flame retardancy of textiles: new approaches. In: Horrocks AR, Price D (eds) Advances in fire retardant materials. Woodhead Publishing, pp 9–40
130. Wang D et al (2021) Intumescent flame-retardant and ultraviolet-blocking coating screen-printed on cotton fabric. Cellulose 28(4):2495–2504
131. Liu C et al (2022) Facile synthesis, high fluorescence and flame retardancy of carbon dots. J Mater Sci Technol 104:163–171
132. Saeidi-Javash M et al (2019) 3D conformal printing and photonic sintering of high-performance flexible thermoelectric films using 2D nanoplates. Adv Func Mater 29(35):1901930
133. Gao T et al (2017) Three-dimensional printed thermal regulation textiles. ACS Nano 11(11):11513–11520
134. Saxena S, Raja ASM, Arputharaj A (2017) Challenges in sustainable wet processing of textiles. In: Muthu SS (ed) Textiles and clothing sustainability: sustainable textile chemical processes. Springer Singapore, Singapore, pp 43–79

135. Arputharaj A, Raja ASM, Saxena S (2016) Developments in sustainable chemical processing of textiles. In: Muthu SS, Gardetti MA (eds) Green fashion, vol 1. Springer Singapore, Singapore, pp 217–252
136. Vílchez S et al (2021) Eco-friendly printing paste replacing urea-based formulations in cotton printing. J Text Inst 112(7):1046–1054
137. Majeed H et al (2023) Bulk industrial production of sustainable cellulosic printing fabric using agricultural waste to reduce the impact of climate change. Int J Biol Macromol 253:126885

Chapter 13
Digital Printing and Applications in the Textiles

Chet Ram Meena and Janmay Singh Hada

Abstract Digital textile printing has transformed the industry by providing an environmentally friendly alternative to traditional dyeing and printing methods, which were labor-intensive, used hazardous chemicals, and caused water pollution. Digital printing eliminates harmful chemical discharge, reduces waste, and offers faster processing with more design flexibility. Different types of textile printers are available, each with unique features, including printing methods like thermal, piezoelectric, and drop-on-demand. Specialized inks are used depending on the fabric's composition, supporting diverse applications. This technology supports sustainable practices and creates opportunities for designers and markets. While digital printing faces some limitations, it is valued for being faster, more economical, and eco-friendly. It is projected to make up a significant portion of the textile printing market, contributing to sustainable development and innovation in the industry.

Keywords Digital printing · Dyeing · Inks · Printing · Printer · Sustainability · Textiles etc.

13.1 Introduction

The word "printing," which emphasizes the basic process of applying pressure to a surface during printing, comes from the Latin word for "pressing." Textile printing is an important method for raising the worth and attractiveness of textile items [1]. This process involves producing multiple copies of graphics or designs, which can vary in complexity and often incorporate a wide range of colors. Unlike dyeing techniques, textile printing involves the precise application of color to specific areas dictated by the design of the image. The use of screens is necessary to accomplish

C. R. Meena (✉) · J. S. Hada
Department of Textile Design, National Institute of Fashion Technology (Ministry of Textiles, Governmet of India), Jodhpur, Rajasthan, India
e-mail: chetram.meena@nift.ac.in

J. S. Hada
e-mail: janmay.hada@nift.ac.in

M. Shahid et al. (eds.), *Advancements in Textile Coloration*,
https://doi.org/10.1007/978-981-96-5091-0_13

precise color placement, and good print registration is crucial to the process's efficacy to guarantee that every color layer aligns as intended. In industrial settings, printing is extensively utilized to efficiently replicate intricate textile designs on a large scale, enabling the mass production of creative patterns while maintaining design integrity. The intricacy of the pattern and the selection of colorants, whether dyes or pigments, affect the printing process's efficiency and length in addition to the printed textile's aesthetic appeal. The equipment used to move the print paste and its application to the material is frequently used to classify printing procedures. Among the common methods are digital printing, inkjet printing, copper roller printing, flat screen printing, and rotary screen printing. Each method offers distinct advantages and is suited to different types of textile applications, allowing for a broad spectrum of creative possibilities. Moreover, technological advancements have continuously expanded textile printing capabilities, enabling more detailed and vibrant designs while optimizing production speeds. This evolution ensures that the textile industry can meet the growing demands for diverse and high-quality printed fabrics, catering to aesthetic and functional requirements. Through these various printing methods, the industry can balance creativity and efficiency, reflecting the dynamic nature of modern textile production [2]. Applying ink to textiles using techniques like screen printing and digital inkjet printing is known as textile printing. The three-color silk prints from China, which date back to 220 BCE, are the first documented examples of printed textiles [3]. Woodblock printing, also originating in China, is the oldest direct printing technique, where designs are transferred onto fabric using carved wooden blocks. This traditional method is still practiced in various forms today [4]. The printing style determines how color is applied to the textile in a certain pattern or design. The textile printing styles can be divided into three basic ways [2, 5, 6], as mentioned in Table 13.1.

The growth of printing has closely mirrored that of textiles, reflecting human history's cultural and material progress. Textiles and printed decorative patterns indicate the technological and material resources available at the time and place of production. Significant advancements were made in textile materials and printing techniques, particularly during the nineteenth century, when modernizations transformed these processes. Currently, industrial printing methods continue to develop in response to the increasing demands of businesses. Introducing new fibers and creative

Table 13.1 Style of textile printings [2, 5, 6]

Printing styles	Description
Direct	It is the most often used printing style. Printing is done directly onto colored or white cloth generally by using screen, roller, and block printing methods
Resist	First, resist paste is used to print the pre-treated substrate, preventing the dye from penetrating the fabric. The intended design remains when the resist material is removed
Discharge	It is predicated on the original dyeing of the printed region being chemically destroyed, where the substrate needs to be dischargeable

designs powered by new technology has expanded the possibilities in textile production. The rising awareness of environmental concerns has driven manufacturers to develop eco-friendly materials that respect the planet and people. Table 13.2 outlines the various textile printing methods widely utilized in the textile industry [2, 7].

In most cases, creating printed designs on textiles requires a prior design. However, it is possible to proceed with printing without a pre-established design, where spontaneity and immediacy are key elements. This direct approach offers a vast space

Table 13.2 Methods of textile printing commonly used in textile industry [2, 7]

Printing methods	Descriptions
Block	This is the most ancient printing technique. The color is uniformly applied to the wooden block, featuring various designs and patterns that are pressed into the fabric
Roller	It is the most cost-effective and rapid printing method. This method is used for printing extensive lengths of cloth with the same pattern. Engravings of designs are applied to the printed roller for each color, which is then pressed into the fabric
Screen	It resembles the photography process. The pattern is executed by forcing the print paste through a nylon screen fabric. It is executed manually or by an automated machine
Heat transfer	Printed paper designs are put on the fabric and then passed through a heated calendar, transferring the pattern from one to the other
Digital	It is new technology. It is an emerging, non-contact technique that drops minute droplets into designated areas of the textile surface
Sublimation	Involves transferring a design onto polyester (PET) fabric using heat and pressure. The ink turns into a gas and bonds with the fabric, resulting in vibrant and permanent prints
Stencil	Involves cutting designs into a stencil and applying dye through the cut areas onto fabric. It's a basic method and often used for simple designs or one-off pieces
Flock	A method where the adhesive is applied to fabric in a specific pattern, and short fibres (flock) are sprinkled onto the adhesive by flock printing machine, creating a soft, velvet-like texture on the design
Foil	A layer of adhesive is applied to the fabric, followed by pressing a metallic foil onto it using heat. The foil sticks to the adhesive areas, creating shiny, reflective patterns
Burn-out	Involves applying a chemical to blended fabrics (usually P/C blend fabric), which destroys the natural fibres, leaving sheer areas where only synthetic fibres remain
Silicone	Silicone-based inks are used to create a soft, stretchable, and durable print on high-stretch fabric. It is applied and cured using heat, forming a layer. Widely used in sportswear and activewear because of their elasticity and resistance to cracking or fading
Puff	It incorporates a heat-reactive ingredient into standard plastisol ink, resulting in expansion and puffing when subjected to heat during the curing process. This produces a raised, three-dimensional impression on the cloth, resulting in a soft and textured pattern

for exploration and creativity. The selected printing technique influences the design, shaping the final outcome. Thus, design and technique are integral to the overall creative process [8].

13.2 Digital Printing [9–12]

The evolution of printed textile design begins with the discovery of materials suitable for creating marks and the innate human desire to express ideas visually, often reflecting emotions. The creative process has consistently utilized available tools and natural resources. To fully grasp the development of digital printing in textiles, it's essential first to explore the key materials and techniques involved. Understanding the origins of these processes and how they transformed over time is crucial to appreciating their progression and the reasons behind their evolution (Fig. 13.1).

Digital printing emerged in the 1970s, pioneered by Milliken for carpet manufacturing. The process involved using air to propel ink droplets onto the carpet's surface. At the same time, punched paper tape evolved from Jacquard's punch card system, patented in 1801 for complex loom weaving. These developments laid the groundwork for digital textile printing. The technology gained significant interest, and in 1995, Stork introduced its Fashion Jet at ITMA. From 2000 onwards, software and hardware developers focused on overcoming the key challenges preventing digital printing from matching conventional methods, such as print speed, color costs, and overall performance.

Four key factors influence the characteristics of digital printing:

- The visual appeal of the original image.
- The printing production process.
- The mechanics involved in reproduction.
- The type of base fabric on which the image is printed.

In digital printing, images can be reproduced quickly, easily, and inexpensively, especially in fast fashion. To regain control over authenticity and the creative process, it becomes essential to re-establish a connection between the artifact and its creator, whether an individual or a group. Increasingly, digital textile printing showcases human input through handcrafted elements, embedding skill and uniqueness into the work. This emphasis on craftsmanship is appreciated across various fields that utilize digital printing, from letterpress to textile design. Consequently, contemporary designers are naturally drawn to handcrafting techniques to redefine the message they

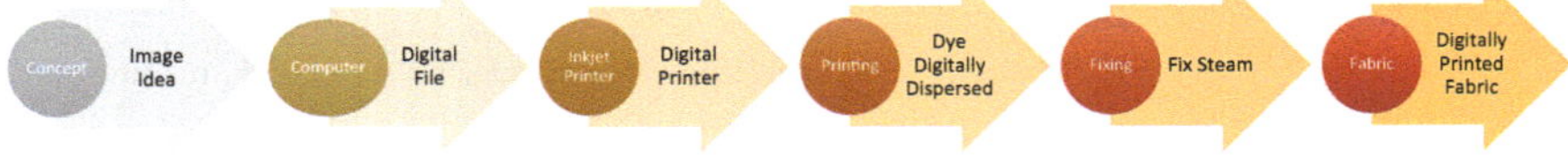

Fig. 13.1 Systematic representation of digital textile printing process

convey through the digital printing medium, infusing their creations with a sense of artistry and originality [13].

Textile digital printing provides a more comprehensive approach than conventional screen printing, which is constrained by many restrictions such as inadequate pattern registration, stains, pinholes, color discrepancies, smearing, dye migration, burning, and inappropriate curing. These challenges might adversely affect print quality. Digital printing offers considerable benefits, including time and space efficiency, accelerated printing rates, and the capability to use both reactive inks (for materials such as wool, cotton, and silk) and disperse inks (for polyester/Nylon). A key benefit is the ability to switch between these ink types in just 10 min, compared to the approx.120 min required in traditional methods [14, 15]

13.2.1 The Evolution of Digital Textile Printing [16, 17]

Textile printing has a rich history spanning centuries, with early examples like a block-printed tunic from the fourth century CE. The craft expanded through global trade networks such as the Silk Road and the spice trade. The shift from traditional methods to digital technologies has revolutionized the industry, driving significant innovation.

13.2.1.1 Digital Carpet Printing

The 1970s marked the advent of digital carpet printing. Milliken of Spartanburg, South Carolina, pioneered the industry with the Milliken Millitron in 1975, followed by Zimmer's carpet printer in 1976. Today, most commercial carpets are produced using advanced digital printing methods.

13.2.1.2 Sublimation Printing

In 1973, RPL Supplies Inc. of Saddle Brook, New Jersey, developed a digital sublimation transfer process, enabling the embedding of digital video images into fabric. This innovation significantly advanced textile design and production.

13.2.1.3 Thermal Inkjet Printing

Canon's Endo discovered the thermal inkjet principle in 1977, later refined by Canon and Hewlett-Packard (HP). Canon introduced the 'Bubble Jet,' while HP launched the first commercial inkjet printer, the Thinkjet, in 1984. These breakthroughs laid the groundwork for inkjet textile printing.

13.2.1.4 Seiren's Digital Printing Innovations

In the early 1980s, Seiren, Japan's largest textile printer, explored inkjet fabric printing. By 1989, it had established a digital textile printing facility, further propelling industry growth.

13.2.1.5 Large-Format Printing Expansion

Digital large-format printing emerged in the 1980s for billboards and banners. Gerber Scientific's innovations in 1987 led to Metro Media Technologies (MMT) becoming a global leader. In 2002, MMT introduced the MegaDrum, a massive inkjet billboard printer.

13.2.1.6 Significant Industry Exhibitions and Achievements

- **FESPA 1996 & 1999**: Introduction of TIJ textile printers and large-format printing advancements.
- **ITMA 1999**: Stork debuted the Amethyst printer for reactive and acid dyes.
- **Drupa 2000**: DPS and Aprion launched Magic PIJ print head technology, later adopted in the Reggiani DReAM printer.
- **Heimtextil 2001**: DuPont introduced the Artistri 3210 for home decor applications.
- **DPI 2001**: Leggett & Platt unveiled the Virtu series UV-cure printers.
- **ITMA 2003**: Ichinose, Reggiani, Robustelli, and Zimmer showcased digital textile printing innovations.
- **Drupa 2004 & SGIA 2004**: Kornit Digital introduced durable inkjet garment printers.
- **FESPA 2005**: Advanced garment and textile printing techniques, including T-shirt printing machinery from the US Screen Printing Institute.

Ongoing advancements in high-speed, efficient, and eco-friendly printing technologies promise an exciting future for digital textile printing. These innovations ensure endless creative possibilities and increased production efficiency, shaping a dynamic and evolving industry [16].

13.3 Types of Inks in Textile Printing

Traditional printable inks generally have four elements: a dye or pigment, a binder, a solvent, and an additive. The principal element, the colorant, may be a dye or a pigment, which must be uniformly distributed in a medium composed of binders, solvents, and additives. Collectively, these factors dictate the ink's viscosity and flow properties [7]. In contrast, functional inks contain an electrically active component, either in place of or alongside the pigment. These types of inks demonstrate features like electromagnetic, thermal, chemical, or optical activity and are categorized as conductive, semi-conductive, dielectric, and resistive inks [7]. Ink can be categorized into two primary types: dye-based and pigment-based. The choice of ink significantly influences the durability of prints. Pigment-based inks offer greater resistance to fading compared to dye-based inks. However, prints exposed to harsh conditions like direct sunlight, frequent washing, or abrasion have a reduced lifespan, regardless of the ink type used [18].

13.3.1 Digital Printing Inks [9, 19–21]

The initial advancement of inkjet textile printing mostly used dye-based inks, encompassing acid dyes for silk, nylon, and wool, dispersion dyes for polyester, and reactive dyes for cotton and rayon. Nonetheless, attaining sufficient wash resistance required intricate pre- and post-treatments, which were often detrimental to the environment and partially negated the benefits of digital printing, including convenience and quick customization. Conversely, pigment-based inks provide enhanced adaptability for different kinds of fabric and need just basic dry heat post-treatment. Nonetheless, they pose unique obstacles in contrast to dye-based inks. Pigment inks include crystalline particles ranging from 50 to 150 nm inside a colloidal environment, while dye inks are homogeneous molecular solutions. The wash and crock resistance of dye-based inks is contingent upon several chemical interactions: acid colors adhere via acid–base interactions, dispersion dyes penetrate polyester fibers, and reactive dyes establish chemical connections with the fabric. Industry leaders in pigment-based digital textile inks include Rohm and Haas, DuPont, BASF, Ciba, Trident, and Sensient. A standard pigment ink formulation has several critical components: a pigment dispersion for coloration, a polymeric binder for resilience, water as a carrier medium, and a co-solvent for solubility and adherence. Supplementary components include surfactants for wetting and stability, humectants to inhibit desiccation, anti-foaming agents, viscosity modifiers, penetrants for expedited drying, and biocides to avert spoiling. These elements collaborate to guarantee superior inkjet printing efficacy, fabric adherence, and durability, making pigment inks a viable option for digital textile printing.

13.3.2 Eco-Friendly Inks in the Printing Industry [22, 23]

Traditional petroleum-based inks contribute to air pollution by releasing volatile organic compounds (VOCs) and are challenging to recycle. In response, eco-friendly alternatives like soy-based and water-based inks are becoming increasingly popular due to their lower environmental impact and safer composition. Soy-based inks, derived from renewable soybeans, emit fewer VOCs, enhance air quality, and produce vibrant colors. Water-based inks, known for their non-toxic properties and minimal emissions, are a sustainable choice for environmentally conscious printing. Additionally, innovative startups are advancing eco-friendly ink solutions. The Egyptian startup Edge Colours produces eco-solvent-based inks that are non-toxic, odorless, and offer superior durability, adhesion, and UV resistance. Similarly, Israeli startup Nur Ink Innovations develops water-based inks using advanced 3D cross-linking chemistry, catering to diverse applications such as direct-to-garment (DTG), direct-to-textile (DTT), and roll-to-roll printing for indoor and outdoor advertising. These advancements highlight the shift towards sustainable printing solutions that benefit the environment and human health.

13.4 Fundamentals of Textile Digital Printing

The most widely used technology in digital textile printing today is inkjet printing. This method is a computer-to-print process in which ink is sprayed onto the fabric through nozzles. Unlike traditional printing methods, no master printing plate is required; instead, the image is transferred directly onto the substrate. Inkjet printing methods are primarily categorized into two types: Continuous Inkjet (CIJ) and Drop-on-Demand Inkjet (DOD) [24, 25].

13.4.1 Continuous Inkjet (CIJ)

Continuous Inkjet (CIJ) technology encompasses two subcategories: binary deflection and multi-deflection. It operates by producing a continuous stream of ink droplets, with only a portion applied to the fabric based on picture data. Excess ink droplets are collected and reused, ensuring efficient use. Originally developed in the mid-1960s by Sweet at Stanford University, CIJ involves forcing ink through nozzles to create a high-frequency stream of uniform droplets, driven by piezoelectric crystal vibrations. These droplets naturally break off after exiting the nozzle. A periodic excitation prevents random droplet formation, typically achieved using a piezoelectric transducer to maintain control over droplet size and frequency. The droplets gain an electrical charge corresponding to the image being printed upon exiting the nozzle. They then pass through an electric field where they are precisely deflected, ensuring accurate ink placement on the fabric [25, 26].

13.4.2 Drop-on-Demand (DOD)

Drop-on-demand (DOD) printing differs from Continuous Inkjet (CIJ) because it only releases ink droplets when needed for the image. DOD includes thermal, piezo-electric, and electrostatic methods, all of which eject ink only when triggered by the image data. This approach reduces ink waste by using only the necessary amount. Both CIJ and DOD are widely used in textile printing because they offer accuracy and flexibility, making them popular choices for digital fabric printing [15, 25].

13.5 Digital Textile Printers [9, 12, 27–29]

The most recent innovation in textile printing is the digital textile printer. These printers are compact, cost-effective, and offer an extensive color range, providing the ability to create unique designs with every print. Digital textile printers are perfect for short-run production, offering rapid turnaround times while maintaining exceptional quality. This flexibility enables businesses to introduce new designs to the market swiftly. Engineered to print directly on both natural and synthetic fabrics, our cloth printing machines combine advanced technology with efficient Indian engineering to meet the dynamic needs of the textile industry. As the industry increasingly adopts digital solutions to enhance quality standards, the demand for digital textile printers continues to rise. One of their key advantages is the ability to print on diverse fabrics, making them a versatile choice for modern textile production. While pigment ink works across various fabric types, reactive ink has become increasingly popular for cotton and cotton-blend fabrics, driving the demand for cotton fabric printers. Disperse ink is commonly used for polyester-based fabrics, while acid ink is preferred for silk fabrics. Digital fabric printers offer versatility, catering to an extensive range of fabrics, including viscose crepe, cotton satin, modal, cotton poplin, organic cotton, rayon, chanderi, polyester georgette, polyester crepe, silk, lace silk, linen, and many others. The variety of ink types available, whether pigment, reactive, disperse, or acid, ensures that these printers can deliver vibrant color options on different fabric types.

Digital textile printers enhance operational uptime, allowing firms to expedite the introduction of new designs to the market. The progression of Direct-to-Fabric printing technology is transforming the textile printing sector, enhancing innovation, customisation, and efficiency. As the industry continues to evolve, these innovations are shaping the future of textile printing, offering unparalleled opportunities for both quality and speed in production. Artificial intelligence (AI) and automation in digital textile printing: as digital textile printing technology continues to evolve; solution providers are making significant investments in intelligent systems that streamline technology management. These advanced systems allow for real-time monitoring and optimization of printer performance, reducing downtime, cutting operating costs, and minimizing waste. As high-speed, complex printing systems become more widely adopted, software applications are playing a critical role in automating system calibration and fleet management, enabling businesses to scale print capacity effectively. Technology providers are developing digital workflow solutions that enhance order management and automate repetitive tasks, such as file processing, to facilitate broader adoption. Table 13.3 presents the latest advancements in textile digital printing technology. These innovations are pivotal in improving efficiency and fostering growth within the digital textile printing industry.

Digital textile printing is poised to revolutionize the manufacturing of bedding, home furnishings, and other textile goods. This technology enables rapid, cost-effective production of small print runs, making it ideal for customized designs.

Table 13.3 Latest innovation in textile digital printers [28]

Sr. No	Printer	Features
1	Colorjet	The Earth series digital printer offers an eco-friendly pigment printing option, using water-based pigment ink with binder fixation. This procedure obviates the need for pre- and post-treatment, considerably reducing water and energy use. Functioning at 160 °C, it guarantees exceptional color retention and great print quality. The Metro NXT industrial printer, including 32 Kyocera heads, produces accurate prints with exceptional efficiency, reaching a production capacity of up to 9,000 sqm daily
2	Konica Minolta	The Nassenger SP-1e is an ultra-high-speed, single-pass printer designed to effectively process over 10,000 m, providing a contemporary substitute for conventional rotary and automatic screen printers. Its advanced print head module and ink control technology enable flexible drop sizes, achieving screen printer-level quality with fine detail, excellent gradation, and impressive speed
3	Kornit digital	Kornit's MAX technology provides unparalleled retail-quality applications by integrating 3D, vinyl, screen printing, dye sublimation, and embroidery effects into a single complete solution. Kornit's comprehensive digital manufacturing process revolutionizes sectors like clothing, sportswear, home décor, and high-fashion textiles, guaranteeing efficient, lucrative, and on-demand fulfillment. By establishing streamlined, integrated workflows, Kornit ensures exceptional quality and durability with a minimal physical footprint, meeting diverse market demands efficiently
4	SPG prints	SPG prints introduced the DART, a robust entry-level digital printer, to support Indian customers entering digital textile printing. Expanding its portfolio, SPG Prints also launched three new high-performance printers: Rose, Jasmine, and Magnolia. Rose provides roll-to-roll paper sublimation at a resolution of 1200 dpi, guaranteeing exceptional print quality. Jasmine, a direct-to-fabric method, improves productivity, whilst Magnolia, with 1200 dpi and advanced Archer + technology, sets a new standard in print quality
5	Embee	The Acumen and Signature rotary printing machines have remarkable output speeds of up to 90 m/min, ensuring outstanding print quality. Equipped with an automated repeat function, a unique head-lifting mechanism, and several repeat printing heads, these machines provide choices for closed-bearing operation (Acumen) and open-bearing (signature) configurations to suit diverse production needs

Digital printing systems are especially conducive to customization, enabling firms to effortlessly alter graphics and fine details using variable data printing (VDP). Due to improved print speed and quality, digital printing emerges as the superior option for organizations seeking to effectively modify designs to accommodate varied client preferences and produce distinctive, high-quality goods.

13.6 Important Drivers Behind the Growth of Digital Printing Technology (DPT)

DPT is revolutionizing the worldwide decorated garment sector with the automation of printing, facilitation of personalization, and minimization of energy usage and manufacturing lead times. Notwithstanding its transformative influence, the core comprehension of the nascent DPT market remains largely unexamined owing to its recent emergence. Multiple critical elements have facilitated the rapid expansion of DPT. From a supply chain standpoint, DPT employs digital files, obviating the need for screen-making and significantly decreasing the time needed for garment printing. This efficiency enables designers to quickly adapt to market trends and consumer demands, providing a significant competitive advantage in fast-paced industries. Additionally, DPT removes the need for large inventories, as it does not require minimum order quantities, thus lowering associated risks and costs. In terms of marketing, DPT allows for on-demand printing of complex, personalized designs, aligning with the growing consumer demand for customization. Environmentally, it is more sustainable compared to traditional printing methods, as it uses less water and energy, helping to address the increasing demand for eco-friendly production solutions. From a business perspective, DPT can be operated from nearly any location with low barriers to entry, making it an attractive option for small businesses and e-commerce ventures. The technology supports short production runs, cost-effective multi-color prints, and requires minimal capital investment, offering significant opportunities for entrepreneurs. Despite its rapid growth and numerous advantages, the deeper dynamics of the DPT market remain largely untapped and underexplored [30, 31].

13.7 Adoption of Digital Printing by Designers

Designers are increasingly adopting digital printing for its versatility, efficiency, and ability to produce intricate, customized designs. This technology allows for faster production, real-time trend adaptation, and reduced waste, empowering designers to experiment with innovative patterns and materials. Digital printing offers a sustainable and creative solution for modern design needs.

13.7.1 Dolce and Gabbana

Dolce & Gabbana introduced the Collezione Genesi NFT collection, seamlessly integrating digital fashion with high-end luxury textiles. This collection showcases cutting-edge printing methods, including metallic and holographic effects, to achieve distinctive textures and visual appeal. Each physical item is paired with a digital NFT, providing buyers with a ground-breaking way to own both a tangible garment and its digital counterpart, redefining the concept of fashion ownership [32].

13.7.2 Julia Koerner's

Julia Koerner's ARID Collection addresses the critical issue of water scarcity through pioneering textile printing. The collection features lifelike patterns inspired by arid landscapes using eco-friendly digital printing techniques that minimize water and chemical use. Comprising 38 distinct textile pieces that can be assembled into a single garment, the collection highlights aesthetic appeal and design adaptability, merging sustainability with innovative fashion [33].

13.8 Challenges in Digital Textile Printing

The rapid advancements in industrial digital printing, particularly inkjet technology, have transformed textile production over the past decade. However, despite its potential, digital textile printing faces several technological and market-related challenges that hinder widespread adoption Fig. 13.2. The future of digital textile printing is promising, with ongoing research into pigment-based digital printing and improved binding mechanisms. Developing specialized inks, advanced pretreatment, and innovative printing techniques will be crucial in overcoming these limitations and ensuring broader adoption across the textile industry [34, 35].

13.9 The Future Prospects of Digital Textile Printing

Digital textile printing is growing daily, driven by continuous advancements in printing technology, material science, and process optimization. Innovations such as augmented inkjet heads, specialized color ink formulations, and enhanced color depth improve print precision, consistency, and vibrancy. As technological barriers are overcome, digital printing is expected to evolve into a more efficient, versatile, and widely adopted solution across various sectors. Beyond traditional inkjet methods, digital textile printing incorporates multiple processes contributing to a seamless and

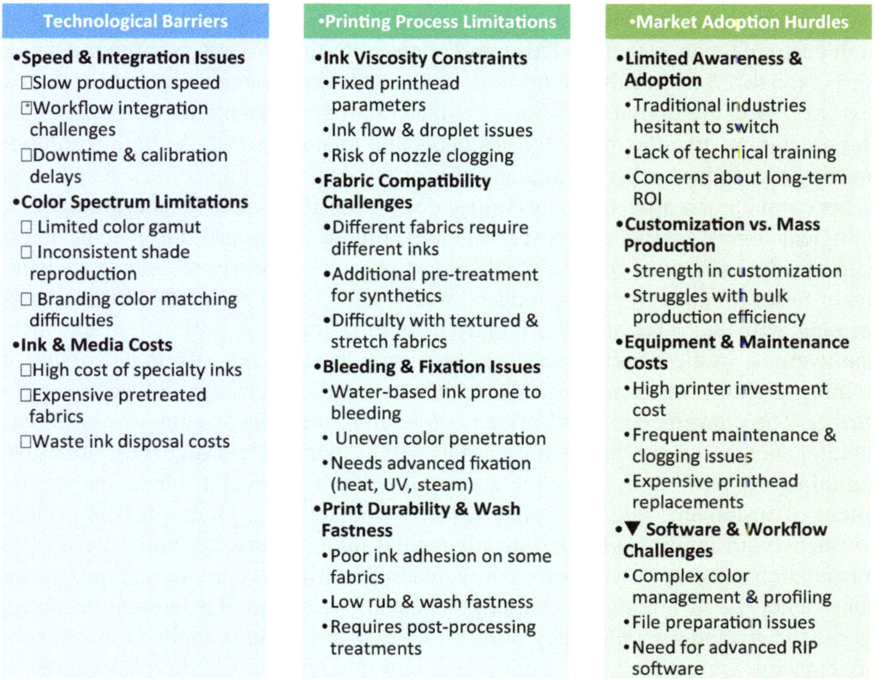

Fig. 13.2 Digital textile printing challenges

adaptable production framework. The integration of high-speed printheads, advanced software for color management, and automated workflow solutions will enable higher printing speeds while maintaining superior resolution and detail. To fully compete with traditional screen printing, digital printing must achieve higher throughput while ensuring reliability and cost-effectiveness. One key focus for the future is increasing printing speed without compromising quality. Current digital printing methods face limitations in terms of output volume, necessitating multiple printers to meet high production demands. Next-generation systems aim to optimize productivity, allowing manufacturers to streamline operations while reducing waste and production downtime. As the textile industry moves towards more sustainable practices, digital printing offers significant environmental benefits over conventional methods. Unlike traditional dyeing and screen printing, digital printing requires minimal water usage, reduces chemical waste, and lowers energy consumption. Eliminating screens and plates further reduces material waste, making it an eco-friendlier option. However, to ensure widespread adoption, cost optimization remains a crucial factor. Ink prices, equipment costs, and post-processing treatments are currently major expenses in digital printing. Future advancements will focus on developing more cost-effective ink formulations, increasing ink efficiency, and improving the durability of printed textiles to enhance market competitiveness. One of the greatest advantages of digital textile printing is its ability to cater to small-batch, on-demand production with high customization potential. This trend aligns with the growing consumer demand for personalized and limited-edition designs in fashion, home décor, and technical textiles. Adopting digital workflows enables rapid prototyping, shorter lead times, and greater creative flexibility for designers and manufacturers. As high-resolution imaging capabilities improve, digital textile printing will offer more intricate patterns, richer color combinations, and innovative design possibilities. With continuous technological improvements and cost reductions, digital textile printing is expected to gain broader market acceptance across industries. As businesses recognize the benefits of faster production cycles, reduced waste, and greater design flexibility, digital printing will likely become the preferred method for mass production and niche applications. In the coming years, research and development efforts will focus on refining digital print technologies to enhance speed, efficiency, and sustainability further. The convergence of AI-driven automation, machine learning for predictive maintenance, and cloud-based design tools will further revolutionize the industry. As digital textile printing continues to evolve, it will play a crucial role in shaping the future of sustainable and innovative textile manufacturing. Digital textile printing advances with innovations such as augmented inkjet heads, specialized color ink formulations, and improved color depth, leading to higher reproduction quality and consistency. As technological challenges are addressed, digital printing is becoming more efficient and versatile, integrating seamlessly into various applications. Beyond inkjet technology, digital printing involves multiple processes that create a cohesive and adaptable framework for widespread adoption. To replace traditional screen printing, digital methods must achieve significantly higher speeds while maintaining high-quality output. Enhancing productivity and optimizing costs, particularly for ink and equipment, are essential for broader acceptance. The industry also focuses on high-resolution imaging and small-batch manufacturing to expand market opportunities. Lowering operational costs, including expenses related to printers, ink, and treatments, will further strengthen the competitiveness and appeal of digital textile printing [36, 37].

13.10 Conclusion

As digital printing technology continues to evolve, it promises even more groundbreaking advancements with significant implications for the textile industry. By combining functionality, customization, and sustainability, digital printing is ushering in a new era of creative possibilities for designers and consumers. The rise of sustainable fast fashion offers a compelling business model that balances ecological, social, and economic considerations, paving the way for more responsible production practices. In contrast to conventional screen printing, digital printing offers a significant benefit by decreasing production time, facilitating rapid adaptability to trends, and lowering water use, thus promoting a beneficial environmental effect. Designers are now empowered to harness this innovative technology, and their willingness to embrace it will play a pivotal role in shaping the industry's future. If adopted strategically, digital textile printing holds the potential to become a key player in India's textile sector and globally, transforming the way textiles are designed, produced, and consumed.

Declaration The authors declare that they have no known competing financial interests or personal relationships that could have appeared to influence the work reported in this chapter. Generative AI and AI-assisted technologies have been used in the writing process to improve the readability and language of the work only. After using this tool, the authors carefully reviewed and revised the content to ensure accuracy and quality. The authors take full responsibility for the final version of this publication.

References

1. Neelima (2009) Printing and washing of textile. Sonali Publications
2. Khalil E, Sarkar J, Rahman MM, Shamsuzzaman M, Das D (2023) Innovative textile printing technology. In: Rahman MM, Mashud M, Rahman MM (eds) Advanced technology in textiles. Springer, Singapore, pp 139–140
3. Choudhury AKR (2022) Principles of textile printing. CRC Press
4. Tariq S (2024) Resist printing on fabric: 3 DIY methods-SewGuide. SewGuide. https://sewguide.com/resist-printing/. Accessed 20 Oct 2024
5. Ahmed T, Toki GFI, Mia R, Islam SR, Fahad MAA, Adeel S, Alpar T (2024) New trends in printing applications of natural dyes and pigments. In: Renewable dyes and pigments. pp 139–163. https://doi.org/10.1016/B978-0-443-15213-9.00003-X
6. Mahapatra NN (2024) Textile printing. CRC Press
7. Islam MR, Afroj S, Yin J, Novoselov KS, Chen J, Karim N (2024) Advances in printed electronic textiles. Adv Sci 11(6):2304140
8. Barron's (2012) Textile printing. Barron's Educational Series
9. Ujie H (ed) (2006) Digital printing of textiles. Woodhead Publishing
10. Tyler DJ (2005) Textile digital printing technologies. Text Prog 37(4):1–65
11. Tippett BG (2002) The evolution and progression of digital textile printing. Digital Printing Systems Orangeburg, New York, USA. https://www.brookstippett.com/docs/Print2002-BGT.pdf
12. Wang H, Memon H (eds) (2023) Digital textile printing: science, technology and markets. Woodhead Publishing
13. Carden S (2016) Digital textile printing. Bloomsbury Academic
14. Koseoglu AU, Gungor Y, Arık YD (2019) Innovations and analysis of textile digital printing technology. Int J Sci Technol Society 7(2):38–43
15. Babar AA, Gianchandani PK, Jhatial AK (2023) Overview of different digital textile printing machines. In: Wang H, Memon H (eds) Digital textile printing: science, technology and markets. Woodhead Publishing, pp 41–55
16. Cahill V (2006) The evolution and progression of digital printing of textiles. In: Ujie H (ed) Digital printing of textiles, Woodhead Publishing, pp 1–15
17. Kumelachew DM, Wagaye BT, Adamu BF (2023) Digital textile printing innovations and the future. In: Wang H, Memon H (eds) Digital textile printing: science, technology and markets. Woodhead Publishing, pp 241–259
18. Sheppard R (2011) New EPSON complete guide to digital printing. Pixel Communications
19. Madhukar Thakker A, Sun D (2023) Inks for digital printing of textiles. J Text Inst 115(12):2377–2390

20. Nie L, Xu X, Chen Y, Dong Y, Chang G, Li R (2023) Development of binder-free pigment inks for direct inkjet printing on cotton fabric without pretreatment. Langmuir 39(17):6266–6275
21. Fu Z (2006) Pigmented ink formulation. In: Ujie H (ed) Digital printing of textiles. Woodhead Publishing, pp 218–232
22. StartUs Insights (2023) Top 10 printing industry trends in 2024. https://www.startus-insights.com/innovators-guide/printing-industry-trends/. Accessed 20 Oct 2024
23. Robert T (2019) Printing Inks From Renewable Resources. In: Höfer R, Matharu AS, Zhang Z (eds) Green chemistry for surface coatings, inks and adhesives: sustainable applications. The Royal Society of Chemistry. pp 339–367
24. Shah MA, Lee DG, Lee BY, Hur S (2021) Classifications and applications of inkjet printing technology: a review. IEEE Access 9:140079–140102
25. Kašiković E, Vladić G, Novaković D (2016) Textile prinitng–past, present, future. J Chem Technol Environ 12(2):35–46
26. Gupta S (2001) Inkjet printing-a revolutionary ecofriendly technique for textile printing. Indi J Fibre Text Res 26:156−161
27. Chapman LP, King KM (2023) Digital printing innovations. Textile World. https://www.textileworld.com/textile-world/features/2023/11/digital-printing-innovations-2/. Accessed 10 Oct 2024
28. Admin (2023) The revolution in textile printing. Indian Text J. https://indiantextilejournal.com/the-revolution-in-textile-printing. Accessed 25 Sept 2024
29. Carnie BW (2010) Design management for the direct digital textile print industry: the development of a framework for measuring the conditions in managing change (Doctoral dissertation, UNSW Sydney)
30. Yanan Y, Chapman LP, Moore MM (2022) Profiling digital printing technology adoption in the fashion industry: a new approach to exploring innovation diffusion. Int J Innov Technol Manag 19(2). https://doi.org/10.1142/S0219877022500043
31. Yu Y, Parillo-Chapman L, Moore M (2020) Fashion printing technology diffusion: big data analytics. In: International textile and apparel association annual conference proceedings, vol 77, no 1. Iowa State University Digital Press
32. Artdaily (2021) Dolce & Gabbana just set a $6 million record for fashion NFTs. https://artdaily.com/news/139973/Dolce-Gabbana-just-set-a--6-million-record-for-fashion-NFTs. Accessed 21 Oct 2024
33. Koerner J (2020) Arid collection, 3D fashion by Stratasys. https://3dprintedart.stratasys.com/fashion/arid-collection. Accessed 15 Oct 2024
34. Work RA (1999) Textile printing. In: Recent Prog Ink Jet Technol II. Society for Imaging Science and Technology, pp 545–547
35. Hasan MA (2023) Technological barriers to digital printing in textiles: a study. In: Wang H, Memon H (eds) Digital Textile Printing: science, technology and markets. Woodhead Publishing, pp 157−183
36. Kobayashi H (2006) Industrial production printers–Mimaki's Tx series. In: Ujie H (ed) Digital printing of textiles. Woodhead Publishing, pp 98−122
37. Gianchandani PK, Jhatial AK, Babar AA, Yesuf HM (2023) Color management and design software for textiles. In: Wang H, Memon H (eds) Digital textile printing: science, technology and markets. Woodhead Publishing, pp 57−72

Chapter 14
Employing Sustainable Technologies for Mitigation of Textile Manufacturing Industry Effluent

Nilendu Basak, Shashwata Mainak, Ankita Chatterjee, and Atif Aziz Chowdhury

Abstract The textile manufacturing industry generates significant environmental concerns due to the discharge of toxic dyes, heavy metals, detergents, and highly alkaline or acidic wastewater during fabric production. These pollutants harm living organisms and the environment, directly or indirectly impacting human well-being. Although widely used, traditional remediation methods often produce secondary toxic by-products that worsen the problem. Recognizing these challenges, sustainable and eco-friendly solutions are becoming critical. Approaches such as microbial remediation and agro-waste use have shown promising results in effectively treating textile effluents. These methods aim to mitigate the environmental impact while avoiding the generation of additional harmful by-products. This chapter provides an overview of the environmental issues caused by textile effluents and examines strategies for their mitigation. It emphasizes sustainable remediation practices and the need to adopt environmentally conscious approaches. The aim is to consolidate existing knowledge and inspire future research to develop effective solutions for managing toxic effluents in the textile industry.

Keywords Textile dyes · Dye toxicity · Biodegradation · Biogenic nanoparticles · Bioremediation · Sustainability

N. Basak
Department of Microbiology, University of Kalyani, Nadia, West Bengal, India

S. Mainak
Government College of Engineering and Textile Technology, Serampore, West Bengal, India

A. Chatterjee
Department of Biotechnology, School of Applied Sciences, REVA University, Bangalore, Karnataka, India

A. A. Chowdhury (✉)
Faculty of Agricultural, Environmental and Food Sciences, Free University of Bozen-Bolzano, Piazza Università, 1, Bolzano, Italy
e-mail: atifaziz.chowdhury@unibz.it

M. Shahid et al. (eds.), *Advancements in Textile Coloration*,
https://doi.org/10.1007/978-981-96-5091-0_14

14.1 Introduction

Apparel is one of the basic three needs of humankind till the pre-historic era. But traveling through the path of history while coming to the present time of fast fashion trend, it has been seen that apparel and textiles are not only being used to ward off shame but through the ingenious hands of human beings, they exhibit the most colorful, intricate and delicate culture of different timelines, clans, kingdoms, habitats and more. In the present time, to meet the growing population's demands, textile industries are using enormous amounts of harsh chemicals and synthetic dye substrates for apparel production. While serving as a significant contributor to the world economy, the textile industries have gained a negative impression for producing large amounts of wastewater [1]. Chandanshive et al. have stated approximately 10% of annually produced synthetic dyes are released into the environment without any treatments [2]. However, the three main sectors of textile effluent generation are pre-treatments of fabrics, dyeing-printing, and finishing of fabrics [3]. Wastewater from these fabric preparation steps contains heavy metal(loids) (HMs), reducing agents, surfactants, detergents, dioxin, phenols, and more [4]. Wastewater has been shown to considerably contaminate the ecosystem [5], harming surface water, groundwater, soil, and the atmosphere. Moreover, many textile and dyeing companies are in underdeveloped nations, and wastewater is frequently improperly treated [6]. Releasing this toxic wastewater without any treatment causes various negative effects on nature and human health. For example, in India, textile industries from Kanpur release a huge amount of contaminated wastewater into canals and drains, which open into the Ganga River and finally end up mixing with the water of the Bay of Bengal [7].

Previously, some conventional chemical and physical wastewater treatments were used by the textile industries before the effluent was released into the nature. Depending upon the pH, total dissolved solids (TDS), chemical oxygen demand (COD), and biochemical oxygen demand (BOD), traditional wastewater treatments are partially efficient in reducing the toxicity of the released water. Because of significantly different concentrations and composition of contaminants present in it, these treatments produce secondary byproducts, which can be more toxic than the primary effluent in some cases [8]. Extensive research on textile wastewater treatment is ongoing, and researchers have introduced some eco-friendly, green, and sustainable mitigation processes like coagulation-flocculation, use of photo-catalysts, electro-Fenton process, microbial decolorization, coagulation-flocculation combined with advanced oxidation processes (AOPs) and more. Di Mauro et al. stated that zinc oxide (ZnO) in the form of nanofibers, nanoparticles, or thin films shows great efficiency as a photocatalyst in wastewater dye degradation in a wide range [9]. In another study, Mcyotto et al. stated that coagulation-flocculation is one of the most effective and promising pre-treatment processes for the removal of solid materials and dyes from wastewater [10]. Also, Dotto et al. have mentioned that the coagulants in coagulation-flocculation processes show maximum efficiency in an acidic medium [11]. Other than these methods, for reducing the turbidity of around 99% of textile wastewater, López-López et al. have combined coagulation-flocculation processes as pretreatment with AOPs to improve the efficiency [12]. Additionally, for sustainable effluent mitigation, microbial fuel cells (MFCs) also have gained the attention of researchers [13]. This chapter will discuss the sources, types, and toxicity levels of wastewater released from the textile industries and the very recent mitigation processes that are eco-friendly, sustainable, and green.

14.2 Characterization of Toxic Effluents from the Textile Industry

The yarn and clothing manufacturing sectors in the textile industry are the primary sources of solid waste. In industries, two methods were used to make yarn: continuous filament fibers like silk, nylon, etc., or staple-length fibers spun. Before spinning, carding and combing are performed on staple-length fibers (such as cotton, jute, polyester, etc.) to remove any shorter fibers from the yarn, resulting in a stronger yarn with a glossy, smooth surface [14]. Thus, those shorter fibers thrown away are either recycled to create a non-woven fabric or disposed of in a landfill. Most waste fibers, even natural fibers, take a long time to break down. For example, synthetic fibers can take more than 200 years to do so in a landfill [15]. In the current scenario, it has been seen that polyester is the most consumed fiber in textile industries, and cotton comes after polyester [14]. After being used by the consumers, when the clothes are discarded and thrown away, around 73% of textile waste is either burned or ends up in landfills. Approximately 12% of the total waste is recycled, and less than 1% is used to produce new clothing [15]. However, this whole post-consumer-based textile waste does not only contain fabrics made up of various kinds of cellulosic, protein, or synthetic fibers; it also contains metal zippers, buttons made up of petroleum-based derivatives, metal curtain holders, and more. It is predicted that by the year 2050, the purchase of textile products will exceed 160 million tons, and the most alarming concern will be the increased number of throwaway unused products [16]. The properties of the wastewater produced at various stages of textile manufacturing are detailed in Table 14.1.

Textile industries generate massive amounts of hazardous, toxic surplus laden with dyes and other chemicals with different heavy metals, including arsenic, cobalt, chromium, copper, cadmium, mercury, and nickel. Since the development of the first synthetic dye in 1856, over 10,000 different varieties of textile dyes have been marketed globally, with an estimated annual production of 8,105 metric tons. A considerable proportion of these poisonous dyes eventually end up in waterways, producing environmental concerns [21]. The amount of poisonous and hazardous colors found in the wastewater effluent is in a range of 5–10% of the total used chemicals [22]. The wastewater generated from textile industries comprises high temperature, BOD, COD, and TDS with altered pH. Reports suggest that all over the world, cotton dyes (2,80,000 tons) are discharged annually, and the most alarming part is that up to 10,00,000 tons of dyes are produced per year [17]. However, various

Table 14.1 Natures of wastewater generated due to the different processes of textile manufacturing industries [17–20]

Process categories	Chemical or agents	Characteristics of generated wastewater
Desizing	Hydrogen peroxide, hydrochloric acid and different enzymes (especially amylase)	Generate 50% of wastewater with high BOD and COD, water temperature approximately 70–80 °C
Scouring for removal of hydrophilic and non-cellulosic components	Phenol ethoxylates, sodium hydroxide, pecrinase, surfactants (wetting agents), and soaps	BOD, high pH and temperature (~70–80 °C) with dark color includes different wax and fatty oils
Bleaching for removal of natural color of fabrics	Ethoxylates, chlorine, hydrogen peroxide, ozone, sodium metasilicate, sodium carbonate and hydroxide	High BOD, COD, pH, and TDS
Mercerization for enhancement of strength and dye adaptability	Sodium hydroxide and acetic acid	High BOD, pH, and TDS
Dying with different textile dyes	3600 dyeing agents (mainly azo dyes) used where they comprise different heavy metals	High toxic dyes (80–90%) with increased BOD, COD, TDS and pH
Printing for the dyed in the specific location of fabrics	Urea, coloring agents (including different metals), rubber and binder materials	High BOD, COD, TDS, pH, and dyes with carcinogenic chemicals
Finishing for improvement and increasing the standard	Chlorinated compounds, resins, waxes, softeners, and antimicrobial agents	Low alkalinity, BOD, high toxicity,

sources of dye are encountered in the environments, either from direct contamination or via passive contamination. During the textile industry's production process, various hazardous elements, both liquid and solid, are produced at different stages. This type of waste is commonly referred to as pre-consumer waste [23]. According to several reports, gallons of water are used during textile manufacturing. A concise review article stated that 2,900 L of water is required to dye one pair of denim jeans while 2,700 L for one cotton t-shirt [16]. During the dying process, a wide range of toxic dyes come out as effluent. However, dyes such as azo, reactive, acidic, and metal-complex dyes have been found in wastewater. Along with the dyes, a survey report around the textile industries in Delhi NCR, India found that effluent contains several metals such as copper (0.17–0.28 mg L^{-1}), chromium (0.11–0.21 mg L^{-1}), cadmium (0.01 mg L^{-1}), lead (0.02–0.10 mg L^{-1}), and nickel (0.11–0.22 mg L^{-1}) [24, 25].

14.3 Toxicity of the Textile Industry Effluent

The significant threats associated with textile industry effluents reach far beyond the harm to the environment and microbial communities, leading to severe alterations in ecological functioning. The chemicals, coloring substances, HMs, and other pollutants within these effluents can harm aquatic ecosystems and humans as well. Around the world, 2,80,000 tons of textile dyes are discharged yearly as industrial wastewater during dyeing [26]. 10–15% of these end up in the environment during the dyeing of cosmetics, fibers, plastics, leathers, and paper-like substrates [27]. Moreover, textile effluents exhibit acute toxicity to various organisms, including zebrafish and *Daphnia* sp., indicating potential health risks to humans by entering into the food chain caused by the contaminated water [28].

14.3.1 Environmental Toxicity

Managing wastewater poses a challenge for maintaining environmental practices. The discharge of textile wastewater is particularly harmful to the environment because it contains poisonous and mutagenic substances that are hard to break down [4]. The textile industry uses around 2,000 distinct types of chemicals for various purposes [6]. Dyes are one of the most common toxins in this sector's wastewater [29]. These harmful substances can pollute water and deteriorate the ecosystems. Toxic dyes, like azo dyes, are especially damaging as they limit penetration in water, impacting photosynthesis in plants and disrupting the food chain [30].

Additionally, the presence of contaminants in textile wastewater can prove fatal for microorganisms that break down organic pollutants naturally. These microbes play a vital role in organic pollutant degradation. When these communities are compromised, the self-purification capacity of the environment is reduced, leading to an accumulation of pollutants and further ecological damage [31]. For example, as a result of textile dye containing wastewater decreased overall microbial diversity and richness, with an increase in taxa involved in aromatic degradation and denitrification as evidenced by advanced sequencing techniques [32]. Biyada et al. reported reduced diversity and abundance of *Actinobacteria*, *Firmicutes*, and *Chloroflexia* in exposure to higher concentrations of textile waste [33]. Additionally, toxic textile effluent can cause a decline in microbial populations and hinder the breakdown of organic matter, leading to increased pollution and a decline in water quality. Moreover, they can alter ecological functioning by disrupting nutrient cycles, reducing biodiversity, and impairing ecosystem services. The health of animals and human can be affected by the consumption of these toxic compounds in their food. The bioaccumulation of heavy metals and toxic compounds in the food chain can have long-term effects on wildlife and human health [30].

14.3.2 Toxicity to Aquatic Life

Wastewater is exceedingly problematic to manage due to its high concentration of dyes and fluctuation. The colors of textile dyes can impact the clarity, pollution levels, acidity, and temperature of the water sources [34]. These harmful substances can harm the living by hindering their growth, reducing photosynthesis, and causing toxin build-up [35]. Ecotoxicological studies using organisms such as algae, small shellfish, and fish have confirmed the nature of these effluents [28, 36]. Approximately 2% of the textile dyes are released straight as effluent, while 10% are lost during the dyeing process [37]. Dye-containing water systems absorb and reflect sunlight, causing most harm to the environment. Light absorption alters algal photosynthetic activity, hence affecting the food chain [37]. The release of hazardous compounds into the different environments caused the decreased level of light transmittance and oxygen uptake. This has a deleterious impact on photosynthesis and aquatic life [38]. Furthermore, textile dyes and effluents have been found to affect seed germination, cause hemolysis, and exhibit toxicity to *Artemia salina* [36].

14.3.3 Toxicity to Human

The toxicity profiles of synthetic dyes have been extensively documented across several studies. The persistent organic dyes exhibit carcinogenic properties, which significantly increase the risk of various cancers among exposed populations when ingested through contaminated water supplies or via dermal contact during industrial processing or laundering activities at home [39]. Vulnerable groups, especially children under five, are highly susceptible to environmental toxins. Chemical contamination from textile industries is associated with severe health effects in young children. These effects include developmental disorders and chronic illnesses, which may arise from maternal exposure during pregnancy or breastfeeding [40]. Apart from having a bad aesthetic effect, these dyes can harm organisms by being carcinogenic and mutagenic [41]. According to Bae and Freeman, 130 out of 3,200 azo dyes have the potential to produce carcinogenic aromatic amines through reduction and degradation [27]. Azo dyes can cause skin irritations, lung and digestive issues, splenetic sarcomas, hepatocarcinomas, and bladder cancer in humans. These dyes can enter the body through the digestive tract and harm hemoglobin and DNA molecules. The chemicals can cause cancer in both humans and animals. Various colors affect the respiratory tract, allergies, and neurological disorders and also share contact with leukemia. There have been reports of kidney, liver, and bladder cancer in textile workers [42].

14.4 Conventional Mitigation Processes

Considering the heavy load of dyes, dissolved solids, and toxic chemicals and their consequences to the environment and the ecology, many methods, including physical, chemical, and biological mitigation methods, have been developed. These have shown quite a bit of success in terms of the management of textile waste effluents. A brief description of the conventional methods is described below.

14.4.1 Physical Treatment

The first step of wastewater treatment is physical treatment, which involves the removal of solid particles and suspended matter. Using a mass transfer mechanism, this cheap method has 85–99% dye removal efficiency from textile wastewater [43]. Adsorption is a commonly used method for removing dyes from textile wastewater, which is efficient and economical, as it relies on the transfer of dissolved dyes to the surfaces of solid or porous materials. Different processes can take place through physisorption or chemisorption, and they are generally driven by hydrophobic interactions and electrostatic forces. The most commonly used adsorbents are silica gel, alumina, zeolites, and activated carbon, the latter being more popular because of its high efficiency in dye removal. Although it would be possible to use less expensive alternative adsorbents such as clay or agricultural wastes, these would often be less effective than activated carbon and require larger volumes [17, 35].

Membrane filtration is one of the advanced techniques used to treat textile wastewater, which would efficiently eliminate color, COD, and salinity. Microfiltration is specially used to treat pigment dye baths and to eliminate suspended and colloidal particles. The membranes are made of various polymers and materials, such as ceramics. Ultrafiltration removes macromolecules and particles, but it does not remove dyes as effectively; therefore, often, these processes need to be combined with other treatments, such as reverse osmosis, for better results [44]. Micellar-enhanced ultrafiltration enhances pollutant elimination by the use of surfactants that encapsulate the contaminants. However, until now, its application in the textile sector remains limited [17].

Nanofiltration is a relatively new membrane technique that combines the aspects of ultra-filtration and reverse osmosis for effective dye removal, among other pollutants [43]. The nanofiltration membranes are made of cellulose acetate and ceramics and operate under size exclusion and electrostatic repulsion. Reverse osmosis is more effective in removing ionic substances, colorants, and salts, which can produce high-quality wastewater; however, this system may suffer from fouling and often requires intensive pre-treatment. The two membrane technologies can be combined with physiochemical techniques to achieve significant benefits in terms of reduced freshwater intake, lower treatment costs, and improved plant performance [43, 45].

14.4.2 Chemical Treatment

Chemical processes involved in dye removal from textile wastewater are often expensive and complex, requiring a great deal of energy and specialized equipment. Coagulation-flocculation is one commonly used method that helps settle dye particles through the use of chemical coagulants like cationic organic polymers. Furthermore, natural coagulants like *Moringa oleifera* are effective against wastewater decolorizing [46]. Another process is electrocoagulation, which generates metal hydroxide flocs by electro-dissolution, and such methods could attain high dye removal rates. An Electro-Fenton process couples oxidation with coagulation using hydroxyl radicals to degrade pollutants [43]. However, scaling up this process for industrial applications remains challenging. Anodic oxidation is quite effective but very expensive and unstable. Ozonation is an efficient oxidation process that degrades dyes but yields dangerous by-products. It is interfered with by several factors, such as pH. In the AOP, strong oxidizing agents like hydroxyl radicals are formed, and dyes are efficiently removed. One common AOP is the Fenton process, which uses a combination of iron salts and hydrogen peroxide to efficiently degrade dyes. Photocatalysis, mediated by TiO_2, is also applied to break down dye molecules in wastewater [44].

Both physical and chemical methods have shown effectiveness in treating textile wastewater; they are often expensive and may generate toxic sludge. Biological approaches and adsorption methodologies are regarded as more sustainable due to their economic efficiency and capacity to decompose various classes of dyes [47].

14.4.3 Hybrid Treatment Technology

Hybrid treatment technologies incorporate more than one process to enhance the efficiency of dye removal from wastewater, thereby offering economies of cost and time as well as reduced operational space. Membrane bioreactor (MBR) technology was developed in the 1960s and produces a very stable water quality. It is considered efficient in removing organics and metals. However, membrane fouling remains a big challenge. In the treatment of wastewater generated from the various textile processes, aerobic MBR (AeMBR) systems have been widely used and are usually very effective in the removal of COD and dyes, especially with supplementary treatments like UV disinfection and granular activated carbon (GAC) filtration [48]. On the other hand, anaerobic MBRs (AnMBRs) have shown energy efficiency, production of less waste, biogas production, and lower operational cost compared with the AeMBRs, hence becoming viable alternatives in treating textile wastewater [44].

Photocatalytic membrane reactor (PMR) technology employs photocatalysts to cleave complex dye molecules into simpler compounds, hence affecting the degradation of toxic substances in wastewater. When it comes to the effectiveness of PMRs, certain factors need to be considered, such as the kind of photocatalyst, the kind of membrane material, how light behaves, and some environmental conditions like temperature and pH. Generally, PMRs use TiO_2-based photocatalysts because they are cheap, stable, and do not harm the living. PMRs can be arranged as photocatalysts immobilized on the membrane surface or suspended in liquid, in which case the catalyst is kept there and recovered afterward [49].

Combining MBR with PMR technologies will increase efficiency in treatment while addressing membrane fouling problems, thus enhancing the removal of pollutants as a whole. Physical cleaning, biological cleaning, or chemical cleaning can improve the performance of an MBR system, with added operational costs. For instance, new hybrid system developments like combining MBRs with extra treatment units have improved their performance in treating complex textile wastewater. Hence, hybrid techniques ensure economic and operational efficiency in the process while guaranteeing a very high level of dye removal [43, 49]. A comparison of the conventional technologies employed in the treatment of textile wastewater is presented in Table 14.2.

14.5 Mitigation Strategies via Sustainable Approaches

14.5.1 Mitigation of Textile Dyes by Microbes

In recent decades, microorganisms have risen as sustainable and emerging approaches to the mitigation of toxic textile dyes. Microbes encompass bacteria, fungi, algae, and yeast altogether. They act as biodegradation agents, and their secondary metabolites and enzymes also remove and degrade the toxic dyes. Microbes-mediated mitigation approaches are advantageous in many ways, as they are more cost-effective and have a less toxic footprint on the environment [45].

14.5.1.1 Bacteria

Living or dead bacterial biomass or their by-products, such as biosurfactants, biogenic nanoparticles, and secondary metabolites (enzymes), can remove or degrade toxic dyes from the soil, wastewater, or other environment [51]. The scientific community is showing their arguments towards the role and efficiency of the indigenous or the non-native bacteria, which is more convenient. Several reports suggest that indigenous bacteria have more potential in the direction of dye removal from any environment [52]. An indigenous bacterial isolate, *Bacillus* sp. AS2 isolated from the soil contaminated with textile dye exhibited the ability to degrade multiple dyes and the potential to degrade a mixture of dyes. *Bacillus* sp. AS2 showed approximate degradation of Congo red (92%), methylene blue (97%), malachite green (94%), and mixed dyes (77%) [53]. Rahayu et al. showed that four ligninolytic bacteria have the potential to reduce the BOD as well as decolorize the azo dyes [54]. However, among them, *Bacillus paramycoides* K7.2 (L11) exhibited the maximum dye decolorization of up

Table 14.2 Comparison between textile wastewater treatment methods [30, 35, 43, 45, 50]

Method type	Mechanism of Process	Advantages	Disadvantages
Physical methods			
Adsorption	Dyes adsorb onto the surface of an adsorbent material, like activated carbon, for removal from wastewater	• Effective for a wide range of dyes • Can achieve high removal efficiencies • The adsorbent is regenerative and can be reused	• Huge amounts of adsorbent materials are consumed • Over a period, adsorbents get saturated • Spent adsorbents may have an expensive disposal
Filtration	Dyes are physically separated from wastewater using membrane filtration techniques like microfiltration, ultrafiltration, or nanofiltration	• Can remove dyes with high-efficiency • Compact and easy to operate • No chemical addition required	• Membranes can be fouled by suspended solids and dyes • Requires regular cleaning and maintenance • Concentrated dye solution requires further treatment
Sedimentation	Dyes are removed by allowing them to settle out of the wastewater due to gravity	• Simple and cost-effective • No chemicals or energy required • Suitable for large volumes of wastewater	• Limited effectiveness for soluble dyes • Requires large settling tanks • Generates sludge that requires disposal
Chemical methods			
Oxidation	Dyes are broken down into simpler, less colored compounds using strong oxidizing agents like ozone, hydrogen peroxide, or chlorine	• Effective for a wide range of dyes • Can achieve complete mineralization of dyes • Rapid reaction times	• May produce hazardous by-products • Requires careful handling of chemicals • High operating costs due to chemical consumption

(continued)

Table 14.2 (continued)

Method type	Mechanism of Process	Advantages	Disadvantages
Reduction	Azo dyes are cleaved into smaller, less colored fragments using reducing agents like sodium hydrosulfite or hydrogen gas	• Effective for azo dyes • Can be combined with other methods • Relatively simple to implement	• Limited effectiveness for non-azo dyes • Requires careful control of pH and redox potential • May produce toxic aromatic amines as by-products
Coagulation/ Flocculation	Dyes are destabilized and aggregated into larger particles using coagulants like alum or ferric chloride, which are then removed by sedimentation or flotation	• Effective for a wide range of dyes • Can remove suspended solids and colloids • Relatively simple to implement	• Generates large volumes of sludge that require disposal • Effectiveness depends on pH and coagulant dosage • May not remove soluble dyes effectively
Hybrid methods			
Adsorption + oxidation	Dyes are first adsorbed on an adsorbent, followed by regeneration of the exhausted adsorbent through oxidation which leads to the destruction of the attached coloring agents	• Combines the advantages of adsorption and oxidation • Can achieve high removal efficiencies • Adsorbents can be reused multiple times	• More complex and costly than individual methods • Requires careful optimization of both processes • Disposal of used adsorbents may remain a concern
Membrane filtration + oxidation	Dyes initially are removed by membrane filtration, and concentrated dye solution is further treated by Oxidation to destroy the dyes	• Combining the benefits associated with membrane filtration and oxidation • Can achieve very high removal efficiencies • Reduces the volume of waste requiring oxidation treatment	• More complex and costly than individual methods • Membrane fouling and oxidation by-products are still concerns • Requires careful optimization of both processes

to 82.79% (Congo red), while the other three isolates, namely L1, L7 and L8 (unidentified) decolorized 40.51% (Alizarin Red S), 38.69% (methylene blue) and 30.34% (Remazol Brilliant Blue R; RBBR) respectively [54]. A bacterium, *B. cereus* SSC, isolated from textile effluent, has the potential to decolorize 97.8% of Reactive Green 12 dye [55]. *Staphylococcus caprae* isolated from textile effluent decolorized (90%; Reactive Red-195) dye in stagnant condition [56]. *Serratia marcescens* ED1, isolated from textile effluent, showed the capability of degrading (81.02%) of methyl orange dye within 48 h [57]. Two Gram-positive bacteria, *Brevibacillus brevis* US575 and *Bacillus cereus* F4810/72, isolated from the tannery industry, Bantala, West Bengal, India, showed to degrade the acid Blue 113 dye (60%) within 72 h [58].

However, at the same time, reports show that non-native bacteria have the same results for the removal of dyes. A total of 19 rhizospheric bacterial strains were isolated from tomato plants and showed decolorization of methyl orange under different conditions, such as aerobic (89.5–96.3%) and microaerophilic (96.3–99.9%) conditions [51]. At the optimized condition, five bacteria identified as *Pseudomonas fluorescens* ESR7, *Staphylococcus saprophyticus* ESD8, *P. veronii* ESR13, *P. parafulva* ESB18, and *Stenotrophomonas maltophilia* ESR20 showed the highest removal. Moreover, the presence of azoreductase and laccase enzymes confirms the degradation pathways [51]. A halotolerant *Pseudomonas aeruginosa* JKAK decolorized Bismarck brown dye in the presence of NaCl (90%) and metal ions (Pb^{2+} and Zn^{2+}) (86.4%) [59]. Another report suggests that more than 90% decolorization (azo dyes) by the *Pseudomonas putida* was noticed within 72 h [60].

14.5.1.2 Fungi

Fungi-based degradation and decolorization of toxic dyes have prevailed in recent years due to their economic friendliness and ability to be a better alternative than a conventional process. The fungal isolates produces an array of enzymes that help to break the textile dyes. However, not only enzymes but fungal biomass also absorb and accumulate different textile dyes. White rot fungi, especially *Candida* sp., are the most studied species for the mitigation of toxic textile dyes [61]. A saprophytic fungi *Pleurotus sajor caju* F-2083, is capable of decolorizing 83 and 63% of two textile dyes such as reactive blue 13 and reactive black 5, respectively, within 8 days at optimum conditions [62]. Moreover, up to 90% decolorization (reactive blue 13) was reported within a shorter incubation time (100 min) by the isolated and purified laccase enzyme from *Pleurotus* sp. F-2083 [62]. A report showed that white rot fungi *Ganoderma lucidum* produce LAC-4 laccase enzyme, which can decolorize 37.23% to 99.94% from 7 different types of textile dyes in single or mixed dye solutions [63]. Three indigenous fungal strains, *Aspergillus iranicus*, *A. terreus,* and *Penicillium crustosum*, were isolated from textile wastewater. Later on, studies confirmed that among these three fungal strains, *A. iranicus* and *A. terreus* showed the potential of decolorization of malachite green by 80% and 90%, respectively, and *P. crustosum* can decolorize more than 93% methylene blue [64]. Puronomo et al. reported that a fungal consortium constructed with *Daedalea dickinsii* and *A. oryzae* has the potential to decolorized 64.77% methylene blue [65].

14.5.1.3 Microbial Consortium

Microbes combined with two or more strains deliver diverse advantages over single strains. The secondary metabolites, such as enzyme production, are increased by many folds, thus enhancing the degradation and decolorization of toxic dyes. A study conducted by Tian et al. demonstrated that a concentration-dependent degradation of crystal violet was noticed by using bacteria and fungi mono or in a consortium culture condition [66]. The best result was archived when a consortium of *Pseudomonas fluorescens* L6 and *Aspergillus fumigatus* A3 together showed 89.73 to 100% dye degradation when dye concentration was decreased from 200 mg L^{-1} to 50 mg L^{-1}. However, the authors further investigated the enzyme responsible for these high degradations and the end products. After the degradation, the product was further inspected by using FTIR and GC–MS spectroscopy. Data suggested that the presence of laccase (Lac), manganese peroxidase (MnP), lignin peroxidase (Lip), and NADH–DCIP reductase enzymes confirms the enzymatic degradation of the dye [66]. A bacterial consortium, namely VITPBC6, was built by using six different *Bacillus* sp. isolated from the leather industry and showed dye (nine reactive dyes) degradation of 44.9–89.29%. However, this VITPBC6 consortium also efficiently degrades (48.85–72.98%) a mixture of reactive dyes [67]. Another study conducted by three bacterial consortia (*Bacillus firmus* TSL9, *Brevibacillus laterosporus* TS5, and *Exiguobacterium aurantiacum* TSL7) exhibited 83.78% decolorization (reactive azo dye) [68].

14.5.2 Agro-Waste and Biochar

The biochar process in textile effluent treatment has lately demonstrated huge potential due to its efficacy, eco-friendliness, thermal stability, and short processing time [69]. Biochar technology can be derived from any natural sources [70], but scientific communities are bent towards agricultural waste mediated [71–73] biochar for the green and sustainable approach. The production of biochar (carbonaceous material) is mainly achieved by pyrolysis of any potent materials [73]. Biochar is mainly negatively charged, so the affinity towards the positively charged contaminants is well established due to the dissociation of functional groups (carboxyl and hydroxyl group), and it can be reused multiple times [72, 74]. One of the effective methods to remove dyes and organic pollutants from textile effluent is the adsorption of the dyes into biochar. A drought-resistant and widely available plant, *Moringa oleifera* (the local name in West Bengal is sajina or sajne), is used as a biochar-adsorbent [75]. At a concentration of 6 g L^{-1}, *M. oleifera* derived biochar successfully removed 99% methylene blue from the textile wastewater within 15 min. However, at the same concentration, this formulated biochar also reduces the load of BOD, COD, and electrical conductivity [75]. Another waste material, pine needles (*Pinus roxbergii*) found in Himalayan forests, is tough to degrade; however, Pandey et al. used these needles as biochar and modified them with immobilized laccase enzyme [73]. One hundred percent of dye and 33% of COD removal is reported from the textile wastewater. Moreover, 73.8% and 88.1% of dye (malachite green and Congo red, respectively) removal was noticed in the case of the stimulated experimental setup [73]. Algae-derived biochar can remove 96.14% Congo red from the aqueous solution [76], while the stem of water lilies-derived biochar adsorbs 99.25% malachite green [77]. The widely reported waste generated from sugarcane industries has developed strategies to use this bagasse for biochar production. At the optimized condition, sugarcane-derived biochar showed 97.46% crystal violet removal from wastewater [74]. However, an interesting study conducted by Singh and Srivastava described the transformation of industrial sludge generated during textile dyeing converted into biochar for the removal of the ofloxacin antibiotic [78].

14.5.3 Phytoremediation

Phytoremediation is another promising approach to mitigate textile pollutants from the environment. Also, it is eco-friendly, sustainable, and less toxic. Live plants have roots, leaves, and inhabitants with multiple microorganisms, which together have the potential to remove or absorb the toxic elements and dyes from the wastewater and wasteland simultaneously [79]. The epicenter of the wastewater generated from textile industries can be tackled with aquatic plants. Phytoremediation of dyes can be achieved in several ways such as phytoextraction or rhizo-filtration (where plant parts can absorb the contaminates and store them), phytodegradation (where the entrapped contaminants by the enzymes and break into less toxic form), phytovolatilization or phyto-stabilization (in this process the pollutants loses their mobility and precipitates in soil or root) [79, 80]. Ultimately, phytoremediation helps to clean up the food chain by removing the toxic pollutants or contaminants from the polluted environment. A study reveals that at 60 °C, *Eichhornia crassipes* (generally known as water lily) can absorb 98% (gentian violet) and 85% (phenol red) from aqueous solution [81]. Two azo dyes, methylene blue (99%) and Congo red (25%) removal were recorded by *Lemna minor* (duckweed) after 24 days [82]. *Ceratophyllum demersum* is a widely studied aquatic plant that removes or degrades toxic dyes from the solution. The biomass of 4 g *C. demersum* can remove 58.83% acid Blue 92 within 7 days at the initial concentration of 20 mg L^{-1} [83]. Another study conducted by Shaalan et al. showed aquatic plant *C. demersum* alone removed 78.53% methylene blue within 15 days of incubation. Still, when *C. demersum* was coupled with *Acinetobacter radioresistens*, the removal efficiency was enhanced, and approximately 88.72% removal was noticed, during the same period of incubation time [84]. A recent study on nano-sorbent composed of dry biomass of *Phragmites australis* and Fe_3O_4-monometallic catalyst removed Congo red (97%) and xylenol orange (47%)

from wastewater; moreover, the experiment showed the removal of these two dyes depends upon the pH [85]. *Pistia stratiotes* can degrade mixed dye (500 mg L^{-1}) by 68.2% from the solution (mimic synthetic wastewater) [86]. Kaur et al. reported the dye removal potentials of *Trachyspermum ammi* L. from synthetic wastewater. The ornamental plant *T. ammi* can remove 99% methylene blue (10 mg L^{-1}) and 91% Congo red, but the removal percentage increased when pH was 5.0 (95% for Congo red) [87].

14.6 Conclusion and Future Direction

The textile industry's environmental impact is a critical problem that requires immediate and continuous intervention. The large amount of hazardous wastewater produced, the considerable quantity of water utilized, and the notable carbon emissions call for urgent, much-needed basic changes to the industry's procedures. Traditional mitigation methods have had little success and are often ineffective due to the generation of harmful secondary by-products. Sustainability and innovative development of the textile industry depend on the implantation of clean technologies, including biological mitigation such as bioremediation and hybrid approaches involving the use of AOPs in combination with eco-friendly dyes and materials. In this context, enabling a circular economy approach, supported by recycling, reusing wastewater, and energy-saving technologies, will be decisive in reducing the adverse environmental impacts generated by the industry. Future research should pay a lot of importance to developing and applying new sustainable technologies, including natural dyeing, advanced water treatment systems, and innovative recycling. In addition, strengthening of regulatory frameworks is needed to apply more stringent environmental standards and foster sustainability concerns; schemes such as the Zero Discharge of Hazardous Chemicals might provide models worth replicating across the industry. The development and improvement of sustainable technologies need cooperation among industry stakeholders, researchers, and policymakers, while continuous research into new materials, processes, and treatments has been imperative for sustainability for a long time.

Declaration The authors declare that they have no known competing financial interests or personal relationships that could have appeared to influence the work reported in this chapter. AI tools were used in the writing process only to improve the readability and language of the work. After using this tool, the authors carefully reviewed and revised the content to ensure accuracy and quality. The authors take full responsibility for the final version of this publication.

References

1. Younis SA, Serp P, Nassar HN (2021) Photocatalytic and biocidal activities of $ZnTiO_2$ oxynitride heterojunction with MOF-5 and g-C3N4: A case study for textile wastewater treatment

under direct sunlight. J Hazard Mater 410:124562. https://doi.org/10.1016/j.jhazmat.2020.124562
2. Chandanshive VV, Kadam SK, Khandare RV et al (2018) *In situ* phytoremediation of dyes from textile wastewater using garden ornamental plants, effect on soil quality and plant growth. Chemosphere 210:968–976. https://doi.org/10.1016/j.chemosphere.2018.07.064
3. Omerogullari Basyigit Z, Eyupoglu C, Eyupoglu S, Merdan N (2023) Investigation and feed-forward neural network-based estimation of dyeing properties of air plasma treated wool fabric dyed with natural dye obtained from *Hibiscus sabdariffa*. Color Technol 139:441–453. https://doi.org/10.1111/cote.12665
4. Hubadillah SK, Othman MHD, Tai ZS et al (2020) Novel hydroxyapatite-based bio-ceramic hollow fiber membrane derived from waste cow bone for textile wastewater treatment. Chem Eng J 379:122396. https://doi.org/10.1016/j.cej.2019.122396
5. Abdel-Karim A, El-Naggar ME, Radwan EK et al (2021) High-performance mixed-matrix membranes enabled by organically/inorganic modified montmorillonite for the treatment of hazardous textile wastewater. Chem Eng J 405:126964. https://doi.org/10.1016/j.cej.2020.126964
6. Khan S, Malik A (2014) Environmental and health effects of textile industry wastewater. In: Malik A, Grohmann E, Akhtar R (eds) Environmental deterioration and human health: natural and anthropogenic determinants. Springer, Netherlands, Dordrecht, pp 55–71.https://doi.org/10.1007/978-94-007-7890-0_4
7. Bharagava RN, Mani S, Mulla SI, Saratale GD (2018) Degradation and decolourization potential of an ligninolytic enzyme producing *Aeromonas hydrophila* for crystal violet dye and its phytotoxicity evaluation. Ecotoxicol Environ Saf 156:166–175. https://doi.org/10.1016/j.ecoenv.2018.03.012
8. Jayapal M, Jagadeesan H, Krishnasamy V et al (2022) Demonstration of a plant-microbe integrated system for treatment of real-time textile industry wastewater. Environ Pollut 302:119009. https://doi.org/10.1016/j.envpol.2022.119009
9. Di Mauro A, Fragalà ME, Privitera V, Impellizzeri G (2017) ZnO for application in photocatalysis: from thin films to nanostructures. Mater Sci Semicond Process 69:44–51. https://doi.org/10.1016/j.mssp.2017.03.029
10. Mcyotto F, Wei Q, Macharia DK et al (2021) Effect of dye structure on color removal efficiency by coagulation. Chem Eng J 405:126674. https://doi.org/10.1016/j.cej.2020.126674
11. Dotto J, Fagundes-Klen MR, Veit MT et al (2019) Performance of different coagulants in the coagulation/flocculation process of textile wastewater. J Clean Prod 208:656–665. https://doi.org/10.1016/j.jclepro.2018.10.112
12. López-López C, Martín-Pascual J, Leyva-Díaz JC et al (2016) Combined treatment of textile wastewater by coagulation–flocculation and advanced oxidation processes. Desalin Water Treat 57:13987–13994. https://doi.org/10.1080/19443994.2015.1063013
13. Li S, Ho S-H, Hua T et al (2021) Sustainable biochar as an electrocatalysts for the oxygen reduction reaction in microbial fuel cells. Green Energy Environ 6:644–659. https://doi.org/10.1016/j.gee.2020.11.010
14. Frazier RM, Vivas KA, Azuaje I et al (2024) Beyond cotton and polyester: an evaluation of emerging feedstocks and conversion methods for the future of fashion industry. J Bioresources Bioprod 9:130–159. https://doi.org/10.1016/j.jobab.2024.01.001
15. Shirvanimoghaddam K, Motamed B, Ramakrishna S, Naebe M (2020) Death by waste: fashion and textile circular economy case. Sci Total Environ 718:137317. https://doi.org/10.1016/j.scitotenv.2020.137317
16. Papamichael I, Voukkali I, Economou F et al (2024) Mobilisation of textile waste to recover high added value products and energy for the transition to circular economy. Environ Res 242:117716. https://doi.org/10.1016/j.envres.2023.117716
17. Khan WU, Ahmed S, Dhoble Y, Madhav S (2023) A critical review of hazardous waste generation from textile industries and associated ecological impacts. J Indian Chem Soc 100:100829. https://doi.org/10.1016/j.jics.2022.100829

18. Madhav S, Ahamad A, Singh P, Mishra PK (2018) A review of textile industry: wet processing, environmental impacts, and effluent treatment methods. Environ Qual Manage 27:31–41. https://doi.org/10.1002/tqem.21538
19. Rippon JA, Evans DJ (2020) Improving the properties of natural fibres by chemical treatments. In: Kozłowski RM, Mackiewicz-Talarczyk M (eds) Handbook of natural fibres (2nd edn). Woodhead Publishing, pp 245–321. https://doi.org/10.1016/B978-0-12-818782-1.00008-0
20. Saravanan A, Yaashikaa PR, Ramesh B et al (2024) Microorganism-mediated bioremediation of dyes from contaminated soil: mechanisms, recent advances, and future perspectives. Food Chem Toxicol 185:114491. https://doi.org/10.1016/j.fct.2024.114491
21. Nabil GM, El-Mallah NM, Mahmoud ME (2014) Enhanced decolorization of reactive black 5 dye by active carbon sorbent-immobilized-cationic surfactant (AC-CS). J Ind Eng Chem 20:994–1002. https://doi.org/10.1016/j.jiec.2013.06.034
22. Prasad SS, Aikat K (2014) Optimization of medium for decolorization of Congo red by *Enterobacter* sp. SXCR using response surface methodology. Desalin Water Treat 52:6166–6174. https://doi.org/10.1080/19443994.2013.812525
23. Uddin F (2021) Environmental hazard in textile dyeing wastewater from local textile industry. Cellulose 28:10715–10739. https://doi.org/10.1007/s10570-021-04228-4
24. Bhardwaj V, Kumar P, Singhal G (2014) Toxicity of heavy metals pollutants in textile mills effluents. Int J Sci Eng Res 5(7):2229–5518
25. Velusamy S, Roy A, Sundaram S, Kumar Mallick T (2021) A review on heavy metal ions and containing dyes removal through graphene oxide-based adsorption strategies for textile wastewater treatment. Chem Rec 21:1570–1610. https://doi.org/10.1002/tcr.202000153
26. Zainith S, Sandhya S, Saxena G, Bharagava RN (2016) Microbes an ecofriendly tools for the treatment of industrial waste waters. Microbes Environ Manage 2016:75–100
27. Bae J-S, Freeman HS (2007) Aquatic toxicity evaluation of new direct dyes to the *Daphnia magna*. Dyes Pigm 73:81–85. https://doi.org/10.1016/j.dyepig.2005.10.015
28. Methneni N, Morales-González JA, Jaziri A et al (2021) Persistent organic and inorganic pollutants in the effluents from the textile dyeing industries: ecotoxicology appraisal via a battery of biotests. Environ Res 196:110956. https://doi.org/10.1016/j.envres.2021.110956
29. Nor FHM, Abdullah S, Yuniarto A et al (2021) Production of lipopeptide biosurfactant by *Kurthia gibsonii* KH2 and their synergistic action in biodecolourisation of textile wastewater. Environ Technol Innov 22:101533. https://doi.org/10.1016/j.eti.2021.101533
30. Alsukaibi AKD (2022) Various approaches for the detoxification of toxic dyes in wastewater. Processes 10:1968. https://doi.org/10.3390/pr10101968
31. Patel DD, Bhatt S (2022) Environmental pollution, toxicity profile, and physico-chemical and biotechnological approaches for treatment of textile wastewater. Biotechnol Genet Eng Rev 38:33–86. https://doi.org/10.1080/02648725.2022.2048434
32. Han T, Zheng J, Han Y et al (2021) Comprehensive insights into core microbial assemblages in activated sludge exposed to textile-dyeing wastewater stress. Sci Total Environ 791:148145. https://doi.org/10.1016/j.scitotenv.2021.148145
33. Biyada S, Merzouki M, Dėmčėnko T et al (2022) The effect of feedstock concentration on the microbial community dynamics during textile waste composting. Front Ecol Evol 10:813488. https://doi.org/10.3389/fevo.2022.813488
34. Verma AK, Dash RR, Bhunia P (2012) A review on chemical coagulation/flocculation technologies for removal of colour from textile wastewaters. J Environ Manage 93:154–168. https://doi.org/10.1016/j.jenvman.2011.09.012
35. Al-Tohamy R, Ali SS, Li F et al (2022) A critical review on the treatment of dye-containing wastewater: ecotoxicological and health concerns of textile dyes and possible remediation approaches for environmental safety. Ecotoxicol Environ Saf 231:113160. https://doi.org/10.1016/j.ecoenv.2021.113160
36. Reddy S, Osborne WJ (2020) Heavy metal determination and aquatic toxicity evaluation of textile dyes and effluents using *Artemia salina*. Biocatal Agric Biotechnol 25:101574. https://doi.org/10.1016/j.bcab.2020.101574

37. Mani S, Chowdhary P, Bharagava RN (2019) Textile wastewater dyes: toxicity profile and treatment approaches. In: Bharagava RN, Chowdhary P (eds) Emerging and eco-friendly approaches for waste management. Springer, Singapore, pp 219–244. https://doi.org/10.1007/978-981-10-8669-4_11
38. Holkar CR, Jadhav AJ, Pinjari DV et al (2016) A critical review on textile wastewater treatments: possible approaches. J Environ Manage 182:351–366. https://doi.org/10.1016/j.jenvman.2016.07.090
39. Lellis B, Fávaro-Polonio CZ, Pamphile JA, Polonio JC (2019) Effects of textile dyes on health and the environment and bioremediation potential of living organisms. Biotechnol Res Innov 3:275–290. https://doi.org/10.1016/j.biori.2019.09.001
40. Shittu E, Lakhanpaul M, Vigurs C et al (2023) A rapid systematic scoping review of research on the impacts of water contaminated by chemicals on very young children. Sci Total Environ 891:164604. https://doi.org/10.1016/j.scitotenv.2023.164604
41. Das A, Mishra S (2017) Removal of textile dye reactive green-19 using bacterial consortium: process optimization using response surface methodology and kinetics study. J Environ Chem Eng 5:612–627. https://doi.org/10.1016/j.jece.2016.10.005
42. Islam MdA, Ali I, Karim SMA et al (2019) Removal of dye from polluted water using novel nano manganese oxide-based materials. J Water Process Eng 32:100911. https://doi.org/10.1016/j.jwpe.2019.100911
43. Islam T, Repon MdR, Islam T et al (2022) Impact of textile dyes on health and ecosystem: a review of structure, causes, and potential solutions. Environ Sci Pollut Res 30:9207–9242. https://doi.org/10.1007/s11356-022-24398-3
44. Behera M, Nayak J, Banerjee S et al (2021) A review on the treatment of textile industry waste effluents towards the development of efficient mitigation strategy: an integrated system design approach. J Environ Chem Eng 9:105277. https://doi.org/10.1016/j.jece.2021.105277
45. Castillo-Suárez LA, Sierra-Sánchez AG, Linares-Hernández I et al (2023) A critical review of textile industry wastewater: green technologies for the removal of indigo dyes. Int J Environ Sci Technol 20:10553–10590. https://doi.org/10.1007/s13762-023-04810-2
46. Muralimohan N, Palanisamy T, Vimaladevi MN (2014) Experimental study on removal efficiency of blended coagulants in textile wastewater treatment. Int J Res Eng Technol 2:15–20
47. Kathing C, Saini G (2022) A review of various treatment methods for the removal of dyes from textile effluent. Recent Prog Mater 04:1–15. https://doi.org/10.21926/rpm.2204028
48. Mahmood A, Bilal B, Naeem Z, Iram S (2021) Physical, chemical, and biological remediation techniques for textile effluents in context with developed and developing countries. In: Kumar V, Prasad R, Kumar M (eds) Rhizobiont in bioremediation of hazardous waste. Springer, Singapore, pp 409–441. https://doi.org/10.1007/978-981-16-0602-1_18
49. Zheng X, Shen Z-P, Shi L et al (2017) Photocatalytic membrane reactors (PMRs) in water treatment: configurations and influencing factors. Catalysts 7:224. https://doi.org/10.3390/catal7080224
50. Chowdhury AA, Islam E (2023) Current advances in bio-membrane technology for wastewater treatment. Membrane and membrane-based processes for wastewater treatment, 1st edn. CRC Press, Boca Raton, pp 35–49. https://doi.org/10.1201/9781003165019-3
51. Haque MdM, Hossen MdN, Rahman A et al (2024) Decolorization, degradation and detoxification of mutagenic dye methyl orange by novel biofilm producing plant growth-promoting rhizobacteria. Chemosphere 346:140568. https://doi.org/10.1016/j.chemosphere.2023.140568
52. Jasińska A, Walaszczyk A, Paraszkiewicz K (2024) Omics-based approaches in research on textile dye microbial decolorization. Molecules 29:2771. https://doi.org/10.3390/molecules29122771
53. Santal AR, Rani R, Kumar A et al (2022) Biodegradation and detoxification of textile dyes using a novel bacterium *Bacillus* sp. AS2 for sustainable environmental cleanup. Biocatal Biotransform 42:41–55. https://doi.org/10.1080/10242422.2022.2113518
54. Rahayu F, Mustafa I, Marjani et al (2023) Newly isolated ligninolytic bacteria and its applications for multiple dye degradation. Water Air Soil Pollut 234:359. https://doi.org/10.1007/s11270-023-06377-7

55. Singh S, Gautam RL, Chaudhary DK et al (2024) Biodecolorization and biodegradation of reactive green 12 textile industry dye and their post-degradation phytotoxicity-genotoxicity assessments. Arch Microbiol 206:262. https://doi.org/10.1007/s00203-024-03994-6
56. Yadav M, Singh AL (2024) Decolourization and detoxification of Reactive Red-195 azo dye by *Staphylococcus caprae* isolated from textile effluent. Folia Microbiol. https://doi.org/10.1007/s12223-024-01175-y
57. Pandey A, Pathak VM, Navneet RM (2024) A feasible approach for azo-dye (methyl orange) degradation by textile effluent isolate *Serratia marcescens* ED1 strain for water sustainability: AST identification, degradation optimization and pathway hypothesis. Heliyon. https://doi.org/10.1016/j.heliyon.2024.e32339
58. Bhattacharjee A, Karmakar S, Sengupta T et al (2024) A study on the dye degrading ability of *Brevibacillus brevis* US575 and *Bacillus cereus* F4810/72 isolated from tannery effluents. Biocatal Agric Biotechnol 60:103287. https://doi.org/10.1016/j.bcab.2024.103287
59. Chebet J, Ramesh M, Karegoudar TB et al (2024) Decolourisation and toxicity reduction of azo dye Bismarck Brown by newly isolated *Pseudomonas aeruginosa*, JKAK strain. Int J Environ Stud. https://doi.org/10.1080/00207233.2024.2357950
60. Samuchiwal S, Sahu A, Selvaraju K et al (2024) Unveiling the azo-reductase mechanism in *Pseudomonas putida* for efficient decolorization of textile reactive dyes: an in-silico study. J Biomol Struct Dyn. https://doi.org/10.1080/07391102.2024.2308768
61. Karthika A, Seenivasagan R, Kasimani R et al (2024) Microbial technologies for sustainable textile effluent treatment: a review. J Environ Chem Eng 12:113275. https://doi.org/10.1016/j.jece.2024.113275
62. Kalia S, Samuchiwal S, Dalvi V, Malik A (2024) Exploring fungal-mediated solutions and its molecular mechanistic insights for textile dye decolorization. Chemosphere 360:142370. https://doi.org/10.1016/j.chemosphere.2024.142370
63. Deng W, Ge M, Wang Z et al (2024) Efficient degradation and detoxification of structurally different dyes and mixed dyes by LAC-4 laccase purified from white-rot fungi *Ganoderma lucidum*. Ecotoxicol Environ Saf 279:116450. https://doi.org/10.1016/j.ecoenv.2024.116450
64. Mian AH, Qayyum S, Zeb S et al (2024) Exploring indigenous fungal isolates for efficient dye degradation: a comprehensive study on sustainable bioremediation in the total environment. Environ Technol Innov 34:103615. https://doi.org/10.1016/j.eti.2024.103615
65. Purnomo AS, Solichah Fauzany U, Dwi Rizqi H et al (2024) Biodecolorization and biotransformation of methylene blue using mixed cultures of brown-rot fungus *Daedalea dickinsii* and filamentous fungus *Aspergillus oryzae* : identification of metabolites and degradation pathway. RSC Adv 14:5061–5068. https://doi.org/10.1039/D3RA08544A
66. Tian Y, Wu K, Lin S et al (2024) Biodegradation and decolorization of crystal violet dye by cocultivation with fungi and bacteria. ACS Omega 9:7668–7678. https://doi.org/10.1021/acsomega.3c06978
67. Saha P, Rao KVB (2024) Biodegradation of commercial textile reactive dye mixtures by industrial effluent adapted bacterial consortium VITPBC6: a potential technique for treating textile effluents. Biodegradation 35:173–193. https://doi.org/10.1007/s10532-023-10047-0
68. Palanivelan R, Ramya S, Aradhana S et al (2024) Bioprospective decolourization of reactive azo dyes at pilot scale by a developed bacterial consortium using the RSM and CCD model. Biomass Conv Bioref. https://doi.org/10.1007/s13399-024-05701-3
69. Obayomi KS, Lau SY, Ibrahim O et al (2023) Removal of Congo red dye from aqueous environment by zinc terephthalate metal organic framework decorated on silver nanoparticles-loaded biochar: mechanistic insights of adsorption. Microporous Mesoporous Mater 355:112568. https://doi.org/10.1016/j.micromeso.2023.112568
70. Sutar S, Patil P, Jadhav J (2022) Recent advances in biochar technology for textile dyes wastewater remediation: a review. Environ Res 209:112841. https://doi.org/10.1016/j.envres.2022.112841
71. Athira TM, Sumi S (2024) Agro-based adsorbents for dye removal from aqueous solutions: a review. Water Air Soil Pollut 235:120. https://doi.org/10.1007/s11270-024-06926-8

72. Kainth S, Sharma P, Pandey OP (2024) Green sorbents from agricultural wastes: a review of sustainable adsorption materials. Appl Surf Sci Adv 19:100562. https://doi.org/10.1016/j.apsadv.2023.100562
73. Pandey D, Daverey A, Dutta K, Arunachalam K (2023) Dye removal from simulated and real textile effluent using laccase immobilized on pine needle biochar. J Water Process Eng 53:103710. https://doi.org/10.1016/j.jwpe.2023.103710
74. Kumari S, Chowdhry J, Choudhury A et al (2024) Machine learning approaches for the treatment of textile wastewater using sugarcane bagasse (*Saccharum officinarum*) biochar. Environ Sci Pollut Res. https://doi.org/10.1007/s11356-024-31826-z
75. Hasan M, Al Biruni MT, Azad S, Ahmed T (2024) Adsorptive removal of dye from textile wastewater employing *Moringa oleifera* leaves biochar as a natural biosorbent. Biomass Conv Bioref 14:11075–11091. https://doi.org/10.1007/s13399-022-03196-4
76. Khan AA, Naqvi SR, Ali I et al (2023) Algal biochar: a natural solution for the removal of Congo red dye from textile wastewater. J Taiwan Inst Chem Eng. https://doi.org/10.1016/j.jtice.2023.105312
77. Jabar JM, Odusote YA (2024) Utilization of prepared activated biochar from water lily (*Nymphaea lotus*) stem for adsorption of malachite green dye from aqueous solution. Biomass Conv Bioref 14:5999–6010. https://doi.org/10.1007/s13399-021-01399-9
78. Singh V, Srivastava VC (2024) Transformation of textile dyeing industrial sludge into economical biochar for sorption of ofloxacin: Equilibrium, kinetic, and cost analysis. Biomass Conv Bioref 14:1881–1893. https://doi.org/10.1007/s13399-022-02554-6
79. El-Sadaawy MM, Agib NS (2024) Removal of textile dyes by ecofriendly aquatic plants from wastewater: a review on plants species, mechanisms, and perspectives. Blue Econ 2. https://doi.org/10.57241/2805-2994.1023
80. Hansa A, Devi A, Upadhyay M et al (2024) Toxicological implications of industrial effluents on plants: a review focusing on phytoremediation techniques. Int J Environ Sci Technol 21:2209–2224. https://doi.org/10.1007/s13762-023-05012-6
81. López-Ahumada E, Salazar-Hernández M, Talavera-López A et al (2022) Removal of anionic and cationic dyes present in solution using biomass of *Eichhornia crassipes* as bioadsorbent. Molecules 27:6442. https://doi.org/10.3390/molecules27196442
82. Wibowo YG, Syahnur MT, Al-Azizah PS et al (2023) Phytoremediation of high concentration of ionic dyes using aquatic plant (*Lemna minor*): a potential eco-friendly solution for wastewater treatment. Environ Nanotechnol Monitor Manage 20:100849. https://doi.org/10.1016/j.enmm.2023.100849
83. Eftekhari Z, Norastehnia A, Masoudian Z (2023) Biodecolorization of azo dye Acid Blue 92 (AB92) by *Ceratophyllum demersum L.*: process optimization using Taguchi method and toxicity assessment. Acta Agriculturae Slovenica 119:1–13. https://doi.org/10.14720/aas.2023.119.2.13593
84. Shaalan MH, Almansoory AF, Al-Baldawi IA (2023) Removing of methylene blue dye from wastewater by interaction between *Ceratophyllum demersum* and *Acinetobacter radioresistens*. IOP Conf Ser: Earth Environ Sci 1215:012042. https://doi.org/10.1088/1755-1315/1215/1/012042
85. Abdallah MAM, Alprol AE (2024) Utilization of aquatic biomass as biosorbent for sustainable production of high surface area, nano-microporous, for removing two dyes from wastewater. Sci Rep 14:4471. https://doi.org/10.1038/s41598-024-54539-2
86. Nair VA, Sonali JMI, Kumar PS et al (2024) Nano-phytoremediation approach using *Pistia stratiotes*: biosynthesized copper nanoparticles for textile wastewater treatment and toxicity analysis. Waste Biomass Valor. https://doi.org/10.1007/s12649-024-02560-x
87. Kaur N, Kaushal J, Mahajan P, Mantri A (2024) Phytoremediation of methylene blue dye (triarylmethane) and Congo red (diazo) by *T. ammi* L.: kinetic studies. Int J Environ Sci Technol 21:1697–1714. https://doi.org/10.1007/s13762-023-05037-x

MIX
Papier aus verantwortungsvollen Quellen
Paper from responsible sources
FSC® C105338

If you have any concerns about our products,
you can contact us on
ProductSafety@springernature.com

In case Publisher is established outside the EU,
the EU authorized representative is:
Springer Nature Customer Service Center GmbH
Europaplatz 3, 69115 Heidelberg, Germany

Printed by Libri Plureos GmbH
in Hamburg, Germany